Dictionary of Agriculture

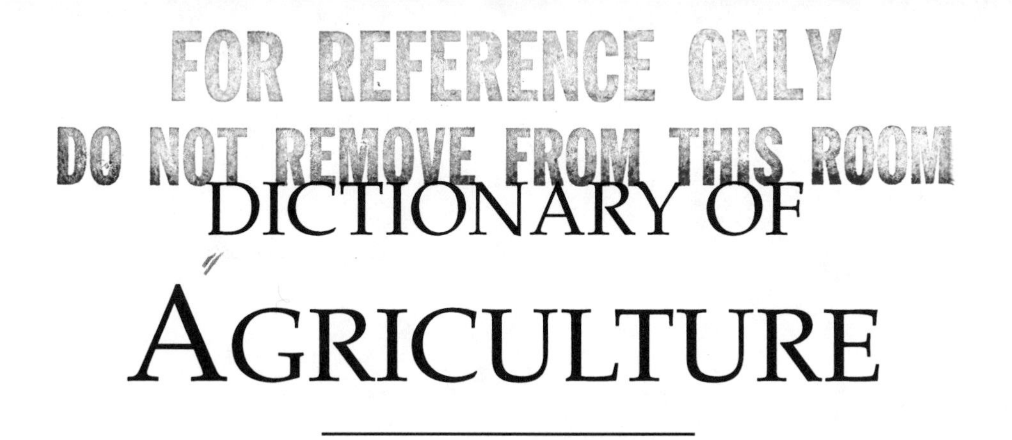

DICTIONARY OF AGRICULTURE

From Abaca to Zoonosis

KATHRYN L. LIPTON

LYNNE RIENNER PUBLISHERS
BOULDER
LONDON

Published in the United States of America in 1995 by
Lynne Rienner Publishers, Inc.
1800 30th Street, Boulder, Colorado 80301

and in the United Kingdom by
Lynne Rienner Publishers, Inc.
3 Henrietta Street, Covent Garden, London WC2E 8LU

Library of Congress Cataloging-in-Publication Data
Lipton, Kathryn L.
Dictionary of agriculture : from abaca to zoonosis /
Kathryn L. Lipton
Includes bibliographical references and subject index.
ISBN 1-55587-523-8 (alk. paper)
1. Agriculture—Dictionaries. 2. Agriculture—United States—
Dictionaries. I. Title.
S411.L55 1995
338.1'03—dc20 94-25260
CIP

British Cataloguing in Publication Data
A Cataloguing in Publication record for this book
is available from the British Library.

Printed and bound in the United States of America

The paper used in this publication meets the requirements
of the American National Standard for Permanence of
Paper for Printed Library Materials Z39.48-1984.

5 4 3 2 1

To Jillian and Brendan

My joy and inspiration

Contents

Acknowledgments, *ix*

Introduction, *xi*

Dictionary of Agricultural Terms 1

Appendix 1: Commonly Used Acronyms 273

Appendix 2: Dictionary Terms by Subject Area 281

Appendix 3: Major U.S. Agricultural and Trade Legislation, 1933–1994 317

Appendix 4: North American Free Trade Agreement (NAFTA) and U.S.-Canada Free Trade Agreement Provisions Affecting Agriculture 325

Appendix 5: Where Major U.S. Crops Are Grown: The Top Producing States, 1991–1993 327

Appendix 6: Standard Weights and Measures with Equivalents 329

Appendix 7: Individual Commodity Weights and Measures 335

References 341

About the Book and the Author 345

Acknowledgments

A special thanks goes to Alden Manchester (Economic Research Service), who encouraged me to expand the glossaries I had developed for previous publications into this dictionary and who shared so many sources of information throughout the preparation of this book. I also want to thank Marlow Vesterby (Economic Research Service) for the hours he spent defining all the terms on farm equipment and buildings.

I gratefully acknowledge the comments and contributions of the many Department of Agriculture reviewers. All of the reviewers listed below are with the Economic Research Service unless otherwise noted: Ed Allen, Mark Ash, Calvin Beale, Mark Bradley (Agricultural Marketing Service), Howard Brooks (Agricultural Research Service), Dick Brown, Tom Capehart, Lee Christensen, E. L. Civerolo (Agricultural Research Service), Jim Cole, Peggy Cooke, Lawrence Duewer, Jim Duke (Agricultural Research Service), John Dyck, Walter Epps, Milton Ericksen, Richard Fallert, Mohinder Gill, Lewrene Glaser, Fred Gray, Verner Grise, Shannon Hamm, Charles Handy, Joy Harwood, Richard Heifner, Fred Hoff, T. Q. Hutchinson, David Isenbergh (Office of the General Counsel Law Library), Doyle Johnson, Artis Jordon (Agricultural Research Service), Judith Kalbacher, Phil Kaufman, Jim Langley (Farm Service Agency), Bill Levedahl, John Link, Janet Livezy, Ron Lord, Gary Lucier, Mary Lisa Madell, Alden Manchester, Margaret Missiaen, Dave Mueller (National Agricultural Statistics Service), Tim Osborne, Victor Oliveira, Dave Peacock, Susan Pollack, Charlene Price, John Price (Agricultural Research Service), Steve Reed, Tanya Roberts, Howard Rosen (Forest Service), Scott Sanford, Dennis Shields, Mark Simone, Bob Skinner, Lewis Smith (Agricultural Research Service), Mark Smith, Jerry Stam, Roger Strickland, John Sullivan, Harold Taylor, Tom Tice, and Leslie Whitener.

I also want to extend my appreciation for the helpful comments of Roy Henwood of the Millers' National Federation, Ted Wilson of the U.S. International Trade Commission, and Barbara Chattin of the Office of the Trade Representative.

Finally, I would like to thank the many researchers in the Economic Research Service who patiently answered my questions in their areas of expertise. Those people are too numerous to name here, but their assistance proved invaluable.

This book is not an official U.S. Department of Agriculture publication. The author is solely responsible for its contents.

—Kathryn L. Lipton

Introduction

From abaca to zoonosis, this dictionary provides an invaluable source of information for everyone from undergraduates to seasoned policymakers. It is designed to serve as a practical, easy-to-use guide to the thousands of terms associated with agriculture. The definitions included here provide important details, but are easily understood by those unfamiliar with farming, food and agricultural policy, trade, conservation, or any other of the myriad of topics covered.

The terms in the dictionary are presented alphabetically. In addition, Appendix 2, "Dictionary Terms by Subject Area," classifies all the terms under 81 topic areas; for example, the reader interested in learning about honey production may look under *bees and beekeeping.* In some cases, a topic may be covered under more than one subject heading. For instance, the terms related to grain are listed under the individual commodity (*wheat, corn, rice,* etc.) and also under *feeds and fodder*; *grain drying*; *milling*; *storage*; *agricultural policies and programs*; *U.S. agricultural policies and programs, foreign*; and *food and feed grain.*

The reader is also guided to other terms defined in the dictionary through the use of italicized words appearing within definitions. The definition for *naps,* for example, includes the italicized words *fibers, gin, cotton,* and *neps,* all defined elsewhere in the dictionary.

Although many words found here have multiple meanings, this dictionary includes only those definitions related to agriculture. Bark, for example, is the outer covering of a tree trunk, the fat covering on a livestock carcass, and the noise a dog makes. Only the first two definitions are included here.

The comprehensive appendixes at the back of the book also include a list of commonly used acronyms, a survey of agricultural and trade-related U.S. legislation, provisions of the North American Free Trade Agreement (NAFTA) and the U.S.-Canada Free Trade Agreement, a listing of the states in which major U.S. crops are grown, standard weights and measures with conversions, and individual commodity weights and measures.

Because the agriculture industry is changing rapidly, this dictionary is by no means exhaustive. However, I have tried to capture the most important terms in each category to provide a basic understanding of agriculture today.

A

Abaca A fibrous product obtained from the leafstalks of a member of the *banana* family. Abaca was widely used for cordage until the development of *man-made fibers* in the mid-1960s. Abaca fibers now are mainly used in the manufacture of high-quality specialty papers, such as porous-plug wrap paper, *tea* bags, stencil-base tissue, meat sausage casings, dust filters, and a number of other applications. The major producing countries are the Philippines and Ecuador, with the Philippines accounting for nearly 85 percent of world output. See also *hard fibers.*

Absolute quotas Specified limitations on *imports* in a definite period. See also *quantitative restriction.*

Absorption The passage of *nutrients* or other substances to the blood, lymph, and *cells,* as from the digestive tract or from the tissues.

Access The availability of a *market* to exporting (see *export*) countries. See also *market access.*

Accession The process of a country becoming a member of an international agreement, such as the *General Agreement on Tariffs and Trade (GATT).* Negotiations determine the specific obligations a nonmember country must meet before it is entitled to full membership benefits of an agreement.

Accession compensatory amount (ACA) A tax or *subsidy* imposed on products entering or leaving a country that is becoming a member (see *accession*) of the *European Union (EU).* ACAs are intended to prevent trade distortions between new and original member countries by compensating for differences in *support prices.*

Acesulfame-k (ace-k) A *low-calorie sweetener* equal to *aspartame* in sweetness. Unlike *aspartame,* it does not lose its sweetness when heated.

Acid rain The atmospheric precipitation composed of the hydrolyzed (see *hydrolysis*) *byproducts* from oxidized halogen, *nitrogen,* and sulfur substances.

Acid soil A *soil* with a *pH* value less than 7.0.

Acorn squash See *squash.*

ACP States See *African, Caribbean, and Pacific States.*

Acquisition The legal transformation through which a company gains ownership and control of another company.

Acre A U.S. and English unit of area equal to 43,560 square feet, 4,840 square yards, or 0.405

hectares. See also Appendix 6.

Acreage allotment An individual *farm's* share, based on the production history, of the total national acreage that the secretary of agriculture determines is needed to produce sufficient supplies of a particular crop. Allotments apply only to *tobacco*.

Acreage base See *crop acreage base*.

Acreage diversion Removing *land* from the *production* of *surplus* crops; shifting land from one crop to another.

Acreage reduction program (ARP) A voluntary *land* retirement system in which participating farmers idle a prescribed portion of their *crop acreage base* of *wheat*, *feed grains*, *cotton*, or *rice*. The base is an historical average of the acreage planted for *harvest* and considered to be planted for harvest. Acreage considered to be planted includes any acreage not planted because of acreage reduction and *acreage diversion* programs during a period specified by law. Farmers are not given a *direct payment* for ARP participation, although they must participate to be eligible for benefits such as *Commodity Credit Corporation (CCC)* loans and *deficiency payments*. Participating *producers* are sometimes offered the option of idling additional land under a *paid land diversion* program, which gives them a specific payment for each *idled acre*.

Acreage reserve A temporary provision of the *Soil Bank* program to reduce *production* and *carryover* of *wheat* and other specific *basic commodities* by taking *acres* out of production on an annual basis.

Acreage slippage A measure of the effectiveness of *acreage reduction programs*. Slippage occurs when the reduction in *harvested acres* is less than the increase in *idled acres*. Slippage may refer to crop acreage (see *acre*), *yields*, or *production*.

Acre foot The quantity of water required to cover 1 *acre* to a depth of 1 foot. This is equivalent to 43,560 cubic feet or 325,851 gallons.

Acres irrigated The acreage (see *acre*) of agricultural *land* to which water is artificially applied by controlled means. Land flooded during high water periods is included as irrigated land only if the water is diverted to the land by dams, canals, or other works. See also *irrigated farms*.

Across-the-board (linear) tariff negotiation A method of *trade negotiation* introduced during the *Kennedy Round* through which uniform percentage reductions may be made in duties (see *tariff*) on major categories of items. It contrasts with the so-called item-by-item type of reduction, which proved difficult to use in the *Dillon Round*.

Acrylic See *synthetic fibers*.

Acute The sudden onset of *symptoms* of a *disease* or the rapid progression of the disease. See *chronic*.

Additionality A requirement that shipments under Title I of *Public Law 480* must be in addition to the normal requirements of the country of destination. The additionality provision is designed to prevent displacement of commercial sales.

Additive An ingredient or combination of ingredients added to a product or parts thereof to fulfill a specific need. In the case of *feed*, for example, additives are usually used in small quantities and require careful handling and mixing. Additives can also be added to *fertilizer* to improve its chemical or physical condition.

Ad Hoc Disaster Assistance An ad hoc aid program providing farmers of virtually all crops with protection from natural disasters. In recent years, a farmer is eligible for a payment if he or she incurs a loss of more than 35 percent (if the farmer has crop insurance) or 40 percent (if the farmer does not have crop insurance). The payment rate depends on whether or not the farmer participates in the annual *commodity* program and on the crop. Farmers do not pay a *premium* for coverage. Between 1988 and 1993, ad hoc disaster assistance has cost an average of $1.5 billion per year. The Federal Crop Insurance Reform Act of 1994 repealed the authority to designate ad hoc disaster programs for crops as "emergency" spending under "pay-go" budget rules, making emergency crop loss aid "on-budget." The *Noninsured Assistance Program (NAP)* is a permanent form of assistance for crops not covered by federal crop insurance.

Adjusted base period A *parity* price concept in which the average *price* received by farmers in the most recent 10 years is divided by the index (1910–1914 = 100) of average prices received (see *prices-received index*) by farmers for all farm products in the same 10 years.

Administrative committee A council of growers, *handlers*, and sometimes a *consumer* or public interest representative responsible for developing *federal marketing order and agreement* policy and administering regulations.

Admission temporaire A customs permit for free importation of products to be reexported after manufacturing or other *processing*. The provision is designed to allow domestic exporting industries to more effectively compete with corresponding *foreign* industries. See *drawback*.

Ad referendum Used during trade or other negotiations (see *trade negotiations*) to indicate that an agreement or *concession* is agreed upon pending final approval by a higher governmental authority.

Adulterated grain *Grain* that contains an added or naturally occurring poisonous or deleterious substance that may make it harmful to health. An example is *aflatoxin*-contaminated *corn*.

Ad valorem equivalent (AVE) A *tariff* assessed on a specific basis and expressed as a fixed percentage of the value of an *import*. See *specific tariff*.

Ad valorem tariff A governmental tax on *imports* assessed as a percentage of the value of the goods cleared through *customs*. For example, 10 percent ad valorem means the *tariff* is 10 percent of the value of the goods.

Advance deficiency payments Payments required by law to be made to crop *producers* when an *acreage reduction program* is in effect and *deficiency payments* are expected to be paid. Advance deficiency payments can range from 30 to 50 percent of expected payments, depending on the crop. Up to 50 percent of the advance payment may be made as *generic commodity certificates*. If total deficiency payments computed after *harvest* are less than the advance amount, producers must refund the excess portion.

Advance recourse loans *Price-support* loans made early in a *marketing year* to enable farmers to hold their crops for later sale. Farmers must repay the recourse loan with interest and reclaim the crops used as collateral. Advance recourse loans have been made to *upland* and *extra-long staple* (ELS) cottonseed *producers* only. However, the secretary of agriculture has authority to make advance recourse loans to producers of other *commodities* if such loans are necessary to provide adequate operating credit to producers.

Advertising The distribution of messages about a product or *commodity* through mass media such as television, radio, newspapers and supplements, magazines, and billboards.

Adzuki bean A variety of *dry edible bean* that is native to Japan. Adzuki beans are an important dry edible bean in the Japanese diet and are cooked mainly with *rice* or used to produce confections. Adzuki beans are not a major crop in the United States.

Aeration, grain The passage of air over or through *grain* to control the adverse effects of excessive moisture, temperature, and humidity. This is usually done by moving air with fans or through ducts.

Aeration, soil The exchange of air in *soil* with air from the atmosphere. The composition of the air in a well-aerated soil is similar to that in the atmosphere; in a poorly aerated soil, the air in the soil is considerably higher in *carbon dioxide* and lower in oxygen than the atmosphere above the soil.

Aeroponics A process where plant roots are suspended in a dark chamber and sprayed with an aerated *nutrient* solution.

A-Fix-A The Brazilian form of the delayed price payment used in the United States. The farmer receives an advance on which interest is paid, until at a day of the farmer's choosing, the final price of the *grain* is set.

Aflatoxin A highly carcinogenic *toxin* produced by a *fungus* which sometimes occurs when crops are stored under warm, dark, and humid conditions. Aflatoxin is most commonly associated with *corn*, *peanuts*, and *soybeans*. *Grain* for *export* is generally rejected if it contains aflatoxin.

African, Caribbean, and Pacific States (ACP States) A group of 69 countries that have special economic relations with the *European Union (EU)* in areas such as trade, *commodities*, and development finance. The areas of cooperation are set forth in the Lome IV Convention of December 1989. Manufactured goods and some agricultural products from these countries receive free entry into the EU.

African Development Bank (AFDB) An organization established to promote economic cooperation among independent African nations and to assist in their economic and social development. Membership is exclusively African.

African swine fever A usually fatal viral *disease* of *swine* that is not transmissible to humans. African swine fever is very similar to *hog cholera*. Trade in live hogs and uncooked pork is restricted from affected areas.

Aftosa See *foot-and-mouth disease*.

Agency for International Development (AID) An agency of the *U.S. Department of State* responsible for assistance programs in *developing countries* friendly to the United States. The agency is also responsible for administration of Title I and Title III programs under *Public Law 480* (P.L. 480) and for overseas execution of all P.L. 480 programs.

Agent 1: An individual or organization acting for another. The latter is referred to as the *principal*. 2: A force or substance that can produce a biological or chemical reaction or change. See *infectious agent*.

Aggregate concentration The share of sales in a sector, such as food retailing, that are made by the largest firms.

Aggregate (of soil) Many fine *soil* particles held in a single mass or cluster, such as a clod or block. Many properties of the aggregate differ from those of an equal mass of unaggregated soil.

Aging A process applied to *cigarette* tobacco whereby the leaf is compressed in *hogsheads* or other containers at a moisture content of 10–13 percent to mildly ferment the *tobacco*.

AgrAbility Project A national program to assist farmers, ranchers, and other agricultural workers with disabilities. Mandated by the *Food, Agriculture, Conservation, and Trade Act of 1990* (P.L. 101-624) (see Appendix 3), the Project is operated by the National Easter Seal Society and the Breaking New Ground Resource Center at Purdue University.

Agreement on Import Licensing Procedures An agreement designed to simplify and harmonize, to the greatest extent possible, the procedures that importers must follow in obtaining an *import license*. Also known as the import licensing code.

Agribusiness The sector consisting of *producers*, manufacturers, and distributors of agricultural goods and services, including *fertilizer* and *farm* equipment makers, food and fiber *processors*, wholesalers (see *wholesaling*), transporters, and retail food and *fiber* outlets.

Agricultural agent See *county extension agent*.

Agricultural association A nonprofit or *cooperative* association of farmers, growers, or ranchers incorporated or qualified under applicable state laws to recruit, solicit, hire, employ, furnish, or transport migrant *farmworkers* or *seasonal agricultural workers*.

Agricultural attaché An individual employed by the *U.S. Department of Agriculture*'s (USDA) *Foreign Agricultural Service* who is assigned to posts throughout the world and is responsible for (1) representing U.S. *agriculture*; (2) reporting on *foreign* supplies of agricultural *commodities*, marketing opportunities, agricultural policies, and related matters; and (3) developing and maintaining foreign markets for U.S. agricultural products.

Agricultural chemicals A broad range of compounds used for fertilizing crops (see *fertilizer*), controlling pests, enriching *feed*, and promoting the health and productivity of *livestock* and *poultry*.

Agricultural commissioner A public employee responsible for enforcing the State Agricultural Code and assuring that the Code is applied to and for *producers* and *consumers*.

Agricultural Conservation Program (ACP) A program to carry out environmental protection and *conservation practices* on agricultural *land*. ACP is a *cost-sharing program* between agricultural *producers*, federal and state agencies, and other groups. Cost sharing is available under annual or long-term agreements.

Agricultural cooperative A business owned and controlled by the people who use its services. These services can include *processing* and marketing members' products and obtaining *production* supplies, such as *feed*, *fertilizer*, petroleum, *seed*, building materials, and equipment. The members finance and operate the business for their mutual benefit. Service cooperatives provide functions related to production and marketing, including credit, electricity, and insurance.

Agricultural Cooperative Service (ACS) A *U.S. Department of Agriculture* (USDA) agency that provided technical and educational assistance to *agricultural cooperatives* to improve organization and operation and to guide further development. ACS also provided management analysis and conducted research, which was published as agency educational materials or in *Farmer Cooperatives*. The agency was merged with USDA's *Rural Development Administration* in April 1994 and later into the *Rural Business and Cooperative Development Service*.

Agricultural Credit Association (ACA) A part of the *Farm Credit System* authorized to make short- and intermediate-term loans and long-term real estate mortgage loans to eligible borrowers. The ACA was created by the merger of a *Production Credit Association* and a *Federal Land Bank Association* or *Federal Land Credit Association*.

Agricultural employment Any labor service or activity related to farming. Services include cultivating (see *cultivate*) and tilling (see *till*) *soil*; dairying; and cultivating, growing, and harvesting (see *harvest*) agricultural, aquacultural (see *aquaculture*), or horticultural *commodities*. Labor associated with handling, planting, drying, packing, packaging, *processing*, freezing, or *grading* prior to delivery for storage of any agricultural or horticultural commodity in its unmanufactured state is also considered agricultural employment.

Agricultural Environmental Quality Council A group established by the *Food, Agriculture, Conservation, and Trade Act of 1990 (P.L. 101-624)* (see Appendix 3) to coordinate and direct all environmental policies and programs of the *U.S. Department of Agriculture* (USDA). The council is composed of top USDA officials. An office operating under the Council coordinates USDA activities.

Agricultural Fair Practices Act A law that enables farmers to file complaints with the *U.S. Department of Agriculture*'s (USDA) *Agricultural Marketing Service* if *processors* refuse to deal with them because they are members of a *producers'* bargaining or marketing association. The Act makes it unlawful for *handlers* to coerce, intimidate, or discriminate against producers because they belong to such an association.

Agricultural Information and Marketing Services (AIMS) A program of the *U.S. Department of Agriculture*'s (USDA) *Foreign Agricultural Service* designed to help expand *exports* by providing a liaison between U.S. companies and *foreign* buyers seeking U.S. food and agricultural products.

Agricultural inputs Components of agricultural *production*, such as *land,* labor, and capital to purchase other *farm inputs*, including *feed*, machinery, *fertilizer*, *seed*, and *pesticides*.

Agricultural marketing The buying, selling, or transferring of agricultural products among individuals, firms, or other organizations.

Agricultural Marketing Agreement Act of 1937 (AMAA) The legislation that authorizes *federal marketing orders and agreements* for selected horticultural crops.

Agricultural Marketing Service (AMS) A *U.S. Department of Agriculture* (USDA) agency that administers and enforces regulatory laws to make *agricultural marketing* more orderly and efficient. AMS's responsibilities include establishing standards for *grades* of *cotton*, *tobacco*, meat, dairy products, *eggs*, fruits, and vegetables; operating *grading* services; and administering *federal marketing orders and agreements*. The agency also operates the *federal-state market news service*, in cooperation with government agencies in 44 States, the District of Columbia, and three territories, reporting up-to-the-minute information on *prices*, *supply*, and *demand* for most agricultural *commodities*.

Agricultural research Research by federal and state institutions to provide new knowledge and technologies to establish low-cost *resources* to enable farmers to produce high-yielding (see *yield*) quality *commodities*, to enhance the *environment,* and to conserve energy and natural resources.

Agricultural Research Service (ARS) A *U.S. Department of Agriculture* (USDA) agency that conducts research aimed at providing the means for solving technical food and agricultural problems of broad scope and high national priority—directed specifically toward meeting objectives in *soil*, water, and sciences; plant and animal *productivity; commodity* conversion and delivery; human nutrition; and systems integration.

Agricultural resource base The natural *resources* (*soil*, water, and *climate*) necessary to produce crops.

Agricultural Resource Conservation Program (ARCP) A conservation program authorized by the *Food, Agriculture, Conservation, and Trade Act of 1990 (P.L. 101-624)* (see Appendix 3) that combines previous conservation programs with new ones. The ARCP contains three main parts: the *Environmental Conservation Acreage Reserve Program (ECARP)*, the *Agricultural Water Quality Protection Program (AWQPP)*, and the *Environmental Easement Program (EEP)*.

Agricultural service Agricultural work performed on a *farm* or *ranch* if the provider of the service is paid on a *contract* basis for materials, equipment, or labor.

Agricultural Stabilization and Conservation Service (ASCS) A *U.S. Department of Agriculture* (USDA) agency responsible for administering federal *price-support programs* and *income-support programs* as well as some conservation and forestry *cost-sharing programs*, environmental (see *environment*) protection, and emergency programs. The agency was abolished in October 1994 and its responsibilities reassigned to the *Consolidated Farm Service Agency.*

Agricultural Stabilization Committee (ASC) A farmer-elected local board responsible for overseeing the federal *price support programs, production* adjustment programs, and conservation programs carried out by the county *Consolidated Farm Service Agency* office. See *county office.*

Agricultural trade and development missions Delegations sent to selected *developing countries* to encourage greater U.S. private sector and *foreign* country participation in U.S. agricultural trade and development programs. Teams are comprised of representatives from the *U.S. Department of Agriculture*, the *Agency for International Development*, the *U.S. Department of State*, private businesses, and nonprofit voluntary organizations.

Agricultural Trade Development and Assistance Act of 1954 See *Public Law 480.*

Agricultural trade office (ATO) Offices established by the Trade Agreements Act of 1979 (P.L. 96-39) to develop, maintain, and expand international markets for U.S. agricultural *commodities.* The offices are located in major centers of commerce throughout the world and serve as centers for *export* sales promotions and contact points for importers seeking to buy U.S. *farm* products.

Agricultural Water Quality Protection Program (AWQPP) A voluntary program authorized by the *Food, Agriculture, Conservation, and Trade Act of 1990 (P.L. 101-624)* (see Appendix 3) and designed to enroll 10 million acres of farmland under agricultural water protection plans by the end of 1995. *Producers* must agree to enroll *cropland* for three to five years in exchange for annual incentive payments and *cost-sharing* with the U.S. government to carry out the wetland and wildlife *habitat* option. The program was implemented as Water Quality Incentive Projects (WQIP) under the *Agricultural Conservation Program.*

Agricultural work Work done on a *farm* or *ranch* in connection with the production of agricultural products, including nursery and *greenhouse* products and animal specialties such as fur farms and *apiaries.* Also included is work done off the farm to handle farm-related business, such as trips to buy *feed* or deliver products to local *markets*.

Agriculture A broad term that encompasses food and *fiber* production, natural *resources*, the agricultural supply sector (including *fertilizer*, chemicals, and equipment manufacturing), service institutions (such as financial, insurance, and communications entities), marketing activities (transportation, *processing*, *wholesaling*, and retailing), public service activities (*market* reporting, *grades* and standards, and *weather* information), and the science and technology that support the agricultural system.

Agriculture Council of America A broad-based membership organization through which food- and *fiber*-related (see *food and fiber system*) interests work together to promote and build support for decisions that will ensure *agriculture*'s profitability and the integrity of the U.S. food system.

Agronomy The branch of *agriculture* dealing with field crop *production* and *soil* management.

Air classification A *grain-milling* process in which swirling air currents in combination with centrifugal force separate larger particles away from smaller, high-*protein* fines, resulting in the separation of low- and high-protein fractions according to size and density.

Air-cured tobacco *Tobacco* that is *cured* in well-ventilated barns under natural atmospheric conditions, usually without the use of supplementary heat. The air-cured class includes *light air-cured tobacco* and *Maryland tobacco* used mainly in *cigarettes* and *dark air-cured tobacco* used mainly in *snuff* and *chewing tobacco*.

Airtight upright silo A sealed, upright, cylindrical tower designed to prevent air from circulating within stored *feed*.

Air drill A machine that plants *seeds* (usually for small *grains*, such as *wheat* and *barley*) by blowing them into *furrows* in the *soil* opened by *disks* or *shanks*. An air drill can also be used to apply *fertilizer* and *pesticides* in the same operation. The machine was developed for *conservation tillage* to minimize field operations and leave maximum *crop residue* on the surface to reduce soil *erosion*.

Air-jet spinning A means of forming *yarn* in which an assembly of *fibers*, after attenuation, passes through air jet(s) that cause a small proportion of the fibers to be wrapped around the remainder to provide a coherent yarn structure. The emergent yarn is wound into a suitable package.

Air planter A machine identical to an *air drill* used for planting row crops, such as *corn* and *soybeans*.

Air seeder See *air drill*.

Alfalfa A *perennial* legume (see *legume*) used as fodder, *pasture*, and as a *cover crop*. Alfalfa is the mostly widely grown *hay* crop in the United States. Major producers include the United States, Canada, Australia, New Zealand, China, Argentina, and the former Soviet Union.

Alkaline soil A *soil* with a *pH* value of 7.0 or more. See *alkali soil*.

Alkali soil A *soil* with a *pH* value of 8.5 or higher. Such soils are so alkaline that the growth of most crop plants is reduced. See *alkaline soil*.

Allergy The reaction, characterized by a variety of clinical *symptoms*, of individuals exposed to a particular substance to which they have a heightened sensitivity (hypersensitivity).

Allocation procedure A set of accounting rules in each *federal milk marketing order* used to determine the amount of producer *milk* that will be priced (see *price*) in each class. Usually, the procedure assigns "other source milk" or unpriced milk from an unregulated *handler* to the lowest class, thereby reserving the *Class I milk* allocation for *producers* within the order.

Allotment A quantity provision in a *federal marketing order* that establishes the amount of a regulated *commodity* that individual *producers* may *market.*

Almond A tree or the edible nut of the almond tree, which is a member of the rose family. Major producers include the United States, Spain, Turkey, Portugal, and Morocco.

Alsike clover See *clover.*

Alternative agricultural product A new use, application, or material that is derived from an agricultural *commodity.*

Alternative Agricultural Research and Commercialization (AARC) Center A *U.S. Department of Agriculture* (USDA) center established by the *Food, Agriculture, and Trade Act of 1990 (P.L. 101-624)* (see Appendix 3) charged with commercializing new industrial nonfood and nonfeed uses for agricultural products. AARC's goals are to create jobs, enhance *rural* economic development, and diversify agricultural *markets.* The center issues grants, *contracts,* and cooperative agreements to carry out its activities.

Alternative farming *Production* methods other than energy- and chemical-intensive one-crop farming (*monoculture*). Alternatives include using animal and *green manure* rather than chemical *fertilizers, integrated pest management* instead of chemical *pesticides,* reduced tillage (see *till*), *crop rotation,* especially with *legumes,* alternative crops, or diversification of the *farm* enterprise.

Amendment Any material, such as *lime,* gypsum, sawdust, or synthetic conditioners, that is worked into the *soil* to make it more productive. The term is used more commonly for materials other than *fertilizer.*

American Agricultural Movement (AAM) A *farm* organization created in the 1970s in response to sharp declines in farm *prices* and *farm income.* In 1978, AAM members drove their *tractors* to Washington, D.C. and camped out for several weeks on the Capitol Mall until Congress acted to raise price supports (see *price-support programs*) above the levels of the 1977 *farm bill.*

American Association of Meat Processors (AAMP) An organization of retail and wholesale (see *wholesaling*) operators of meat *processing* plants, locker plants, frozen food centers, freezer food suppliers, and food plants. See also *International Association of Meat Processors.*

American Bankers Association A trade association of commercial banks designed to enhance their role as providers of financial services. The association has an Agricultural Bankers Division.

American Dry Pea and Lentil Association An Idaho-based organization of U.S. growers, shippers, and dealers of *dry edible peas* and *lentils.*

American-Egyptian cotton See *American-Pima cotton.*

American Farm Bureau Federation The largest *farm* organization with a membership that includes *farm operators,* farmland owners, and others with an interest in *agriculture.* The federation supports *market*-oriented farm policies, unrestricted access to markets, and reduced government regulation.

American Frozen Food Institute An *agribusiness* organization representing frozen food manufacturers.

American Meat Institute (AMI) An *agribusiness* organization representing *meat packers, processors,* sausage manufacturers, meat suppliers, and canners throughout the United States. AMI was founded in 1906 and its activities include marketing, research, improved operating methods and products, conservation, and industrial education.

American-Pima cotton An *extra-long staple (ELS) cotton* formerly known as American-Egyptian cotton in the United States, grown chiefly in the irrigated (see *acres irrigated* and *irrigated farms*) valleys of Arizona, New Mexico, and West Texas. American-Pima cotton represents only 1 percent of the U.S. *cotton* crop and is used chiefly for thread and high-valued fabrics (see *cloth*) and apparel. See also *supima.*

American Seed Trade Association (ASTA) An association of American, Canadian, and Mexican firms that produce and *market* seed for planting and provide services and products to *seed* companies, state and regional seed associations, and crop improvement associations.

American Society of Farm Managers and Rural Appraisers (ASFMRA) An association of professional *farm* managers, appraisers, teachers, researchers, and extension workers (see *county extension agent*) in farm and *ranch* management and *rural* appraisal. The ASFMRA was founded in 1929.

American Soybean Association A national, nonprofit organization founded by *soybean* farmers and extension workers (*see county extension agent*) to promote the crop and increase profits for farmers producing soybeans through *export* and *market* expansion, education, research, and legislative action.

American Sugarbeet Growers Association (ASGA) A federation of state and regional associations of *sugarbeet* growers. Formerly known as the National Sugarbeet Growers Association.

American Textile Manufacturers Institute The national trade organization for the U.S. *textile* industry. Member companies operate in more than 30 states and account for approximately 75 percent of all textile *fibers* consumed by mills (see *milling*) in the United States.

Amino acids *Nitrogen*-containing compounds that are the building blocks of living tissues and the chief components and determinants of the characteristics of a *protein.* Eighteen different amino acids commonly occur in the food supply. Eight are considered essential to ingest because the body cannot make them from other materials.

Ammonia A colorless gas (82.25 percent *nitrogen* and 17.75 percent hydrogen) composed of one atom of nitrogen and three atoms of hydrogen. Ammonia liquefied under pressure is used as a *fertilizer.*

Ammonification The formation of *ammonia* or ammonium compounds in the *soil.*

Amylase Any of a group of *enzymes* that accelerate the *hydrolysis* of *starch* and *gluten.*

Amylopectin A component of *starch* characterized by its heavy molecular weight, its branched structure of *glucose* units, and its tendency not to gel in aqueous solutions. The starch of normal *corn* is made up of amylopectin and *amylose.*

Amylose A component of *starch* characterized by straight chains of *glucose* units and the tendency of its aqueous solutions to set to a stiff gel. The starch of normal *corn* is made up of *amylopectin* and amylose.

Anaerobic Living or functioning in the absence of air or molecular oxygen.

Andean Group A group of Latin American countries formed in 1969 to promote regional economic integration among medium-sized countries in the region. Members include Bolivia, Colombia, Ecuador, Peru, and Venezuela.

Anhydrous Dry, or without water. Anhydrous *ammonia*, for example, is water-free, in contrast to the water solution of ammonia commonly known as household ammonia.

Anhydrous ammonia tank wagon A towed carrier for liquid *nitrogen* fertilizer (see *fertilizer*) and applicator, either mounted on the carrier or pulled behind.

Anhydrous dextrose See *dextrose*.

Anhydrous milk fat (AMF) See *butteroil*.

Animal and Plant Health Inspection Service (APHIS) A *U.S. Department of Agriculture* (USDA) agency that conducts regulatory and control programs to protect animal and plant health. The agency is responsible for the inspection and certification of animals, plants, and certain related products to meet health and/or sanitary requirements. The agency is also responsible for the inspection of *processing* facilities in the United States and in countries that *export* to the United States.

Animal biotechnology Biological or genetic modification of living organisms. See also *biotechnology*.

Animal Damage Control (ADC) Program A program operated by the *U.S. Department of Agriculture*'s (USDA) *Animal and Plant Health Inspection Service* to provide recommendations and direct assistance to government agencies and private individuals to protect *agriculture* from injury and damage caused by wild animals, including mammals, birds, and reptiles. Most activities are conducted on a partnership basis with individuals and local, state, and federal cooperators using matching funds to help pay for assistance.

Animal drug residue See *drug residue*.

Animal fats See *fats and oils*.

Animal protein feeds *Protein* derived from *meat-packing* and *rendering* plants (see *plants or establishments*), *surplus* milk or *milk* products, and marine sources for use as animal *feed*. They include proteins from meat, milk, *poultry*, *eggs*, fish, and their products.

Animal unit A unit of measurement of *livestock* used to make comparisons of *feed* consumption. An animal unit is the equivalent of one mature *cow* weighing 1,000 pounds. Five mature ewes, for example, are considered an animal unit.

Animal unit month The amount of *feed* needed for good growth and *production* of one *animal unit* during a month.

Animal Welfare Act A law that prescribes care and treatment for animals used in research, the wholesale pet trade, zoos and circuses, and while in commercial transportation. The act covers all warm-blooded animals, with certain exceptions (rats, mice, and birds). It specifically excludes domestic farm animals raised for food or *fiber*. The Act is enforced by the *U.S. Department of Agriculture*'s (USDA) *Animal and Plant Health Inspection Service (APHIS)* through a system of licensing and registration of regulated businesses and inspections to ensure licensees and registrants are complying with the standards for proper care and handling of animals covered by the Act.

Annual A plant that completes its life cycle from *seed* to plant, flower, and new seed in one growing season or less.

Annual pasture *Land*, usually *cropland*, on which annually (see *annual*) seeded *grasses* and/or *legumes* are established and used primarily for grazing (see *graze*); also, plant materials grazed from such intensively managed pastureland (see *pasture*).

Ante-mortem condemnations Animals condemned prior to their entry into the *slaughter* plant.

Anthelmintic A product that expels or destroys internal *parasites*.

Anthropic soil A man-made *soil* produced from a natural soil or other earthy deposit that has new characteristics. Examples include deep, black surface soils resulting from centuries of manuring (see *manure*), and naturally acid soils that have lost their distinguishing features due to many centuries of liming (see *lime*) and use for growing grass.

Antibiotic A chemical substance usually synthesized by a living *microorganism* and used to treat bacterial (see *bacteria*) infections. In proper concentration, antibiotics inhibit the growth of other microorganisms in animals and humans.

Antidumping duty A duty (see *tariff*) that attempts to discourage or prevent *dumping*. The usual aim is to levy a duty that equals the difference between the selling *price* of the goods in the country of origin and the selling price in the importing country. For example, Country A may manufacture an item and sell it within Country A for $2 but may *export* the item to Country B for sale within that country for only $1. To prevent dumping of the item, Country B may levy an antidumping duty of $1, which would equalize the selling price of the item in both countries.

Antidumping law A provision (Title VII) of the U.S. Tariff Act of 1930 (P.L. 71-361) that allows the *U.S. Department of Commerce* (DOC) to levy *antidumping duties* equivalent to the *dumping* margins under certain conditions. The DOC must determine that an imported product is being sold at less than its *fair value*, and the *U.S. International Trade Commission* must determine that a U.S. *producer* is thereby being injured.

Antioxidant A substance added to *feed* and food to provide greater stability and longer shelf-life by delaying the onset of oxidative rancidity.

Antivitamin Any substance that inhibits the normal function of a *vitamin*.

Apiarist A person who keeps *honeybees* for hobby or profit. See also *commercial beekeeper*.

Apiary A group of *honeybee* colonies (see *colony*); also the yard or area where bees are kept.

Apiculture The process of raising *honeybees* for hobby or profit (see *apiarist*).

Apis The genus for bees. More specifically, the *honeybee* belongs to *Apis mellifera.*

Apis mellifera The scientific name for the *honeybee.* See also *Apis.*

Apparent consumption A gross estimate of product utilization before deductions for shrinkage, waste, spoilage, or the like.

Apple The firm, fleshy, edible fruit of the apple tree. *Varieties* include the McIntosh, Golden and Red Delicious, Granny Smith, Jonathan, Rome, Stayman, and Winesap. The United States, Canada, Mexico, Europe, China, and Taiwan are the major apple producers.

Applicator A device, such as a sprayer, used to apply *pesticides, fertilizers,* or other *agricultural chemicals.*

Appreciation (depreciation) of currency The increase (decrease) in the value of one currency relative to another. An appreciation implies that one currency becomes more valuable relative to another and hence less of the valued currency is required in exchange for the other.

Appropriations bill Legislation that, upon becoming law, gives legal authority to spend or obligate *U.S. Treasury* funds for specific purposes. Appropriations bills provide part or all of the funds approved in authorization legislation (see *authorization bill*).

Apricot A temperate-zone fruit related to the *peach* and *plum. Varieties* include Royal-Blenheim, Derby-Royal, Tilton, Patterson, Pomo, Katy, Castlebrite, Perfection, and Improved Flaming Gold. Southern Europe, the United States, Argentina, and Australia are major apricot producers.

Aquaculture The production of *aquatic plants* or animals in a controlled environment, such as ponds, raceways, tanks, or cages, for all or part of their life cycle. In the United States, baitfish, catfish, clams, crawfish, freshwater prawns, mussels, oysters, salmon, shrimp, tropical (or ornamental) fish, and trout account for 85 percent of aquacultural production. Other species include alligator, hybrid striped bass, carp, eel, red fish, northern pike, sturgeon, and tilapia.

Aquatic plant A plant that lives in water.

Aquifers Subterranean water-bearing rock formations that can be tapped for wells. Aquifers supply water to about 97 percent of *rural* residents and almost 50 percent of the U.S. population.

Arabica See *coffee tree.*

Arable soils See *land capability.*

Arid climate A dry *climate,* such as found in desert or semidesert regions, where there is only enough water for widely spaced desert plants.

Arid region An area where the potential water losses by evaporation and transpiration are greater than the amount of water supplied by precipitation. In the United States, this area is broadly considered to consist of the dry parts of the 17 Western States, basically west of the 98th line of longitude.

Aromatic rice *Rice* that has a distinctive odor when cooked. Examples include *basmati rice* and jasmine rice.

Arracacha A root crop that is often used in place of *potatoes.*

Arrivals Annual and monthly rail, *piggyback*, truck, and air shipments received at wholesale or *terminal markets* in major U.S. cities.

Artichoke A member of the thistle family that produces edible flowers. Green Globe is the California *variety*, while European varieties include the Thistle, Prickly, and Green French artichoke. The United States, Italy, France, Spain, and Chile are the major artichoke producers.

Artificial insemination The artificial, rather than natural, injection of previously collected *semen* into a female in estrus. Artificial insemination is most widely used with *cattle* and offers several advantages, including greater control over herd characteristics through selection of genetic material. Artificial insemination is more economical than keeping an animal for breeding purposes.

Arugala An edible, leafy green with a spicy, mustardy tang.

Aseptic processing (packaging) A food *processing* technique that uses high temperatures over short periods for storage stabilization. The product is sterilized outside the container and cooled, then the container is sterilized, and the two are brought together in a stable environment.

Asexual reproduction A reproductive process that does not involve the *germ cell* or sexual *cells.*

Ash The nonvolatile residue resulting from the complete burning of organic matter. It is commonly composed of oxides of elements such as silicon, aluminum, iron, calcium, magnesium, and potassium.

Asian pear A fruit that looks like a round *apple* and tastes like a cross between apples and *pears.* They are eaten fresh and in desserts.

Asparagus A member of the lily family that produces edible young shoots. The major asparagus producers are the United States, Mexico, Taiwan, China, and Chile.

Aspartame (APM) The leading *low-calorie sweetener*, with a sweetening power of 180 to 200 times that of *sucrose*. Aspartame is composed of two naturally occurring *amino acids*, phenylalanine and aspartic acid.

Aspergillus flavus The *fungus* that produces *aflatoxin.*

Aspiration The process of removing *chaff*, dust, and other light materials from *grain* using controlled velocities of directed air. The process is used during *wheat* cleaning in preparation for *milling* or *feed* manufacturing.

Assay A technique for examining or determining characteristics.

Assembler's margin The total cost of all services of assemblage rendered by *country elevators*, *subterminal elevators*, or *terminal elevators*, dealers, *brokers*, and others involved in accumulating supplies of raw materials.

Assessment programs Programs requiring *producers* to pay a fee per unit of *production* in order to share program costs with the government. For example, to be eligible for price supports under the 1994 *flue-cured tobacco* program, producers had to agree to contribute 3 cents per pound of *tobacco* marketed to ensure the operation of the program at *no net cost* to taxpayers. Flue-cured tobacco buyers were also required to contribute 5 cents per pound. In addition, both producers and buyers had to contribute 0.7915 cent per pound each for a budget deficit assessment. Also referred to as "check-off" programs.

Assets (current) Cash, as well as crops, *livestock*, and similar assets that can be liquidated quickly to meet current obligations.

Assets (fixed) *Land*, buildings, and other long-lived assets that are difficult to convert into cash quickly to meet current obligations.

Association of American Feed Control Officials Officials of state, provincial, dominion, and federal agencies engaged in the regulation of *production*, analysis, labeling, distribution, and sale of animal *feeds* and *livestock* remedies.

Association of American Pesticide Control Officials An association comprised of state, municipal, and federal officials dedicated to the uniform enforcement of laws relating to proper and safe use of *pesticide* chemicals throughout North America.

Association of American Seed Control Officials U.S. and Canadian officials who administer state and provincial *seed* regulations. The organization is composed of one member from each state, one from Canada, and one from the *U.S. Department of Agriculture (USDA)*.

Association Nacional dos Exportadores de Cereais (ANEC) A Brazilian trade organization that sets the standards for the *grading* of export *grain* in Brazil.

Attainable yield The *yield* expected through the use of known technology. See *yield, economic maximum*.

Auction A *market* near major *production* or metropolitan areas where products, such as *livestock* or *tobacco* leaf, are sold sequentially to the highest bidder (see *bid*). In the case of tobacco, for example, the bidders are buyers for manufacturers, dealers, and *exporters*, as well as independent dealers or speculators.

Auction market A *market* facility that receives *livestock* from the seller and for a fee sells to buyers on an *auction* basis open to the public.

Auger A spiral device that moves *grain* through a tube.

Auger wagon A *wagon* with high sides, enclosed or open, with an *auger* to move *feed* or *grain* into *feed bunks* or *bins*.

Australian Wheat Board (AWB) The most important government institution in the Australian *wheat* industry. The AWB is involved in *variety* control, the establishment of *grain* standards, the administration of *producer* price (see *price*) policy, and *export* grain sales.

Authorization bill Upon becoming law, it establishes or continues operation of a federal program or agency and establishes the limits and duration of its funding. See also *appropriations bill*.

Autotrophic Capable of self-nourishment by using (oxidizing) simple chemical elements or compounds, such as iron, sulfur, or nitrates, to obtain energy for growth.

Auxins Organic substances that cause stems to elongate, leaves and fruit to fall, or cuttings to grow roots.

Available nutrients in soils The part (percentage) of the supply of a plant *nutrient* in the *soil* that can be taken up by plants at rates and in amounts significant to plant growth.

Available water in soils The part of the water in the *soil* that can be taken up by plants at rates significant to their growth.

Avian Refers to birds. In an agricultural context, the term pertains to *poultry* and/or fowl.

Avian influenza A highly infectious viral *disease* of *poultry* that affects growth rate, mortality, and *egg* laying. When an outbreak occurs in chickens (see *poultry*), the entire flock is destroyed and trade in live birds and poultry products is restricted. Mild forms in turkeys (see *poultry*) can be controlled by vaccination (see *vaccine*) in conjunction with strict *quarantines* and improved biosecurity.

Avocado The tropical American fruit of the laurel family. Avocados are 4 or 5 inches long, with thin, dark green skin. Major producers include the United States, Israel, Mexico, and Southeast Asia. California produces 85 to 90 percent of the U.S. domestic crop.

B

Babcock test A test for *butterfat* quantity in *milk* and milk products.

Baby vegetables Miniature *varieties* of vegetables, including *artichokes, beets, carrots, cauliflower, corn, eggplant, lettuce, squash,* and *tomatoes.*

Bacillus thuringiensis See *B.t.*

Backcross The breeding of a first-generation *hybrid* with either parent.

Backgrounding 1: An activity in which grazing (see *graze*) or harvested (see *harvest*) *forages* are the predominant *feeds* used to grow *calves* into *feedlot*-ready *feeder cattle.*

2: The preparation of young *cattle* for the feedlot, habituating them to *confinement* facilities and feeds.

Back hoe A *shovel* mounted on the back of a *tractor* and used to dig trenches or pits in *soil.*

Backward integration Gaining ownership and control over an earlier stage of the *production* or marketing (see *agricultural marketing*) process by a marketing firm.

Bacteria Microscopic, round, rod-shaped, spiral, or filamentous single-*cell* organisms that can be beneficial or cause *disease.*

Bacterial ring rot A highly contagious and infectious *disease* affecting *potatoes.* The disease is caused by a bacterium (see *bacteria*) and can destroy an entire crop.

Bactericide A chemical agent that kills *bacteria.* A biocide or *germicide* may be a bactericide.

Bagasse See *pulp.*

Bake-off format A bakery that makes 95 percent or more of its output from frozen dough that is either purchased or made in the bakery's own plant (*see plant or establishment*).

Baking or frying fats (shortening) According to the *Bureau of the Census,* products that meet all of the following conditions: (1) manufactured from vegetable oils (see *fats and oils*), meat fats (see *fats and oils),* or a combination; (2) deodorized (see *deodorizing*) or hydrogenated (see *hydrogenation*) and deodorized; (3) containing a significant amount of glycerides solid at room temperature; and/or (4) produced and sold entirely or primarily for baking or frying purposes.

Baking quality The performance of *wheat* and *flour* when made into bread. If a wheat flour has the potential of producing a loaf of good volume and good color, grain, and texture, it is said to have good baking quality. Important elements include *protein* content and quality (including *gluten* quality), ash content, dough "strength," and water absorption capability.

Balanced Budget and Emergency Deficit Control Act of 1985 See *Gramm-Rudman-Hollings Deficit Reduction Act* and Appendix 3.

Balanced ration A meal that provides an animal with the proper amounts and proportions of all the required *nutrients* for 24 hours.

Balance of payments A statement of economic transactions showing the relative difference between the inflow and outflow of goods, services, and capital claims and liabilities between a country and its trading partners. The balance of payments includes: (1) *current accounts,* including trade and services; (2) *capital accounts,* including short- and long-term items; (3) official transactions in reserve assets (gold, *special drawing rights (SDRs),* and *foreign* currency holdings; and (4) unilateral transfers of gifts by governments and individuals. See also *creditor nation* and *debtor nation.*

Balance of trade The difference between the value of goods that a nation *exports* and the value of the goods it *imports.* A trade surplus occurs when a country's exports exceed its imports, resulting in a favorable trade balance. Similarly, a trade deficit means that imports total more than exports for a country, producing an unfavorable trade balance.

Balancing The *market* service of distributing *milk* among various uses and *processors* to meet fluctuating needs from varying supplies.

Balancing plant A plant (see *plant or establishment*) in a fluid *milk* market which principally manufactures storable products such as *cheese, butter,* and *nonfat dry milk* powder from *Grade A milk* supplies in excess of needs for *fluid milk products.*

Bale A large, compressed, bound, and often wrapped bundle of a *commodity,* such as *cotton, hops,* and *hay.* See *bale (cotton).*

Bale (cotton) A package of compressed *cotton lint* as it comes from the *gin.* Including bagging and ties, a *bale* weighs about 500 pounds, and its dimensions vary depending on the degree of compression, 12–32 pounds per cubic foot. A bale is the form in which *cotton* moves in domestic and international commerce. However, cotton is bought and sold on a net weight (pound or kilogram) basis. For statistical purposes, cotton is reported in terms of running bales, in 480-pound net weight bales, or in pounds. A running bale is any bale of varying lint weight as it comes from the gin. To maintain comparability, bale weights are commonly converted to 480-pound net equivalents.

Bale pickup loader See *bale wagon.*

Baler, conventional, large, and round A machine that compresses *hay,* straw, or *forage* into a compact unit, either round or rectangular (see *bale),* and wraps and ties it with *bale twine.* Conventional balers make a rectangular bale that weighs about 50 to 100 pounds. *Round balers* and *large-bale balers* make bales that may weigh a half ton or more.

Bale shredder A machine with knives or flails used to remove *hay* from *bales.* The machine may have a blower incorporated to blow *forage* into *feed bunks.*

Bale stacker A flat-bed vehicle used in fields to pick up baled (see *bale*) *hay*, stack it, and unload it.

Bale twine Heavy string made from sisal, *hemp,* or plastic used to hold *bales* of *hay* together.

Bale wagon A unit, usually a flat-bed, used for hauling *bales*. A bale wagon may be self-propelled or a pull-type. Some models may include a device for picking a bale up from the field and stacking it on the *wagon*. Others may have a slanted bed so bales will slide to the rear. See also *bale stacker*.

Bale wrapper A device for wrapping *bales* (usually round) in plastic to prevent *hay* from spoiling.

Bambara groundnut A *drought*-resistent *legume,* cultivated throughout the drier regions of tropical Africa. Also called round bean or earth pea. The crop is not commonly grown in the United States.

Banana A tropical fruit with white, pulpy flesh and a rind that changes from green to yellow as the fruit ripens.

Band A flock of *sheep*.

Band application A method of applying *fertilizer* in bands near plant rows, providing a more efficient use of fertilizer than other methods that apply it over the entire *soil* surface.

Banding (of fertilizers) The placement of *fertilizer* in the *soil* in continuous narrow ribbons, usually at specific distances from the *seeds* or plants. The fertilizer bands are either on or below the soil surface.

Bang's disease See *brucellosis*.

Bank for Cooperatives (BC) *Farm Credit System* institutions that offer a complete line of credit and related financial services to *agricultural cooperatives, rural* utility systems, and other eligible customers.

Three banks, each with a national charter, comprise the BC system. The National Bank for Cooperatives (CoBank), a lending institution authorized to provide loans to and for the benefit of agricultural cooperatives, was created January 1, 1989, by the merger of the Central Bank for Cooperatives and 10 of the 12 District Banks for Cooperatives. CoBank also finances the *export* of U.S. agricultural products and provides international banking services for the benefit of farmer-owned cooperatives. CoBank is headquartered in Denver, Colorado, and has offices throughout the United States.

Two of the former District Banks for Cooperatives elected not to merge into the National Bank for Cooperatives (CoBank). The two banks, however, operate under the same charter as CoBank.

The St. Paul Bank, based in St. Paul, Minnesota, has four regional offices in the upper Midwest. The Springfield Bank for Cooperatives is headquartered in Springfield, Massachusetts, and primarily serves the eight Northeastern States. The three banks participate with one another on larger loans, as needed.

Bank out wagon A bulk *wagon* used to haul *rice* and other crops out of the field to *farm trucks* for highway travel.

Banque de Developpement des Etats de l'Afrique Centrale (BDEAC) See *Central African States Development Bank.*

Bargaining association An organization of *producers* intended to influence *prices* and other terms of trade, usually by negotiating *contract* terms with *processors*.

Bargaining power The ability of a buyer (seller) to negotiate the lowest (highest) *price* that a seller (buyer) is willing to pay for a given quantity of a *commodity.*

Bark 1: The fat (see *fats and oils*) covering on a *carcass.*
2: The outer covering of a woody branch, stem, or trunk.

Barley A *grain* used mainly for livestock (see *livestock*) *feed* and the manufacture of *malt* beverages. Barley is the third leading *feed grain* grown in the United States and is the most important grain product used by brewers. The United States, the former Soviet Union, *European Union,* Canada, Australia, and Spain are the leading producers.

Barn cleaner Machinery used to pull, push, or flush *manure* from a barn.

Barrier to entry Any characteristic of a *market* that gives an existing firm a cost saving or other advantage over a potential entrant.

Barrier to exit A factor that makes it difficult for a firm to exit a *market.* Most often it is because the return from using fixed *resources* in their best alternative uses is low.

Barrow A young castrated male hog (see *swine*). Barrows and *gilts* account for about 95 percent of slaughtered (see *slaughter*) hogs.

Barter A *countertrade* in which goods of equal value are exchanged under a single *contract,* within a specified time period, and without an exchange of money.

Base acreage See *crop acreage base.*

Base period price The average *price* for an item in a specified time period used as a base for an index, such as 1910–1914, 1957–1959, 1967, and 1977. See *prices-paid index.*

Basic commodities Six crops (*corn, cotton, peanuts, rice, tobacco,* and *wheat*) declared by legislation as price-supported (see *price-support programs*) *commodities.*

Basing point A geographical site used to establish fixed rates and/or *prices.* Generally, rates or prices increase according to the distance from the basing point. Basing points are used primarily in the *federal milk market order* program.

Basin irrigation (or level borders) The application of irrigation water (see *acres irrigated* and *irrigated farms*) to level areas that are surrounded by border ridges or levees. Also known as border-strip or check irrigation. See *irrigation methods.*

Basis The difference between a specific cash *price* and a futures price, usually the price for the *futures contract* for the same *commodity* that is nearest to maturity. For example, if the price of *corn* in Omaha was $2.50 per *bushel* and the price of the corn futures contract nearest to delivery was $2.80, the Omaha basis would be $0.30 under.

Basis risk Randomness in *price* relationships that prevents predicting cash-futures price differences with certainty.

Basmati rice *Rice* that has a distinctive odor when cooked. The basmati grains double in length and remain completely separate when cooked. Basmati rice is grown mostly in the Punjab area of central Pakistan and northern India and is mainly bought by higher-income Middle Eastern countries and the United States. Basmati rice is sold at *prices* roughly double those for *long-grain rice.*

Bean The edible *seed* of any *variety* of leguminous plants (see *legume*), including *dry edible beans,* snap or green beans, and *soybeans.*

Bean Market News A weekly statistical release produced by the *U.S. Department of Agriculture*'s (USDA) *Agricultural Marketing Service.* The Bean Market News quotes *prices, market* trends, and *exports* of U.S. *dry edible beans.*

Bear A person who sells *commodities* or other assets in anticipation of a *price* decline. See also *bull.*

Beater See *corn harvester.*

Bed 1: A narrow, flat-topped ridge on which crops are grown, bordered by a *furrow* on each side for irrigation (see *acres irrigated* and *irrigated farms*) drainage of excess water.
2: An area in which seedlings or sprouts are grown before transplanting.

Bedding and garden plants Young flower or vegetable plants that are sold for outdoor or patio use in flower *beds,* borders, patio planters, and home gardens. They are handled as *annuals.* Open acreage vegetable transplants and *herbs* grown for use as garden plants are also classified as bedding and garden plants. These plants are marketed in flats, individual pots, or hanging baskets. Examples are geraniums, impatiens, begonias, petunias, marigolds, pansies, and vegetable plants.

Bedding soil Growing areas made by plowing or grading the surface of fields into a series of elevated *beds* separated by shallow ditches for drainage.

Bed leveler A machine used to flatten seedbeds (see *bed* and *seed*), usually to allow for the uniform application of irrigation water (see *acres irrigated* and *irrigated farms*). See *land plane.*

Bed shaper A machine used to make rows or mounds of hills in a seedbed (see *bed* and *seed*).

Beefalo A *hybrid* bred from *cattle* and American *Bison.*

Beef breaking See *breaking (beef).*

Beef cattle *Breeds* of *cattle* suitable for meat *production.* These animals usually have broad backs and loins and heavy rumps.

Beef cow-calf production An enterprise in which *cows* are bred and maintained for the primary purpose of producing stocker (see *stocker cattle*) or *feeder calves* and/or *yearlings.*

Beef cows Female *cattle,* kept for nondairy purposes, that have calved (see *calving*) one or more times. Also referred to as brood cows.

Bee glue See *propolis.*

Beehive A box or similar structure containing a *colony* of *honeybees* (see *hive*).

Beekeeper See *apiarist.*

Beekeeping associations Local, regional, state, national, and/or international organizations of *apiarists* that meet regularly for educational and social activities.

Bee pasture Plants and trees visited by *honeybees* for *nectar* or *pollen.*

Beeswax A secretion from eight glands on the underside of a *honeybee's* abdomen that is molded to form the *honeycomb* in which the *colony* lives. Bees may consume 8 to 20 pounds of *honey* to secrete 1 pound of beeswax.

Beet A plant whose bulbous root is cultivated (see *cultivate*) as a garden vegetable and a source of *sugar.* See also *sugarbeet.*

Beginning stocks Inventories of a product on hand at the beginning of an accounting period, such as a month or year. See *carryover.*

Belgium-Luxembourg Economic Union See *Common Market.*

Belt horsepower See *horsepower.*

Bench terrace An embankment constructed across sloping (see *slope*) *soils* with a steep drop on the downslope side.

BENELUX See *Common Market.*

Best management practices *Conservation practices* determined by federal or state natural *resource* agencies to be the most practical means of controlling *erosion* or nonpoint-source pollutants (see *nonpoint-source pollution*) at levels compatible with environmental goals.

Bid 1: A *price* offered by a would-be buyer for a product, asset (see *asset, current* and *asset, fixed*) or service—for example, a price offered for *cattle* at an *auction* or a price offered for a *futures contract* on the floor of an exchange (see *commodity futures exchange*).

2: A price at which a would-be seller offers to sell a product, asset, or service—for example, the price at which a food *processor* agrees to supply a standardized food product to an institution. Also referred to as an offer.

Biennial A plant that lives for two years, producing vegetative growth the first year and blooming and fruiting the second year.

Bilateral oligopoly (monopoly) A *market* with relatively few large sellers and few large buyers.

Bilateral trade agreement A trade agreement between any two nations. The agreement may be either preferential, applying only to the two countries involved, or *most-favored-nation,* negotiated between the two countries but extending to all or most other countries.

Bilateral trade negotiations Discussions of trade issues between two countries.

Binder tobacco A class of *tobacco* that was originally used for binding bunched *filler tobacco* into the form and shape of a *cigar.* However, most cigars now use reconstituted sheet for the inner binder. As a result, binder tobacco is principally used for loose leaf *chewing tobacco.*

Bin dryers On-*farm* dryers (see *crop dryer*) that are generally low-capacity, low-temperature systems, capable of producing excellent quality *grain.*

Bin, grain A round or rectangular enclosure made of wood, metal, plastic, or some combination used to store *grain,* such as *wheat, corn,* or *barley.*

Biochemical The product of a chemical reaction in a living organism.

Biochemistry The branch of chemistry that deals with chemical compounds and processes occurring in living organisms.

Biocide See *bactericide.*

Biodegradation The use of *microorganisms* to break down the physical and/or chemical structure of a compound.

Biodiesel See *biofuel.*

Biodiversity A *production* system comprised of multiple plant and/or animal species, versus the specialization associated with *monoculture.*

Biofuels A broad range of biological materials that are used to produce energy. Biofuels include wood, *ethanol,* and diesel substitutes (also referred to as biodiesel).

Biological control of pests The control of some weeds and insect pests using indigenous or imported natural enemies or *diseases* to which the pest is susceptible. For example, biological control may use *Bacillus thuringiensis (B.t.)* against gypsy moths.

Biomass Any living matter that can be converted into usable energy through biological or chemical processes. It includes feedstocks, such as agricultural crops and their residues, animal wastes, wood, wood residues and grasses, and municipal wastes.

Biopulping The pretreatment of wood chips with white-rot fungi (see *fungus*). The technology has a number of benefits, including reduced chemical reactions, increased strength properties, decreased energy and waste treatment costs, and lower capital investment per unit increase in *production* capacity.

Biotechnology The use of technology, based on living systems, to develop commercial processes and products. Includes specific techniques of plant regeneration and *gene* manipulation and transfer.

Bison See *buffalo.*

Bitter melon A *squash* used in Asian cooking that looks like a bumpy *cucumber* and has a bland taste.

Black bean A *variety* of *dry edible bean* that is among the most nutritious because it is high in *protein* and potassium. Black beans are used in making soups, chili, *rice* dishes, and casseroles.

Blackberry An edible berry that, despite its name, actually ranges in color from red (logan berries) to black (Cherokee). Other *varieties* include Boysen, Evergreen, Marion, and Ollabe.

Blackeye pea (bean) A *dry edible bean* named for its dark-colored eye. While the U.S. crop is grown primarily in California, blackeye peas (beans) are a popular food in the southern United States.

Black liquor A term for the *byproduct* of paper *production* that results when sulfur compounds are used to separate *lignin* from *cellulose* pulp (see *pulp*). Black liquor is used as the primary source of carbon disulfide (a chemical intermediate in making other sulfur compounds) and dimethyl sulfoxide (an organic solvent).

Black wool See *shorn wool.*

Bleaching A chemical process used to reduce color. For example, bleaching is done to remove color-producing substances from oils and to further purify *fats and oils*. It is normally accomplished after the oil has been refined (see *refining*). Edible oils are generally bleached with adsorbents such as fuller's earth or activated charcoal, while industrial oils are usually bleached with sulfuric acid or chemical agents. Similarly, bleaching is done to remove the yellowish color of *wheat* and improve the baking quality.

Blended credit A form of export *subsidy* that combines direct government *export* credit and *credit guarantees* to reduce the effective interest rate.

Blending Mixing other ingredients with a *commodity* to produce a product with the unique properties or characteristics of the components of the blend, often providing a superior end product. In the case of *textiles,* for example, other *natural fibers* or *man-made fibers* may be mixed with *cotton.* Similarly, different *wheat* varieties (see *variety*) can be combined into a uniform blend.

Blend price A weighted average *price* based on the proportion of *Grade A milk* in a pool allocated to each of the use classes (see *classes of milk*). *Producers* participating in a pool receive its blend price with adjustments for *butterfat* content and *farm* location. Used primarily in the *federal milk market order* program.

Blight A range of *symptoms* of plant *diseases,* including spotting, sudden wilting, or death of leaves, flowers, twigs, stems, or entire plants.

Bloat A distention of the *rumen,* paunch, or large colon of an animal by the gases of *fermentation.*

Blocks *Feed* compressed into a solid mass cohesive enough to hold its form and generally weighing 30 to 50 pounds.

Blood meal Coagulated packing house blood that has been condensed, flash-dried, and ground (see *grind*) into *meal.* Its crude *protein* content ranges from 80 to 82 percent.

Bloom The bright red color (considered desirable) taken on by meat when exposed to air (oxygen).

Blower; feed, grain, or forage A machine used to move *feed, grain,* or *forage* from one area to another with forced air generated by a large, motor-driven fan.

Blown oils See *boiled, blown, bodied.*

Blowout An area from which *soil* material has been removed by wind. Such an area appears as a nearly barren, shallow depression with a flat or irregular floor consisting of a resistent layer, an accumulation of pebbles, or wet soil lying just above a water table.

Blowroom The area in a textile mill (see *mill, textile*) in which raw *cotton* is opened and cleaned by machines in an "*opening line*" prior to being fed to the card (see *carding*).

Blue bag A severe case of *mastitis* in which the *udder* tissue rots away.

Blueberry An edible, seedless berry native to North America. The United States, Canada, and Europe are the major blueberry producers.

Bluetongue A viral *disease* of *cattle, sheep,* and *goats* that causes respiratory and digestive disorders. The disease, transmitted by insects, is particularly fatal to sheep. Many countries restrict animal *imports* from bluetongue areas.

Boar A mature male hog (see *swine*) that has not been castrated. See *barrow.*

Board for International Food and Agricultural Development (BIFAD) An organization created by Congress to advise officials managing *foreign* assistance programs regarding the use of U.S. *land-grant* and other qualified universities to support agricultural development abroad.

Board of Governors See *Federal Reserve System.*

Bodied oil See *boiled, blown, bodied oils.*

Bog soil *Soil* with mucky or peaty surfaces underlain by *peat.* Bog soils usually have swamp or marsh vegetation and are most common in humid regions.

Boiled, blown, bodied oils Oils (see *fats and oils*) refined by processes used by the paint and varnish industry to enhance the drying qualities of the oils. Boiled oils are made by adding a "drier" (such as oxide of lead, manganese, or cobalt), heating the oil to 125 degrees to 150 degrees centigrade, then cooling it immediately. Air is continually blown through the oil during the heating. Blown oils are prepared in much the same manner as boiled oils except that the air-blowing continues for a longer period and temperatures are kept below 125 degrees centigrade. Bodied oils (originally known as stand-oils) are made by heating the oil to around 300 degrees centigrade for about six hours in the absence of air.

Boiled oils See *boiled, blown, bodied oils.*

Bok choy A vegetable with thick white stalks and dark green leaves. Also called chinese cabbage, bok choy is frequently used in stir-fry and other Chinese dishes.

Boll The *seed* pod of the *cotton* plant.

Bolt The formation of an elongated stem or seedstock, as in *sugar beets.*

Bonded warehouse A warehouse owned by persons approved by the *U.S. Treasury Department* and under bond or guarantee for the strict observance of the revenue laws. A bonded warehouse is used for storing goods until excise taxes and *customs* duties (see *tariff*) are paid or goods are otherwise released. See *foreign trade zone.*

Boned out The process of removing bone from meat or the meat remaining after the bone is removed.

Boners Firms that debone meat; usually from *utility, canner, and cutter* cows (see *cow*) and/or *bulls.* This meat is used mainly for ground beef and sausage products.

Boniato A staple root crop in many Latin and Asian countries that resembles a sweet potato but is drier and not as sweet.

Boning Removing all the bones from the meat *carcass, primal,* or other cut available.

Bonus commodities See *commodity distribution.*

Booking the basis Determining the *price* for a cash sale by applying a specified difference (see *basis*) to a *futures price* to be observed on a later date selected by the seller or the buyer. Similar to *call pricing* for *cotton.*

Border irrigation Irrigation (see *acres irrigated* and *irrigated farms*) in which the water flows over narrow strips that are nearly level and are separated by parallel, low-bordering banks or ridges. See *irrigation methods.*

Border tax adjustment The remission of taxes on exported goods so that national tax systems do not impede exports. The U.S. government makes little use of border tax adjustments because it relies more heavily on income (or direct) taxes than most other governments.

Bovine Pertaining to *cattle.*

Bovine growth hormone (bGH) See *bovine Somatotropin (bST).*

Bovine Somatotropin (bST) A *protein* occurring naturally in *cattle* that controls the amount of *milk* produced. bST has been genetically engineered as a synthetic *hormone* to inject or implant into cows to increase milk *production.* On November 5, 1993, the Food and Drug Administration (FDA) announced approval of the new animal drug SOMETRIBOVE, a recombinant bovine Somatotropin (bST) product for increasing milk production in cows.

bST is also referred to as bovine growth hormone (bGH) and can be used in cattle feed to produce more, and leaner beef. A similar product (porcine somatotropin (pST)) for use in pork production is in the experimental stage.

Boxed beef Beef cut into *primals, subprimals,* or both; vacuum-wrapped and placed in cartons by the *packer.*

Boxed industry The sector of the *meat packing* industry involved with *boxed beef,* including both slaughterers that box their own beef and *processors* that buy *carcasses* to convert to boxed beef.

BPSY (bleachable prime summer yellow) Once-refined (see *once-refined oil*) *cottonseed* oil of bleachable quality that, after *bleaching,* will contain 0.25 percent *free fatty acids* or less, a very small quantity of color bodies, and water and volatile matter not to exceed 0.10 percent.

Brahma A large-boned, heavy *breed* of beef *cattle* originating in Asiatic countries. Brahmas are used primarily for *crossbreeding* with other beef breeds.

Bran The coarse outer covering of the *wheat* or other *grain* kernel (see *kernel*) that is separated from the *endosperm* during the *milling* process.

Branded processed meat operation A plant (see *plant or establishment*) or firm that produces a processed (see *processing*) meat product that it sells using a brand name, such as Armour Star bacon or Smithfield hams.

Brannan plan A *price-support program* proposed by Secretary of Agriculture Charles Franklin

Brannan in 1949. The plan called for *price* supports and *production* payments based on a moving average of prices in 10 of 12 previous years instead of the 1910–1914 basis used for calculating *parity*. Producer eligibility was to be based on compliance with allotments, *quotas,* and conservation provisions. The plan also called for payment limits (see *payment limitations).*

Bread for the World A citizens' movement that lobbies Congress on behalf of the world's poor and hungry. See also *Bread for the World Institute on Hunger and Development.*

Bread for the World Education Fund See *Bread for the World Institute on Hunger and Development.*

Bread for the World Institute on Hunger and Development A nonprofit organization that is separate from, but works with, *Bread for the World.* The Institute conducts policy research, provides educational outreach, and publishes resource materials. The Institute is the successor to the Bread for the World Education Fund, founded in 1975.

Breadfruit A starchy (see *starch*) vegetable used in the Caribbean, many Pacific islands, and Brazil. The fruit is large (often 5 pounds or more) and yellow-green.

Breaker and/or breaking house A firm that breaks (see *breaking (beef)*) dressed beef *carcasses.* These firms normally do not *slaughter.*

Break flour Flour produced as the *bran* is separated from the *endosperm* during the *milling* process. See *break system.*

Breaking (beef) Reducing a beef *carcass* to smaller cuts, such as *primals* or fabricated (see *fabrication*) cuts.

Break system A portion of the milling process in which a series of rotating corrugated rollers separate the majority of wheat kernels' interior *endosperm* from the outer *bran* coat in the *milling* process. See *middlings rolls.*

Breed 1: A group of plants or animals with common ancestry and inherited characteristics that distinguish it from other groups. When plants or animals of the same breed are mated, the offspring have the same characteristics as the parents.

2: The reproduction of plants or animals to obtain offspring with the same or improved characteristics.

Breeding herd *Livestock* used or intended primarily to produce progeny, rather than for direct use for *stock,* feeding, or *slaughter.*

Breeding unit index A measure of a *breeding herd,* including the total number of female animals capable of giving birth, weighted by the *production* per head, in a base period.

Brewer's rice The smallest size of broken *rice* fragments. Brewers' rice is used in making pet foods and as a source of *carbohydrates* in brewing.

Broad-base terrace A low embankment, with such gentle *slopes* that it can be farmed, constructed across sloping *soils* approximately on the contour. Broad-base terraces are used to reduce *runoff* and soil *erosion.*

Broadcasting The random scattering of *seeds* over the surface of the ground. The seeds are then covered with *soil.*

Broadcast seeder See *end-gate seeder.*

Broccoli The undeveloped flower of a plant related to *cabbage.* California accounts for over 90 percent of U.S. fresh *market* output, and Mexico is a major *exporter.*

Broiler A young chicken (see *poultry*) produced for meat. A broiler is 8 to 12 weeks old, usually weighs 4 to 6 pounds *liveweight,* and 3 to 4.5 pounds dressed weight. The terms broiler, fryer, and young chicken are often used interchangeably.

Broiler complex The total or overall organization, facilities, and personnel required to perform all functions of *production,* processing, and marketing (see *agricultural marketing*) to the wholesale (see *wholesaling*) level. A *broiler* complex is usually vertically integrated (see *vertical integration*) through combined ownership and contractual (see *contract*) agreements.

Broken corn and foreign material (BCFM) Any material passing through a 12/64-inch sieve, plus non-*corn* material remaining on top.

Broken kernel A *kernel* separated into two or more pieces, exclusive of insect boring or surface *contamination.*

Brokens *Kernels* of *rice* that are less than three-fourths of the length of the whole kernels. Brokens are used in beer, processed (see *processing*) foods, and pet foods.

Broker An agent who buys or sells products, assets (see *assets, current* and *assets, fixed*), or services for clients, usually on a commission basis. For example, a broker might buy and sell *futures contracts* for a customer. A broker serves as an intermediary but does not take ownership.

***Bromus secalinus* (cheat)** Any of several grasses, especially the common chess. This weed is a major problem for *winter wheat* producers in the central Plains.

Brood *Honeybees* in the *egg,* larval (see *larva*), and *pupal* stages.

Brood chamber The section of a *beehive* in which the *brood* is reared and food is stored for the *colony's* survival.

Brood cows See *beef cows.*

Brooder house A structure equipped with heating units (brooders) in which day-old chicks or *poults* are started and kept for the first six weeks. For *broilers,* brooding generally occurs in an area of the *growout house.*

Brooding Using special, controlled temperature conditions to raise young chickens (see *poultry*) or turkeys (see *poultry*) during the first few weeks of life.

Brown rice Whole or broken *kernels* of *rice* from which only the *hull* has been removed. Brown rice may be eaten as is or may be milled into regular-milled white rice. Cooked brown rice has a slightly chewy texture and a nutty flavor. The light brown color is caused by the presence of seven *bran* layers, which are very rich in minerals and *vitamins*—especially the B-complex group.

Brucellosis A bacterial (see *bacteria*) *disease* of *cattle, swine, sheep,* and *goats* that reduces breeding and *milk* production. The disease can be transmitted to humans as undulant fever through direct contact with infected animals or through unpasteurized (see *pasteurization*) milk products. Infected animals must be slaughtered (see *slaughter*). In the United States, *vaccines* are used to help prevent this highly contagious disease. Trade in live animals from affected areas requires testing of animals prior to *export* and certification concerning herd health status.

Brussel sprout A small, edible head produced by a member of the *cabbage* family. Most of the U.S. crop is grown in California.

Brussels Tariff Nomenclature (BTN) The widely used international *tariff* classification system that preceded the *Customs Cooperation Council Nomenclature (CCCN)*. The CCCN was replaced in 1988 by the *Harmonized Commodity Description and Coding System.*

B.t. A *protein* produced by the bacterium *Bacillus thuringiensis* (see *bacteria*). B.t. is toxic to certain caterpillars. B.t. is routinely sprayed around homes or commercial areas to control insect pests.

Buckwheat An *herb* yielding *seeds* used for fodder and bread and pancake *flour*. The United States is the major producer of buckwheat.

Bud An unexpanded flower or rudimentary leaf, stem, or branch.

Buffalo Several types of wild oxen, the most common being bison (American or European) and water buffalo (Asian and African).

Buffer A substance in a solution or *soil* that makes the degree of acidity resistant to change when an acid or base is added.

Buffer stock Reserve *commodities* held by a government or an international association to ensure adequate supplies and/or stabilize *prices.*

Buffer strip An area of *erosion*-resisting vegetation designed to reduce *runoff* and *soil* erosion. Buffer strips are usually planted on the contour in cultivated (see *cultivate*) fields or along their boundaries. For the purposes of the *Conservation Reserve Program,* a buffer strip is an area of vegetation 66- to 99-feet wide placed along a river, stream, lake, or other body of water to prevent delivery of nonpoint source sediment (see *nonpoint-source pollution*), *nutrients,* and so on. Also known as filter strips.

Bulb A stem or root, including corms, rhizomes, and *tubers,* containing stored food and the undeveloped shoots of a new plant. The bulb is planted underground and can yield *annual* and *perennial* varieties of plant, as well as flowering and nonflowering types (to be used for forcing a plant or flower). Examples are tulips, daylilies, crocus, daffodils, dahlias, cannas, and amaryllis.

Bulgur *Wheat* that has been parboiled, dried, and partially debranned (see *bran*) for later use in either cracked or whole *grain* form.

Bulk cargo Loose cargo or freight in large amounts, numbers, weight, or volume. Examples include oil, *grain,* or ore that is not packaged, baled (see *bale*), bottled, or otherwise packed.

Bulk carrier An ocean-going, single-deck vessel of more than 10,000 tons deadweight. Applies to dry-bulk carriers and liquid-bulk carriers. Also known as tankers.

Bulk curing A *curing* process for *flue-cured tobacco*. Leaf is suspended in the curing atmosphere in bulk (loose armfuls are held in place by racks). Humidity and temperature are controlled by forcing heated air vertically through the tightly packed leaves in a completely closed system.

Bulk milk tank An enclosed tank, cooled by a refrigeration unit, used for storing raw *milk.*

Bulk products Agricultural *commodities* for *export* that are not processed (see *processing*) and/or have a relatively low per-unit value. Examples include raw *grains* and *oilseeds.* See also *high-value products (HVPS).*

Bull 1: A noncastrated male *bovine* animal used for breeding (see *breed*).
2: A person who buys *commodities* or other assets (see *assets, current* and *assets, fixed*) in anticipation of a *price* rise. See also *bear.*

Bull table See *squeeze chute.*

Bulldozer A *tractor* with a heavy steel blade used on a *farm* to move dirt, brush, trash, and similar materials.

Bunker silo See *horizontal silo.*

Bunk feeder system A mechanical configuration of *elevators, augers, bins,* and mixers used to mix, meter, and deliver *feed* to *livestock.*

Burdock A long, thin, brown root used in Japanese cooking.

Bureau of Agricultural Economics A *U.S. Department of Agriculture* (USDA) agency that was the predecessor of the *Economic Research Service.*

Bureau of Labor Statistics (BLS) An agency of the *U.S. Department of Labor* that collects, processes, analyzes, and disseminates data relating to employment, unemployment, and other characteristics of the labor force; *prices* and *consumer* expenditures; wages, other worker compensation, and industrial relations; productivity and technical change; economic growth and employment projections; and occupational safety and health.

Bureau of Reclamation An agency of the *U.S. Department of Interior* that is responsible for building dams and canals and providing water to local water districts. These districts, in turn, sell the water to agricultural *producers.*

Bureau of the Census An agency of the *U.S. Department of Commerce* that collects, tabulates, and publishes a wide variety of statistical data about the people and the economy of the United States. The bureau conducts censuses of population, housing, *agriculture,* state and local governments, manufacturers, mineral industries, distributive trades, construction industries, and transportation. The bureau also compiles current statistics on U.S. *foreign* trade, including data on *imports, exports,* and shipping.

Burley tobacco The major type of *air-cured tobacco.* Burley is light in body and neutral in flavor, with a low sugar content and high alkaloid content. It is used chiefly in *cigarettes.* The United States, Brazil, Italy, China, Malawi, and Spain are the major producers of burley tobacco.

Burned lime See *lime, agricultural.*

Bushel A unit of dry measure containing 2,150.42 cubic inches and equivalent to 8 gallons of liquid. The weight of a bushel varies with the density and bulk of the *commodity.* For example, a bushel of *oats* weighs 32 pounds; *barley,* 46 pounds; and *corn,* 56 pounds.

Bush or forest fallow *Land* left idle under woody vegetation between crops.

Butter A solid emulsion of *milkfat,* air, and water made by churning *milk* or cream. Major producers include the former Soviet Union, India, France, and the United States.

Butterfat See *milkfat.*

Buttermilk The product left after *butter* is extracted from cream. Cultured buttermilk is skimmed or partially skimmed milk (see *milkfat content*) that has been cultured with lactic acid *bacteria,* packaged, and sold to *consumers.*

Butteroil Melted and clarified *butter* from which most of the water has been removed. Also known as anhydrous milk fat. Major producers include the *European Union,* India, the former Soviet Union, and New Zealand.

Buy-back A *countertrade* whereby the seller of a product, such as machinery, equipment, or technology used as input in the *production* of another product, agrees to accept the resulting products as full or partial payment.

Buying broker An agent who arranges terms of purchase on behalf of a buyer (wholesaler (see *wholesaling*), retailer, institution, or *foodservice* establishment) for a fee paid by the buyer.

Bypass protein The portion of *protein* that remains undegraded by the *rumen* microbes, as opposed to *fermentable protein.*

Byproduct A secondary product resulting from the *processing* of a primary product. For example, byproducts from the *slaughter* of *livestock* include blood and bone *meal* used as *fertilizers* and *leather* for shoes.

C

Cabbage One of the oldest vegetables known. *Varieties* include head (green, red, and Savoy) and nappa (also called celery or Chinese cabbage). The United States, Mexico, and Canada are the major cabbage producers.

Cacao A tropical tree whose dried (see *drying*), partly fermented (see *fermentation*), fatty *seeds* are used to make chocolate. Major producers are Cote d'Ivoire, Brazil, Ghana, Malaysia, Indonesia, Nigeria, and Cameroon. The United States is among the top five cocoa bean *importers.*

Cactus pads The large round pads of the *prickly pear* cactus used in Mexican cooking.

Cactus pear See *prickly pear.*

Cairns Group A group formed in 1986 at Cairns, Australia. The group seeks the removal of *trade barriers* and substantial reductions in *subsidies* affecting agricultural trade in response to depressed *commodity* prices (see *price*) and reduced *export* earnings stemming from subsidy wars between the United States and the *European Union.* The members account for a significant portion of the world's agricultural exports. The group includes major food *exporters* from both developed and *developing countries*: Argentina, Australia, Brazil, Canada, Chile, Colombia, Hungary, Indonesia, Malaysia, New Zealand, the Philippines, Thailand, and Uruguay.

Calabaza A large, round, *pear*-shaped *squash* that resembles a pumpkin.

Calf The offspring of a *cow* until it is a year old or sexually mature.

Calf loss rate The ratio of the number of calves (see *calf*) lost because of death, theft, and other causes during a calendar year per 100 calves born alive during that calendar year.

Calf table See *squeeze chute.*

Call option The right, without obligation, to buy a *futures contract* at a specified *price* during a specified time period.

Call pricing Agreeing to determine the price for a cash *cotton* sale by applying a specified difference (see *basis*) to a futures price to be observed on a future date selected by the buyer (buyer's call) or by the seller (seller's call). Call pricing is essentially the same as *booking the basis.*

Callus A hard or thickened layer at the base of unorganized tissue and formed from organized plant tissue.

Calving Giving birth to a *calf.*

Calving rate The ratio of the number of calves (see *calf*) born alive during the production year per 100 sexually mature female *cattle* (*cows* plus breeding-age *heifers*) exposed to a *bull* or artificially inseminated (see *artificial insemination*) approximately nine months earlier.

Camplobacteriosis A *disease* of humans frequently associated with consumption of undercooked or improperly handled *poultry. Symptoms* range from a day or two of mild diarrhea and vomiting, to hospitalization for dehydration and diarrhea, blood poisoning, and sometimes death. The severity of the symptoms depends, in part, on how many *bacteria* are consumed and how well the body fights the bacteria. See also *salmonellosis.*

Canadian Wheat Board (CWB) The largest *marketing board* in Canada in terms of sale values, and the sole authority for interprovincial and international trade of Canadian *wheat* and *barley* (*oats* were dropped from the CWB in 1989). CWB activities include: setting *prices* to *producers*, regulating producer deliveries through *quotas*, and organizing *grain* handling and transportation. The CWB is mandated by the Canadian Parliament to achieve three main objectives through its operations: (1) to *market* as much grain as possible at the best price that can be obtained; (2) to provide price stability to grain producers; and (3) to ensure that each producer obtains an equitable share of the available grain market.

Candling Holding or passing an *egg* before a bright light to check the shell and interior contents for quality and defects.

Canola A *variety* of *rapeseed* that yields oil (see *fats and oils*) and *meal.* Canola meal is a high-quality feedstuff (see *feed*) with a *protein* content of 37–38 percent. Canola oil contains the lowest level of *saturated fat* and the highest level of *unsaturated fat* of vegetable oils. Canola is the fastest-growing *oilseed* crop in the world because it can be produced in areas of the world not suitable for *soybeans.* Kentucky, Illinois, Indiana, Ohio, and Tennessee are the major U.S. producing states.

Cantaloupe A *muskmelon. Varieties* include Topscore, Ambrosia, and Saticoy. The United States, Mexico, and Honduras are the major producers. California produces 65 percent of the fresh *market,* followed by Texas and Arizona. The United States is a *net exporter* of cantaloupes.

CAP See *Common Agricultural Policy.*

Capacity of feedlot The maximum number of *cattle* that can be fed in a *feedlot* at one time.

Capacity of housing The maximum recommended number of animals to be kept in a specified type of housing at one time, commonly reported in numbers per unit of area.

Capital account Part of a nation's *balance of payments* that includes purchases and sales of assets (see *assets, current* and *assets, fixed*), such as stocks, bonds, and *land.* A nation has a capital account surplus when receipts from asset sales exceed payments for the country's purchases of *foreign* assets. The sum of the capital and *current accounts* is the overall balance of payments.

Capital account surplus See *capital account.*

Capital budgeting Decisions made by individuals and firms regarding how much and where *resources* will be obtained and spent, setting standards for project acceptability, evaluating individual projects, and determining the source of capital to be used.

Capital expenditures As used in the *farm income* accounts, it is the value of farmers' medium-term gross investments in durable *resources*, such as *tractors* and buildings. This investment capital includes tractors, trucks (see *farm truck*), and other machinery; buildings; and *land* improvements. It does not include investment in land.

Capital replacement The value of machinery, equipment, and breeding stock used up during the year, plus the cost of returning these items to the same levels of quality and quantity at which they began the period.

Capon A *cockerel* chicken (see *poultry*) that has been castrated to encourage larger mature size, with more tender meat with a higher fat (see *fats and oils*) content. A capon is usually 7–9 months old with a *dressed weight* near 10 pounds.

Capper-Volstead Act A federal law passed in 1922 that permits farmers to act together in associations to collectively *market* their products, which in the absence of such enabling provision could result in antitrust actions against them. The act assigns to the secretary of agriculture responsibility for determining if such associations unduly enhance *price* through monopolization (see *monopoly*) or restraint of trade.

Capsaicin The major active ingredient in red pepper and used in prescription medications for such things as arthritis and herpes zoster (shingles).

Captive bakery A bakery owned by a supermarket (see *foodstore*) *chain*.

Carambola A small, yellow, oval-shaped fruit also known as starfruit because it is star-shaped in cross section. The fruit is native to the Malay Archipelago but is grown commercially in southern Florida.

Carbohydrate Various neutral compounds containing *carbon*, hydrogen, and oxygen, most of which are formed by green plants. Examples include *sugar*, *starch*, and *cellulose*.

Carbohydrate-based fat substitute A *fat substitute* that is sometimes mixed with water to replace part of the fat (see *fats and oils*) in salad dressing, *margarine*, and frozen desserts. Examples include N-OIL, a tapioca dextrin, and maltodextrins made from corn *starch*.

Carbon One of the most common chemical elements, occurring in lampblack, coal, and coke in varying degrees of purity. Compounds of carbon are the chief constituents of living tissue.

Carbon dioxide A gaseous compound that is formed when *carbon* combines with oxygen.

Carbon disulfide See *black liquor*.

Carbon-nitrogen ratio The ratio of the weight of organic *carbon* to the weight of total *nitrogen* in a *soil* or in an organic material.

Carcass The body of a slaughtered (see *slaughter*) meat animal with the *hide or skin*, feet, head, tail, and most internal organs removed. The definition varies slightly for each *species*. For example, pork *carcasses* include the feet.

Carcass cut-out The percentage of retail cuts that can be sold from a *carcass*. In other words, it is the amount of meat versus fat (see *fats and oils*), bone, and waste.

Carcass proportion The relative amounts of each cut (*primal, subprimal*, or retail cut) obtained from a beef *carcass*.

Carcass trim (Y4 to Y3) The practice of upgrading (see *grading*) a *carcass* hindquarter from Yield Grade 4 to Yield Grade 3 by partial trimming of internal fat (see *fats and oils*).

Carcass weight, dressed For *cattle*, *sheep*, and *goats*, the animal's weight after *slaughter* and removal of the *hide or skin*, feet, head, tail, most internal organs, genito-urinary organs (except kidneys), and slaughter fats (see *fats and oils*). For hogs (see *swine*), part removal includes internal organs, genito-urinary organs (including kidneys), and slaughter fats.

Carcass weight equivalent (CWE) The weight of meat cuts and meat products converted to an equivalent weight of a dressed *carcass* (see *dressed animal*). Includes bone, fat (see *fats and oils*), tendons, ligaments, and inedible trimmings.

Carcinogen A substance, either natural or man-made, that causes cancer when ingested, inhaled, or otherwise introduced into the body.

Carding A process in *yarn* manufacturing in which *fibers* are separated from tufts, allowing cleaning and blending, and their partial alignment in the form of *sliver*.

Cardoon A thistlelike relative of the *artichoke* mostly used in southern Europe. Cardoon resembles an extra-large *celery* head.

Cargo Preference Act A U.S. law that requires a certain portion of goods or *commodities* financed by the U.S. Government to be shipped on U.S. flag ships. The law specifically states "whenever the United States contracts for, or otherwise obtains for its own account, or furnishes to or for the account of any foreign nation without provision for reimbursement, any equipment, materials, or commodities," the United States shall ship in U.S. flag vessels, to the extent that they are available at fair and reasonable rates, at least 50 percent of the gross tonnage involved. The law has traditionally applied to *Public Law 480* and other *concessional* financing or donation programs.

Cargo ton See *measurement ton*.

Caribbean Basin Economic Recovery Act (CBERA) **(P.L. 98-67)** More popularly known as the Caribbean Basin Initiative (CBI), this 1983 act eliminates duties (see *tariff*) on *imports* of products from designated Caribbean countries until September 30, 1995. The CBI also provides for import relief to U.S. industries injured or threatened by increased imports from CBI countries.

Caribbean Basin Initiative (CBI) See *Caribbean Basin Economic Recovery Act*.

Caribbean Community and Common Market (CARICOM) The successor to the Caribbean Free Trade Association (CARIFTA). CARICOM was organized by the Treaty of Chaguaramas in 1973 by the British Commonwealth Nations and Territories in the Caribbean Basin to promote political cooperation and integration, *free trade*, and economic development among its members. The members are Antigua and Barbuda, The Bahamas, Barbados, Belize, Dominica, Grenada, Guyana, Jamaica, Montserrat, Saint Kitts and Nevis, Saint Lucia, Saint Vincent and the Grenadines, Trinidad, and Tobago. Observers include Anguilla, Bermuda, British Virgin Islands, Dominican Republic, Haiti, Netherlands Antilles, and Suriname.

Caribbean Development Bank (CDB) An organization established in 1969 to promote economic development and cooperation. Regional members include Anguilla, Antigua and Barbuda, The

Bahamas, Barbados, Belize, British Virgin Islands, Cayman Islands, Columbia, Dominica, Grenada, Guyana, Jamaica, Mexico, Montserrat, Saint Kitts and Nevis, Saint Lucia, Saint Vincent and the Grenadines, Trinidad and Tobago, Turks and Caicos Islands, and Venezuela. Canada, France, and the United Kingdom are nonregional members.

Caribbean Free Trade Association (CARIFTA) See *Caribbean Community and Common Market (CARICOM).*

Carotene A fat-soluble (see *fats and oils*) pigment widely distributed in nature and synthesized in plants. Carotene is the precursor of Vitamin A (see *vitamin*).

Carrier 1: An edible material to which ingredients are added to facilitate uniform incorporation of the ingredients into *feeds*. The active particles are absorbed, impregnated, or coated into or onto the edible material.

2: A person or animal that harbors and possibly transmits an infectious *agent*, such as a *pathogen* or *parasite.* A carrier often does not display the symptoms of the *disease* the agent causes.

Carrot An orange-gold root vegetable high in Vitamin A (see *vitamin).* Major producers include the United States, Canada, and Mexico.

Carrying capacity The number of animals that a *pasture* can adequately feed for a certain period.

Carry-out Prepared food purchases bought but not consumed on the premises. Carryout foods can include purchases from drive-through lanes at *fast food restaurants*, hot sandwiches purchased at convenience stores, and salads bought at supermarket (see *foodstore*) salad bars and service delicatessens.

Carryover Existing supplies of a *farm* commodity at the beginning of a new *harvest* for a *commodity* (end of a designated *marketing year*). It is the remaining *stock* carried over into the next year.

Cartel An alliance or arrangement of independent sellers in the same field of business organized to function as a *monopoly* with respect to *production* or marketing (see *agricultural marketing*) of the *commodity.*

Carter mill A machine with a series of *disks* with different sized notches, pockets, or indentations used to separate weed *seeds*, *chaff*, dirt, and other debris from *wheat, barley*, *oats*, and other *grains.*

Casein A *milk* product composed of the major portion of milk *protein.* It is manufactured from skim milk and is usually marketed in dry form. Food grade casein is used in processed (see *processing*) foods, and industrial-*grade* casein is used in making glue, paint, and plastics.

Case of eggs A container of 30-dozen shell *eggs*, packed in cartons or 2 1/2 dozen cardboard flats.

Cash and carry wholesaler A *wholesaler* with a warehouse where the retail customers buy orders and transport merchandise back to their stores. Cash-and-carry wholesalers usually stock a limited line of fast-moving products, have no outside sales staff, and offer no additional services. Many regular wholesalers operate cash-and-carry departments as part of their regular wholesaling activity. See also *wholesale club store.*

Cash basis See *basis.*

Cash commodity A *commodity* traded outside the rules of a *commodity futures exchange.*

Cash contract An agreement negotiated between a buyer and seller outside the rules of a *commodity futures exchange* that defines rights and obligations to deliver, exchange ownership, and pay for a product or service on one or more future dates. Such agreements normally specify quantity, quality, time, and place of delivery, and *price* or formula for determining price (see *formula pricing*).

Cash crop Any crop that is sold off the *farm* for ready cash.

Cash expenses Short-run, out-of-pocket variable (see *variable costs*) and fixed *production* costs (see *fixed costs*). Cash expenses are equivalent to the minimum break-even crop or *livestock* values needed to maintain an average *acre* or livestock unit in production.

Cash Export Certificate Program See *Wheat and Feed Grain Export Certificate Programs.*

Cash forward contract See *cash contract.*

Cash grain farm A *farm* on which *corn*, *sorghum*, *oats*, *barley*, other small *grains*, *soybeans*, or field *beans* and *peas* account for at least 50 percent of the value of the products sold.

Cash grains *Grains* commonly produced for sale rather than for animal *feed* or *seed*. Cash grains include *barley, corn, oats, sorghum, soybeans, wheat,* and field *beans* and *peas*.

Cashing out Providing food assistance in cash rather than in *commodities* or food stamps (see *Food Stamp Program*). Eligible residents of Puerto Rico have received cash instead of food stamps since July 1982 under the *Nutrition Assistance Program.*

Cash market Trading that is not under the rules of a *commodity futures exchange.* Such trading may be for *spot* or *deferred delivery.*

Cash-out option for generic certificates The option provided to original holders of *generic commodity certificates* to redeem the certificate at its face value for cash from the *Commodity Credit Corporation* (CCC) instead of exchanging it for *commodities*. To encourage exchange of certificates for *surplus* commodities, generic certificates cannot be redeemed for cash until five months after the issue date. Those who purchase or trade the certificates from original holders are not permitted to cash-out the certificates. Certificates issued under the *Export Enhancement Program* cannot be cashed out.

Cash receipts The gross income earned by farmers through marketing (see *agricultural marketing*) crop and *livestock* commodities, plus the receipts from placing eligible *commodities* under *nonrecourse loan.* Cash receipts include sales from such commodities as *corn, soybeans, cattle,* and *milk.*

Cash settlement Final settlement of a *futures contract* by a payment from the seller to the buyer of an amount equal to the value of the *commodity* in the *cash market.* This method is used in lieu of physical delivery to settle maturing *feeder cattle* futures contracts at the Chicago Mercantile Exchange.

Castormeal A toxic material used as a *mulch* in specialized applications. See also *castoroil* and *castorseed.*

Castoroil A viscous fatty oil (see *fats and oils*) used for industrial purposes, including cosmetics, waxes, nylons, plastics, coatings, and lubricants. See also *castormeal* and *castorseed.*

Castorseed The poisonous *seed* of the castor plant used to produce *castoroil* and *castormeal.* Brazil, India, and China rank as the major castorseed producers.

Catalo The *hybrid* animal produced by breeding American Bison (see *buffalo*) with domestic *cattle.* See also *beefalo.*

Catalyst An *agent* (such as an *enzyme* or a metallic compound) that initiates or facilitates a chemical reaction but is not itself changed during the reaction.

Catch crop A crop seeded in rotation either with or between the growing periods of regular crops to increase *production* and potential income.

Catching frame A self-propelled, plastic- or rubber-covered metal frame that unfolds around the base of a tree to catch fruit or nuts shaken off during harvesting (see *harvest*).

Catfish Farmers of America A trade association of U.S. catfish and trout farmers.

Cat skinner An experienced *tractor* operator.

Cattle Domesticated *bovine* animals raised for beef and dairy. Major producers are India, the former Soviet Union, the United States, Brazil, and Argentina.

Cattle and calves on feed Animals for the *slaughter* market (see *market*) fed on a *ration* of *grain* or other *concentrates* that are expected to produce a *carcass* graded Select or better.

Cattle cycle The repetitive pattern of *cattle* production (see *production*) and *price* movements. A period of approximately 10 years in which the number of beef cattle is alternatively expanded and reduced for several consecutive years in response to perceived changes in the profitability of beef products. When prices (and profits) are cyclically low, for example, cattlemen reduce their herds, and they increase cattle numbers when prices are favorable.

Cattle disposition *Cattle* sale, or *feedlot* placement without ownership transfer.

Cattle feeder A wood, metal, or plastic container to hold *feed* or *hay* for feeding *livestock.*

Cattle feeding The feeding of *grain* and other *concentrate* feedstuffs (see *feed*) to produce *slaughter* cattle (see *cattle*) grading (see *grade* and *grading*) Good or better.

Cauliflower The undeveloped flower of a plant related to *cabbage.* The United States, Mexico, and Guatemala are the major producers. California is the single largest supplier in the U.S. fresh *market.* New York is the second major U.S. growing area.

CCC commercial credit A general term for the short- and intermediate-term commercial *credit guarantee* programs operated by the *Commodity Credit Corporation (CCC)*; GSM-102 guarantees repayment of private, short-term credit (up to three years); GSM-103 is an intermediate-term program that covers credit extended for 3 to 10 years.

Celeriac See *celery root.*

Celery A member of the *carrot* family cultivated for its edible stalk. The United States, Canada, and Mexico are the major producers.

Celery knob See *celery root.*

Celery root The brown, knobby root of the wild *celery* plant. The root has creamy, white flesh with a mild flavor. Also known as celeriac or celery knob.

Cell The smallest structural unit of a living organism able to grow and reproduce independently.

Cell culture The growth of *cells* under laboratory conditions.

Cellulose A chemically complex *carbohydrate* making up most of the structural parts of a plant, such as the *cell* walls. *Cotton* fibers (see *fiber*), for example, are almost pure cellulose. Paper is mainly cellulose separated by chemical processes from wood or other plant remains.

Cellulosic fibers All *fiber* of plant or vegetable origin. These fibers include *natural fibers* such as *cotton, linen,* and *jute,* and *man-made fibers* of wood *pulp* origin, such as rayon and acetate.

Census of Agriculture A survey taken by the *Bureau of the Census* every five years (in years ending in 2 and 7) to determine the number of *farms,* farm acreage (see *acre*), crop acreage and *production, livestock* numbers and production, production expenses, farm facilities and equipment, farm tenure, value of farm products sold, farm size, type of farm, and so forth. Data are reported for states and counties.

Center for Science in the Public Interest (CSPI) A nonprofit nutrition advocacy organization founded in 1971 by three scientists, Albert Fritsch, James Sullivan, and Michael Jacobson, to conduct research and public education programs on food safety, nutrition, and alcohol abuse.

Center pivot irrigation A large, mechanical device for crop irrigation (see *acres irrigated and irrigated farms*) that pumps *groundwater* from a well through a long pipe elevated on wheels and driven electrically or by water pressure. The pipe pivots around a central point and irrigates a field in a large circular pattern.

Central African Customs and Economic Union An organization of six Central African nations established in 1964 to promote the establishment of a Central African *Common Market.* Member nations are Cameroon, Central African Republic, Chad, Congo, Equatorial Guinea, and Gabon. Also called Union Douaniére et Economique de l'Afrique Centrale (UDEAC).

Central African States Development Bank A financial organization with nine member countries established in 1975 to provide loans for economic development. The members are Cameroon, Central African Republic, Chad, Congo, Equatorial Guinea, France, Gabon, Germany, and Kuwait. Also called Banque de Developpement des Etats de l'Afrique Centrale (BDEAC).

Central American Common Market (CACM) A regional organization formed in 1960 to promote economic development through a *customs union* and industrial integration in the member nations of Costa Rica, El Salvador, Guatemala, Honduras, Nicaragua, and Panama. See also *common market.*

Central breaking Cutting *carcasses* to *primals* or *subprimals,* either by a firm for a retailer or at the retailer's central warehouse (see *chain warehouse*).

Central cutting Cutting meat to retail cuts and packaging before delivery to local *foodstores.* Central cutting can be done by various firms and/or at a retail *chain's warehouse.*

Central farrowing house A nonportable farrowing house with individual spaces for females. See *farrow*.

Centrally-planned economy An economic system in which state directives rather than *market* forces determine the *production*, pricing (see *price)*, and distribution of goods and services. Also referred to as a "nonmarket economy." China and most other communist nations are examples of centrally-planned economies.

Cereals 1: The generic name for grasses that produce edible *seeds*. The term is also used for certain products made from the seeds. Cereals include *wheat*, *oats*, *barley*, *rye*, *rice*, *millet*, *corn*, and *sorghum*.
2: A food prepared from *grain* and traditionally eaten at breakfast.

Certificate of origin A document showing the country producing a shipment. A certificate of origin is required by *customs* officials of importing (see *import*) countries that apply different *tariffs* depending on the country of origin. Some countries, for example, may be eligible for preferential tariff treatment.

Certified ready-to-cook A measure used in reporting federally inspected (see *inspection, federal*) *slaughter*. It is the weight of meat certified wholesome by inspection after *postmortem condemnation* pounds are removed. See also *meat and poultry inspection* and *inspection, meat*.

Certified seed The progeny of foundation or registered *seed* that has been handled to maintain satisfactory genetic purity and that has been approved and certified by the certifying agency.

Certs See *generic certificates*.

Chaff *Glumes*, *husks*, and other *seed* coverings, together with other plant parts, that are separated from seeds during *threshing* or *processing*.

Chain A food retailer, *foodservice* operator, or other company owning 11 or more stores or outlets. See also *independent grocer* and *foodstore*.

Chain warehouse A central plant used by a *chain* retail firm to assemble, store, and distribute a product to local stores.

Chard A type of *beet* that develops lush leaves rather than a fleshy root.

Chayote A pale green, tropical, summer *squash* sometimes referred to as vegetable *pear*. Chayote has a deeply ridged surface and a single flat *seed* in the center.

Check irrigation See *irrigation methods*. Also known as border-strip or basin irrigation.

Checks *Eggs* with cracked shells but with shell membranes and contents intact.

Check-off programs See *assessment programs*.

Cheese A dairy product made by separating milk *curds* from *whey*. Cheese is usually made from cow's *milk*, but it is also produced from the milk of *sheep*, *goats*, water buffalo, camel, and other mammals. The United States, the former Soviet Union, France, Italy, and Germany are the major cheese producers.

Chemigation The application of chemicals, such as *fertilizers*, *herbicides*, *insecticides*, *fungicides*, and *nematocides* to *soils* and plants through irrigation systems (see *irrigation methods*).

Cherimoya A heart-shaped, pale green fruit native to the Andes, but also grown along the California coast. The flavor is similar to a *pineapple* crossed with a *banana*. Also known as starfruit.

Cherry A red to purplish red, heart-shaped fruit that can be sweet or sour. Sweet cherry *varieties* include the dark red- to purple-skinned Bing, Lambert, Black Tartarian, Chapman, and Burlat and the yellow-skinned Royal Ann and Rainier. Sour cherry varieties are Montgomery (the leading variety) and Early Richmond. The United States leads the world in cherry production.

Chess See *Bromus Secalinus (cheat)*.

Chewing tobacco A product made from *tobacco* leaf. Almost all the stems and some of the coarser *fibers* are removed before *processing*. The product consists of irregular fragments or flakes of tobacco leaf, about 1/4 to 1 inch in diameter, and sold in small packages. Three types of chewing tobacco are produced in the United States. These include: (1) **plug**—the leaf is pressed into flat cakes after the stems have been removed; (2) **twist**—the leaf is stemmed and twisted into small rolls; and (3) **loose leaf**—made almost entirely from *cigar*-leaf tobacco.

Chicago Board of Trade The United States' largest futures market (see *commodity futures exchange*) for agricultural *commodities*. Members include merchants, *processors*, and *brokers* engaged in futures (see *futures contract*) trading.

Chicago Rice and Cotton Exchange See *cotton exchange*.

Chicken See *poultry*.

Chick pea See *garbanzo bean*.

Child and Adult Care Food Program (CACFP) A program that provides cash reimbursements and USDA-donated foods or cash equivalents to non-residential child and adult day care facilities based on the number of breakfasts, lunches, suppers, and snacks served.

Child nutrition programs See *National School Lunch Program*, *School Breakfast Program*, *Special Milk Program*, *Child and Adult Care Food Program*, and *Summer Food Service Program*.

Chinese cabbage See *bok choy*.

Chinese gooseberry See *kiwifruit*.

Chinese pea pod See *snow pea*.

Chisel A *soil*-penetrating point that can be drawn through the soil to loosen the subsoil, usually to a depth of 12 to 18 inches.

Chisel plow *Tillage equipment* with metal points pulled behind a *tractor* for tilling (see *till*) and breaking up *soil*.

Chisel point An implement used on *tillage equipment* to *cultivate* or break up the *soil*, kill weeds, conserve moisture, and prepare the ground for seeding (see *seed*). Also known as a spike.

Chitin An animal *byproduct* used to treat sewage effluent and remove metals from waste water. The most common source is the tough outer shell of shellfish.

Chlorophyll The light-absorbing pigment responsible for the green color of plants. Chlorophyll is important in *photosynthesis* in plants.

Chloroplasts Small bodies in plant *cells* where the green pigment *chlorophyll* is concentrated.

Chlorosis A condition in plants resulting from the failure of *chlorophyll* to develop, usually because of a deficiency of an essential *nutrient*. With chlorosis normally green tissues turn yellow.

Cholesterol A component of body *cells* of humans and animals. Cholesterol is needed to form *hormones*, cell membranes, and other body substances. Cholesterol is present in all animal products—meat, *poultry*, fish, *milk* and milk products, and *egg* yolks. Foods of plant origin contain no cholesterol.

Chopper, flail A rotating "barrel" of individual knives that cut *forage* and other crops into fine material.

Chopper, rotary A horizontal rotating blade used to finely chop *forage*, weeds, brush, limbs, and similar material.

Chromosome A thread-like structure contained in the nucleus of a *cell* that carries *genes* conveying hereditary characteristics.

Chronic The gradual onset of symptoms of a disease or the prolonged or continual existence of symptoms that are not necessarily progressive. See also *acute*.

Chute A stationary or portable structure that restricts animals to single-file passage. See *squeeze chute*.

Cigar A *tobacco* product for smoking that consists of *filler tobacco*, *binder tobacco*, and *wrapper tobacco*.

Cigar classes of tobaccos These include *filler tobacco*, *binder tobacco*, and *wrapper tobacco* classified according to their traditional use in *cigars*.

Cigarette The primary product made from *tobacco*. U.S. cigarettes are a blend of *flue-cured*, *burley*, *Maryland*, and Oriental tobaccos.

Cilantro An *herb* widely used for flavoring and garnishing Asian and Latin dishes. See also *coriander*.

Citrus canker A bacterial (see *bacteria*) *disease* of citrus plants that can cause serious damage to trees, including leaf, twig, and fruit spotting, defoliation, and fruit loss. Movement of *citrus fruit* from affected areas is restricted.

Citrus fruit Primarily *oranges*, *grapefruit*, *lemons*, *limes*, *tangelos*, and *tangerines*.

Classification of cattle for type The type scoring of individual animals in a herd by an experienced classifier using a rating system that combines all the characteristics contributing to the animal's value.

Classified pricing The *federal milk marketing order* system under which regulated *processors* pay into the pool for *Grade A milk* according to the class (see *classes of milk*) in which it is used.

Class I differential The amount added to the value of *milk* for manufacturing uses to obtain the *Class I minimum price*. Presently, in most *federal milk marketing orders* the value for manufacturing use is the average *price* paid for *manufacturing grade milk* by plants (see *plant or establishment*) in Minnesota and Wisconsin (see *Minnesota-Wisconsin [M-W] price*). The Class I differentials, which vary by market order, are established at levels that, in conjunction with the dairy *price support program*, ensure present and future supplies of high-quality milk for use in fluid dairy products (see *fluid milk products*) throughout the federal order system.

Class I effective price The total Class I *price* consisting of two parts: the minimum federal order Class I price (see *Class I minimum price*) and, if any, the *over-order payments*. (Note: the over-order payments are not part of a *federal milk marketing order*.)

Class I minimum price The lowest *price* that is paid by a *handler* for producer *milk* used in Class I (see *classes of milk*) *fluid milk products* under a *federal milk marketing order*. Except in the New York-New Jersey order, prices are f.o.b. the plant (see *plant or establishment*) of first receipt; that is, at the plant where the milk is first delivered.

Classes of milk A system to divide producer *milk* in a *federal milk marketing order* depending upon its intended use. The *prices* for each class of *milk* are different. Some federal milk marketing orders have four classes and others only three. The products included in Class I and Class II are the same in each Federal order. Class III is nearly the same in each order.

Class I is milk used in *fluid milk products*, such as *buttermilk*, whole, lowfat, skim, and flavored milk. See also *milkfat content*.
Class II is milk used in cream or "soft" manufactured products, such as sour cream, cottage cheese, ice cream, and *yogurt*.
Class III is milk used in other *hard manufactured dairy products*, such as *butter*, *cheese*, and canned milk.
Class III-A is skim milk used to make *nonfat dry milk*.

In orders with only three classes, Classes III and III-A are combined into Class III.

Cleaning (grain) Removing dockage, insects, and to a degree, shrunken and broken *kernels* from *grain* by means of mechanical screening and scalping devices. Cleaning practices vary from country to country. See *precleaning*.

Clearing accounts A *countertrade* whereby two countries agree to purchase specific amounts of each other's products over a specified period of time. If an imbalance of trade between the two countries exceeds a preestablished level, trade is stopped, and the imbalance is corrected by a cash payment. Also referred to as switch trading.

Clearinghouse An adjunct to, or division of, a *commodity futures exchange* that settles trades, guarantees *contracts*, and oversees delivery procedures.

Clear flour The portion of *flour* remaining after the "patent" cut of flour (see *patent flour*) has been taken off. Clear flour is normally higher in ash and *protein* than patent flour and of lower *market* value.

Climate The atmospheric and meteorological influences, principally temperature, moisture, wind, and evaporation, that characterize a region and influence the nature of its *soils*, vegetation, and *land* use.

Clod buster A machine used to break up large chunks of dry *soil* to help prepare the ground for seeding (see *seed*).

Clone An organism derived by asexual reproduction from a single parent. Because genetic material is not combined as in sexual reproduction, clones are genetically identical to the parent. In the case of plants, clones are derived from a single plant by means of vegetative propagation such as the rooting of cuttings or slips, budding, or *grafting*.

Close grown crops Small *grains*, such as *wheat*, *barley*, and *oats* that are seeded in rows too close together to *cultivate* with a row *cultivator*. Rows are usually about 2–8 inches apart.

Cloth A *textile* product obtained by *weaving*, *knitting*, braiding, felting, bonding, or fusing *fibers*. Cloth is synonymous with fabric.

Clover An *annual* or *perennial* legume (see *legume*) that ranks second to *alfalfa* as a *forage* crop. The most important *species* are red clover, alsike clover, white clover, and sweet clover.

Club wheat Usually *white wheat* varieties, either winter or spring habit.

Coarse breaks Separate rolls that *grind* the larger particles in a *break system* into "coarse" by size. See also *fine breaks*.

Coarse grains *Corn*, *barley*, *oats*, *grain sorghum*, and *rye*. *Millet* is also included in the statistics of some *foreign* nations. These are generally considered *feed grains*.

CoBank See *Bank for Cooperatives (BC)*.

Cockerel A male chicken (see *poultry*) less than one year old.

Cocoa See *cacao*.

Coconut A tropical fruit with a hard, round, brown shell. The flesh and juice are used.

Codes of Conduct A trade agreement specifying acceptable standards for implementing national trade legislation and cooperating with other governments on trade matters.

Codex Alimentarius See *Codex Alimentarius Commission*.

Codex Alimentarius Commission Created in 1962 by the *Food and Agriculture Organization (FAO)* and the *World Health Organization (WHO)* to negotiate agreements from member countries on international standards and safety practices for foods. The Codex standards set minimum quality, safety, and hygiene standards that countries voluntarily apply to their *exports* and *imports* of *commodities* for human consumption. The standards are published in a listing called *Codex Alimentarius*. Approximately 145 countries are members.

Coefficient of equivalence A coefficient used in calculating daily *levies* on *grain* imported into the *European Union (EU)*. The EU official *prices* relate to grain of a standard EU quality, and the quoted values of grain offered are adjusted according to quality by the addition or subtraction of these coefficients.

Coffee, Sugar, and Cocoa Exchange, Inc. A trading organization where cocoa (see *cacao*), *coffee*, and world and domestic raw cane *sugar* futures and options contracts (see *futures contract*) are traded daily. The cocoa *contract* is traded in 10 metric ton lots (2,204.6 pounds) of cocoa beans with *prices* quoted in dollars per metric ton. The coffee contract is traded in 37,500 pound lots of washed arabica coffee beans with prices quoted in cents per pound. The world sugar price is the *No. 11 contract price* for raw cane sugar (f.o.b. Caribbean), while the domestic price is the *No. 14 contract price* for raw cane sugar (c.i.f., duty/fee paid, New York). Both the world and domestic sugar contracts are traded in 50 long ton lots (112,000 pounds), with prices quoted in cents per pound.

Coffee tree A tropical tree whose *seeds* are roasted and ground (see *grind*) to make coffee. *Varieties* include Arabica coffees, which are higher-quality and preferred for ground coffees, and robustas, which are lower-quality types used for instant coffee. A large percentage of the world's coffee is produced in South America. Major producers include Guatemala, Costa Rica, El Salvador, Honduras, and Indonesia.

Coir A *fiber* obtained from the *husk* of *coconuts* that is highly resistant to rot. Coir *yarn* is used mainly for the manufacture of floor coverings, such as door mats, matting, and rugs. Commercial coir *production* is centered in India and Sri Lanka.

Cold confinement barn An open-sided shelter building in which *cattle* are confined without access to an outside *lot.*

Cold wall storage tank A *milk* storage tank in which a refrigerant is circulated in the walls.

Colic A digestive disturbance (infectious or otherwise) that causes pains in the abdomen.

Colony A social community of several thousand worker *honeybees*, usually containing a *queen bee* with or without *drones.*

Color additives See *additives.*

Colorado River Salinity Control Program A program that provides financial and technical assistance to identify salt source areas in the Colorado River Basin; to install *conservation practices* to reduce salinity levels in the Colorado River; to carry out research, education, and demonstration activities; and to carry out monitoring and evaluation activities. Several *U.S. Department of Agriculture (USDA)* agencies cooperate in this effort.

Colt A young male *horse* under four years of age. See also *filly.*

Colostrum The *milk* secreted by mammalian females for the first few days following the delivery of offspring. Colostrum is high in antibodies and laxative.

Comb See *honeycomb.*

Combination bakery A bakery that uses both *bake-off format* and *scratch/mix format.*

Combination dryer On-*farm* dryer (see *crop dryer*), mainly used for *corn*, that combines the best characteristics of *bin dryers* and *nonbin dryers* (i.e., high quality and high capacity).

Combine (self-propelled, pull-type) A large machine used to thresh (see *threshing*) *grain*, *corn*, *beans*, *rice*, *seed*, and other crops. A combine may be a pull-type, but most are now self-propelled.

Combines have a platform, called a *header*, that cuts and forces the crop into a system of beaters, sieves, and fans that separate the seed from the rest of the plant, or *chaff*, which is spread behind the combine.

Combing The process by which short *fibers* are removed from either *cotton* or animal fibers, while improving the alignment of the fibers within the *sliver*, in order to provide (ultimately) *yarns* of higher quality, in terms of evenness, luster, fineness, and strength. Combing is essential for the production of worsted yarns (see *worsted system).*

Combless package See *package bees.*

Commercial beekeeper A person who keeps *honeybees* full-time for income, producing *package* and *queen bees, honey*, or wax or other products, and pollinating services. The industry generally considers commercial beekeepers as maintaining 300 or more *colonies*.

Commercial feedlot An enterprise where *cattle* are fattened in *lots*. In some cases, the cattle may be custom fed (see *custom feeding*) for others. *Feed* usually is purchased, and labor is hired. Commercial feedlots usually have the capacity for 1,000 or more head of cattle.

Commercialization The process of placing a product into the marketplace (see *market*) with the goal of generating profits.

Commercial (meat) production A reporting measure of all *slaughter* and meat *production* in federally inspected (see *inspection, federal, meat and poultry inspection,* and *federally inspected slaughter*) and other plants (see *plant or establishment*), excluding animals slaughtered on *farms* (see *slaughter, farm*).

Commercial slaughter The term applied to all animals slaughtered (see *slaughter*) under state or federal inspection (see *inspection, federal, meat and poultry inspection*, and *federally inspected slaughter*). The term does not include animals slaughtered on *farms* (see *slaughter, farm*).

Commercial supplement A high-*protein* feedstuff (see *feed*) manufactured or formulated to provide extra *nutrients* to *feed grains*.

Commercial treaty An agreement between two or more countries outlining the conditions under which business among countries may be transacted. The treaty, for example, may outline *tariff* privileges, terms on which property may be owned, and the manner in which claims may be settled.

Commission merchant A *receiver* who handles more than half of his volume on consignment (see *consign*) from growers and shippers.

Committee on Surplus Disposal (CSD) A subcommittee of the *Food and Agriculture Organization*'s (FAO) Committee on *Commodity* Problems that regularly monitors and consults on food aid transactions to ensure that *surplus* disposal does not interfere with the normal *production* and trade patterns. The CSD monitors FAO member countries' compliance with the FAO Principles of Surplus Disposal, adopted in 1954.

Commodity A product offered by a number of sellers in essentially the same form, especially an agricultural or mining product that can be transported.

Commodity Credit Corporation (CCC) A federally owned and operated corporation within the

U.S. Department of Agriculture. The CCC was created to stabilize, support, and protect *farm income* and *prices* through loans, purchases, payments, and other operations. The CCC functions as the financial institution through which all money transactions are handled for agricultural *price-support programs, income-support programs,* and related programs. The CCC also helps maintain balanced, adequate supplies (see *supply*) of agricultural *commodities* and their orderly distribution. The CCC does not have any operating personnel or facilities, although its activities are handled by the *Foreign Agricultural Service* and the *Consolidated Farm Service Agency.*

Commodity distribution Direct donation of food products by the federal government to needy persons, schools, and institutions. *Commodities* are either entitlement or "bonus." Bonus commodities can be received when they are available from *surplus* stocks (see *stocks*) purchased by the *Commodity Credit Corporation (CCC)* under its *price support programs.*

Commodity futures exchange An organization that establishes and enforces rules and provides facilities and services for trading (see *futures trading*) standardized *forward contracts.*

Commodity Futures Trading Commission (CFTC) The independent federal agency that regulates U.S. *futures trading.*

Commodity Import Programs (CIPs) Grants and loans made by the *Agency for International Development (AID)* to countries judged important to U.S. *foreign* policy objectives. By making dollars available, CIPs help these countries finance purchases of U.S. agricultural *commodities* or other inputs needed to meet their development objectives and also provide *balance-of-payments* support to countries with very limited foreign exchange.

Commodity option A *contract* traded on a *commodity futures exchange* that gives the buyer the right without obligation to buy or sell a *futures contract* over a specified time period. The *price* paid for a commodity option is called the *premium,* and the price at which the option can be exercised by the option is called the *strike price.* The seller is called an *option grantor* and must make and maintain a *margin* deposit with a *broker* to ensure performance. The advantages of options for *hedging* include the ability to limit losses from price changes in one direction while gaining from price changes in other directions, absence of *margin calls* for option holders, and the wide range of choices available. See *call option* and *put option.*

Commodity price and income support programs Federal programs designed to support crop *prices* and *farm income.*

Commodity Supplemental Food Program (CSFP) A program to provide *commodities* to supplement the diets of low-income infants, children up to age six, women during pregnancy and up to six weeks postpartum, breastfeeding women up to 12 months postpartum, and persons 60 years of age and over. Under the CSFP, the *U.S. Department of Agriculture (USDA)* donates foods, such as juice, *egg* mix, and canned fruits and vegetables, and distributes them through state and local agencies. Participants in the CSFP cannot also participate in the *Special Supplemental Food Program for Women, Infants, and Children (WIC).*

Common Agricultural Policy (CAP) A set of regulations by which member states of the *European Union (EU)* seek to merge their individual agricultural programs into a unified effort to promote regional agricultural development, fair and rising standards of living for the *farm* population, stable agricultural *markets,* increased agricultural productivity, and methods of dealing with food supply security. The three principal elements of the CAP are the *variable levy,* intervention purchases, and *export subsidies.*

Common external tariff The *tariff* schedule applied by members of a *customs union*, such as the *European Union*, to *imports* of nonmember countries.

Common market A regional grouping of countries that levies common external duties (see *tariff*) on *imports* from nonmember countries but that eliminates tariffs, *quotas*, and other miscellaneous government restrictions on trade among member countries. Also referred to as a tariff union. The *European Union* is probably the best known example of a common market. Others include the Belgium-Luxembourg Economic Union, BENELUX (Belgium, the Netherlands, and Luxembourg), the Central African Customs and Economic Union, the East African Community, the *West African Economic Community*, and the Central American Common Market. See also *customs union.*

Communaute Economique des Etats de l'Afrique Centrale (CEEAC) See *Economic Community of Central African States.*

Communaute Economique de l'Afrique de l'Ouest See *West African Economic Community.*

Community Nutrition Institute (CNI) A nonprofit organization founded in 1969 to serve as an advocate for *consumer* protection, food program development and management, and sound federal diet and health policies.

Companion crop A crop grown with another crop, usually a small *grain* with which *alfalfa*, *clover*, or other *forage* crops are sown.

Company A legal entity that controls and operates, and usually owns, one or more *plants or establishments.*

Comparative advantage An economic theory that in international trade it is more advantageous for a country to devote its *resources* to the goods and services that it can produce and *export* more efficiently than other nations. At the same time, the theory implies that it is more advantageous for a country to *import* those goods and services that can be produced more efficiently elsewhere.

Compensatory duty A duty (see *tariff*) sometimes levied on imported (see *import*) manufactured articles to offset the increased costs of a domestic manufacturer of similar articles when such costs are attributable to a tariff on the raw materials used by the domestic manufacturer. For example, at one time the U.S. duty on *wool* was 3 cents a pound. Because it takes about 4 pounds of wool to produce 1 pound of cloth, a compensatory duty of 12 cents a pound was placed on imported woolen cloth. At that point, the domestic producer of wool cloth could compete on even terms with *foreign* manufacturers.

A compensatory duty also may be placed on an imported *commodity* in order to offset an excise tax on the same commodity produced in the importing country.

Compensatory payment A payment or assessment under *federal milk marketing orders* designed to place all *handlers* selling in a regulated area on the same pricing basis with respect to the cost for the *milk* used in fluid sales (see *fluid milk products*). The compensatory payment is equal to the difference between the order's Class I (see *classes of milk)* price and its Class III *price* in some situations and between the order's Class I price and its *blend price* in other situations.

Compensatory tax A tax imposed by the *European Union* on certain imported (see *import*) fruits, vegetables, wine, and fish when the *price* of the imported product falls below the *reference price*.

Competing brand A product packaged under a *packer's* or *processor's* second brand name and often used by small retail operations to compete with other retailers' *private label brands*.

Competitive advantage A situation where one *producer* can produce a particular *commodity* more cheaply than another producer.

Competitive imports Imported (see *import*) products, such as vegetables, *sugar*, and meat, that are also produced domestically. Also referred to as supplementary imports.

Competitive market An industry with many relatively small firms, no insurmountable barriers against firms to enter (see *market entry*) or exit (see *market exit*) the industry, and relatively small costs for firm entry or exit. In a competitive *market*, no one firm is able to impact market prices (see *price*) by its actions. See *conditions of entry* and *conditions of exit*.

Complementary imports See *noncompetitive imports*.

Complete feed A nutritionally adequate *feed* for animals. It is a specific formula compounded to be fed as the sole *ration* and is capable of maintaining life and/or promoting *production* without any additional substance being consumed except water.

Complete fertilizer A *fertilizer* containing sufficient amounts of *nitrogen*, phosphorous, and potassium to sustain plant growth.

Compost Rotted organic matter made from waste plant residues. Inorganic *fertilizers*, especially *nitrogen*, and a little *soil* are usually added.

Compound tariff A "mixed *tariff*" composed of an *ad valorem tariff* plus a *specific tariff* on the same imported item.

Computer assisted trading (CAT) A bidding (see *bid*) and trading procedure using a central computer and computer terminals.

Conasupo An acronym for the Compania Nacional de Susistencias Populares, S.A., which is the official Mexican government agency responsible for ensuring adequate food supplies (see *supply*) for the country. Until 1985, the Conasupo was the sole *importer* of basic food and feedstuffs (see *feed*) for Mexico. Since then, the Mexican government has allowed private companies to *import* basic *grains*, *oilseeds* and products, and *tallow*, although they still maintain control through a system of prior *import licensing* for these and other basic products.

Concentrate Feedstuffs (see *feed*) that are high in digestible *nutrients* and energy and low in crude *fiber* (under 18 percent), including *barley*, *oats*, *cottonseed*, and *beet* pulp (see *pulp*).

Concentration One measure used to determine the degree of competitiveness in an industry or *market*. It measures the combined sales of the leading firms—typically the four largest—as a share of total market or industry sales. See *concentration ratio*.

Concentration ratio The *market* share held by the top-ranked firms in an industry. The *concentration* ratio may be measured by sales, productive capacity, or other criteria and is often examined as a 4-firm, 8-firm, or 12-firm grouping.

Concession A *tariff* reduction, tariff *binding*, or other agreement to reduce *import* restrictions. In negotiations, a country may offer to reduce its own tariff and *nontariff trade barriers* to induce other countries to reciprocate.

Concessional sales Credit sales of a *commodity* in which the buyer is allowed more favorable payment terms than those on the open *market*. For example, Title I of *Public Law 480* provides for financing sales of U.S. commodities with low-interest, long-term credit.

Concurrent-flow dryer An off-*farm* commercial dryer (see *crop dryer*) in which *grain* and air flow vertically. The gentle drying and cooling methods used in these dryers result in grain of superior quality. Their main disadvantage is their high initial cost.

Condensed Reduced to denser form by removal of moisture.

Condensery A plant (see *plant or establishment*) or factory producing *condensed* or evaporated *milk*.

Conditional Most-Favored Nation The according of *most-favored nation (MFN)* treatment subject to compliance with specific terms or conditions. All members of the *General Agreement on Tariffs and Trade (GATT)*, including the United States, accord unconditional MFN treatment to most other GATT members. The United States, however, confers annually renewable MFN treatment to a limited number of countries conditional on their compliance with the terms of Title IV of the Trade Act of 1974 (see Appendix 3). See also *Jackson-Vanik Amendment*.

Conditioner See *hay conditioner*.

Conditioner (of fertilizer) A material added to a *fertilizer* to prevent caking and to keep it free-flowing.

Conditioning See *tempering*.

Conditions of entry The ease or difficulty of entry by a new firm, commonly measured by the cost advantage that an existing firm has over the most favored potential entrant. See *market entry*.

Conditions of exit The ease or difficulty of exit from the *market* by an existing firm. See *market exit*.

Conduct Firm-level strategies in pricing (see *price*), advertising, research and innovation, and plant (see *plant or establishment*) investment.

Confinement A partially or completely solid-floored and enclosed or covered facility in which *livestock* are kept in *dry-lot* for maximum year-round *production*.

Confinement production system The *production* of *swine* in nonportable buildings with or without access to *lot* space. *Pasture* is used only for breeding animals, if at all.

Confinement rearing A method of raising *poultry*, typically turkey *poults*, which are not permitted to run on *range* but are confined within a house or small fenced apron alongside the house.

Conifers See *softwood*.

Conselho Nacional do Comercio Exterior (CONCEX) A Brazilian association of private traders and government agencies that acts in an advisory capacity on *exports* of *grain*. It is not a government agency.

Conservation compliance provision A provision authorized by the *Food Security Act of 1985 (P.L. 99-198)* (see Appendix 3) that requires farmers with *highly erodible cropland* to have obtained an approved *conservation plan* by 1990. The plan must be completed by 1995 to maintain eligibility for federal *farm* program benefits.

Conservation district Any unit of local government established under state law to carry out a program for the conservation, use, and development of *soil*, water, and related *resources*. There are over 3,000 conservation districts in the United States.

Conservation easement Rights to *land*, using a reserved interest deed, where the grantee acquires all rights, title, and interest in a property, except those rights that might run with the land expressly reserved by a grantor.

Conservation plan A combination of *land* uses and practices to protect and improve *soil* productivity and to prevent soil *erosion*. A conservation plan must be approved by local *conservation districts* for acreage (see *acre*) offered in the *Conservation Reserve Program*. The plan sets forth the conservation measures and maintenance that the farm owner or *farm operator* will carry out during the term of the *contract*.

Conservation practices Methods to reduce soil *erosion* and to retain *soil* moisture. Major conservation practices include *conservation tillage*, *crop rotation*, *contour farming*, *stripcropping*, *terraces*, and *diversion terraces*.

Conservation Reserve Corn Bonus Program A program effective only for the 1987 *crop year* and designed to encourage farmers to place highly erodible (see *highly erodible cropland*) *corn* base acreage (see *crop acreage base*) into long-term *conserving uses*. Bonus payments to participants were paid entirely in *generic certificates*.

Conservation Reserve Program (CRP) A program first authorized by the *Food Security Act of 1985 (P.L. 99-198)* (see Appendix 3) designed to reduce *erosion* on 40–45 million *acres* of *cropland*. Under the program, *producers* who sign *contracts* agree to convert *highly erodible cropland* to approved *conserving uses* for 10–15 years. In exchange, participating producers receive annual rental payments and cash or *payments-in-kind* to share up to 50 percent of the cost of establishing permanent vegetative cover. Under the *Food, Agriculture, Conservation, and Trade Act of 1990 (P.L. 101-624)* (see Appendix 3), the program was broadened and renamed the *Environmental Conservation Acreage Reserve Program*.

Conservation, soil A combination of *land* use and practices to protect and improve *soil* productivity and to prevent soil deterioration from *erosion*, exhaustion of plant *nutrients*, accumulation of toxic salts, excessive compaction, or other adverse effects.

Conservation tillage Any of several farming methods designed to reduce *erosion* of *soil* caused by water or wind while maintaining crop *yields* and quality. These conservation tillage (see *till*) methods provide for *germination* of *seeds*, plant growth, and weed control yet maintain effective ground cover throughout the year and disturb the soil as little as possible. To reduce soil erosion by water, a minimum of 30 percent of the soil surface is kept covered by residue after planting. Where wind erosion is a problem, at least 1,000 pounds per *acre* of flat, small-*grain* residue equivalent are maintained on the soil surface during the critical erosion period. *No-till* is the most restrictive (soil-conserving) form of conservation tillage. Other practices include *ridge-till*, *strip-till*, and *mulch-till*.

Conservation tillage equipment *Farm* cultivation (see *cultivate*) and seeding (see *seed*) machinery designed to leave the previous year's *crop residue*, or *mulch*, on the surface in order to slow *erosion*. In order to work the *soil*, machines usually have *coulters*, which are specially designed to cut through stubble, to prevent the machine from clogging. Sometimes called mulch tillage equipment.

Conserving uses *Land* idled from *production* and planted in a *soil*-conserving crop, such as *annual*, *biennial*, or *perennial* grasses. Uses exclude acreage (see *acre*) (1) devoted to a crop of *rice*, *upland cotton* or *extra-long staple cotton*, *feed grains*, *wheat*, *soybeans*, *peanuts*, other *program crops*, or approved *nonprogram crops*; (2) required to be taken out of *production* under an *acreage reduction program*; and (3) designated under the *Conservation Reserve Program* or other conservation programs.

Consign To turn over the sale of a shipped product to an *agent* who normally charges a fee for services rendered.

Consolidated Farm Service Agency (CFSA) A *U.S. Department of Agriculture (USDA)* agency created in October 1994 by merging the *Agricultural Stabilization and Conservation Service (ASCS)*, the *Farmers' Home Administration (FmHA)*, and the *Federal Crop Insurance Corporation (FCIC)*. The FSA provides *commodity* price and income support, farm loans, and crop insurance.

Consolidated Metropolitan Statistical Area (CMSA) A large concentration of metropolitan population composed of two or more contiguous *Metropolitan Statistical Areas (MSA's)* which together meet certain criteria of population size, *urban* character, and social and economic integration. Each CMSA must have a population of 1 million or more.

Constant dollars See *real dollars*.

Consultative Group on International Agricultural Research (CGIAR) A group sponsored by the *World Bank*, the United Nations (UN) Development Programme, and the *Food and Agriculture Organization* of the UN to coordinate and fund international agricultural research centers, such as the International *Rice* Research Institute in the Philippines and the International *Maize* and *Wheat* Improvement Center in Mexico.

Consumer A person who buys and uses products and/or services and does not resell them.

Consumer Expenditure Survey (CES) A survey conducted by the *Bureau of Labor Statistics*. The diary survey collects data on small, frequently purchased items such as food and beverages over two consecutive weeks. Information on household characteristics is also collected at the end of the second week. The interview study collects information on larger expenditure items.

Consumer Price Index (CPI) A measure of the average *price* level of a *market basket* of goods and services at the retail level for a specific period compared against a benchmark period. The CPI is published by the U.S. Department of Labor's *Bureau of Labor Statistics*. It is the most widely accepted measure of change in *consumer* prices.

Consumer subsidy equivalent (CSE) A measure of the value or cost to *consumers* of government food and *agriculture* programs. CSEs estimate the amount of *subsidy* consumers would need to maintain their economic well-being if all agriculture programs were discontinued.

Contamination The introduction of *microorganisms*, disease *agents*, or chemicals that can cause human *disease* into foods.

Continental climate The *climate* typical of great *land* masses where wide ranges in temperature and other *weather* conditions occur because the area is not greatly influenced by nearness to the sea.

Continuing Survey of Food Intakes by Individuals (CSFII) A survey of food consumption by individuals conducted by the *U.S. Department of Agriculture*'s (USDA) Agricultural Research Service (ARS). The CSFII collects information about foods eaten at home and away from home during three consecutive days. The data provide estimates of food actually ingested for individuals classified by sex, age, income, race, and region. See also *food disappearance data* and *Consumer Expenditure Survey*.

Continuous crop *Land* that is cropped year after year as opposed to *cropland* that is occasionally left idle. See *summer fallow*.

Contour farming A field operation, such as *plowing*, planting, cultivating (see *cultivate*), and harvesting (see *harvest*) on the contour, or at right angles to the natural *slope*, to reduce *erosion*, protect *soil* fertility, and use water more efficiently.

Contour furrows *Furrows* plowed at right angles to a *slope* to intercept and retain *runoff* water.

Contract 1: An agreement between two parties that defines rights and obligations to deliver and pay for a designated *commodity* or service on one or more future dates. See *cash contract, futures contract, minimum-price contract,* and *production contract.*

2: The rules under which futures are traded for a particular commodity on a particular exchange—for example, the live *cattle* futures contract at the Chicago Mercantile Exchange.

3: To enter into a written or oral agreement calling for delivery and payment for a product or service on one or more future dates.

Contract farming The contractual (see *contract*) specification of the *market*-quantity, *grade*, size, inspection, timing, or pricing (see *price*) to both the grower and the *processor* or shipper before production begins. Contract farming is one method of *vertical integration.*

Contract grower (poultry) A grower who typically provides housing, equipment, labor, litter, power, and heat, and cares for the *broilers* during growout. The contract grower does not own the birds but receives a fee for these services, as specified in a *contract* with a *poultry* or *feed* processing (see *processing*) company.

Contracting party (CP) A country that has signed an agreement, such as the *General Agreement on Tariffs and Trade (GATT)*, and has accepted its specified obligations and benefits.

Contract of affreightment A *contract* to hire a ship, or part of it, to carry goods.

Contract production See *production contract* and *contract farming.*

Control To prevent regeneration of a pest or other hazard by treating *imports* for exotic pests or *diseases* to eliminate the hazard or eradicate the *pathogen.* Cold treatment of *citrus fruit* imports to kill Mexican fruit flies is an example of such a method. See *sanitary and phytosanitary regulations.*

Controlled brand See *private label brand.*

Controlled exchange rates Relative values of currencies established and maintained by government central banks. U.S. *exchange rates* were fixed until 1973. Also known as fixed exchange rates.

Controlled lighting Artificial lighting in *poultry* housing to increase or decrease the number of hours of light in order to control sexual maturity, fertility, molt, and timing of feeding and activity levels.

Conventional tillage A method that inverts the *soil* by *plowing* or otherwise leaves less than the minimum residue after planting required to qualify as *conservation tillage.*

Conversion factors Factors used to convert one unit of measure to another. For example, conversion factors can be used to convert from metric to English units:

Metric Equivalents

Metric weight	*Equivalent*
1 gram	0.035274 ounce
1 meter (m)	39.37008 inches
1 hectare (ha)	2.471044 acres
1 liter (l)	.9080830 dry quarts
	1.056688 liquid quarts
1 kilogram (kg)	2.204623 pounds
1 quintal (100 kgs)	220.4622 pounds
1 metric ton (1,000 kgs)	2,204.622 pounds

Conversion factors also indicate the number of pounds of a *commodity* per *bushel* or the number of bushels per ton.

Pounds Per Bushel

Wheat	60
Corn (shelled)	56
Oats	32
Barley	48
Rye	56
Sorghum grain	56
Rice, rough	45
Soybeans	60
Flaxseed	56

Bushels Per Metric Ton

Wheat	36.7437
Corn (shelled)	39.36825
Oats (32-pound bushel)	68.89444
Barley	45.929625
Rye	39.36825
Sorghum grain	39.36825
Rice, rough	48.991
Soybeans	36.7437
Flaxseed	39.36825
White potatoes	36.7437

Commodity Conversion Factors

100 pounds of:	*Can be obtained from:*
White flour	2.252 bushels of wheat
Cornmeal (degermed)	3.16 bushels of corn
Cornmeal (nondegermed)	2.00 bushels of corn
Rye flour	2.23 bushels of rye
Milled rice	152 pounds of rough rice
Soybean oil	549 pounds of soybeans
Cottonseed oil	588 pounds of cottonseed
Linseed oil	277 pounds of flaxseed

For more details, see Appendices 6, 7, and U.S. Department of Agriculture *Weights, Measures, and Conversion Factors for Agricultural Commodities and Their Products.* Econ. Res. Serv., AHB-697, June 1992.

Converted wetlands *Wetlands* that have been drained or otherwise manipulated to produce agricultural *commodities.*

Convertible currency The currency of a nation that may be exchanged for another nation's currency without restriction. Also referred to as hard currency.

Convertible local currency credit A *Public Law 480* credit sale in which installments are paid in dollars or, at the option of the U.S. government, in currencies that can be converted into dollars.

Cooperative See *agricultural cooperative.*

Cooperative arrangement An arrangement among small firms to gain efficiency and improve their competitive position. The arrangement may include cooperative purchase of supplies, joint management services, and use of a common brand name to obtain effective brand promotion and reduce advertising expenditures.

Cooperative Extension Service Educational work for people outside of classrooms carried on by the states, usually through the resources of the *land-grant universities* and colleges in cooperation with the *U.S. Department of Agriculture* (USDA). The *Cooperative State Research Education and Extension Service* represents USDA in conducting cooperative extension work. See also *county extension agent.*

Cooperative federation An organization formed by *cooperatives* to perform certain functions, such as serving as a marketing (see *agricultural marketing*) or bargaining *agent.*

Cooperative-owned wholesaler A *wholesale* warehouse wholly owned by a group of retailers in order to obtain merchandise at the lowest possible cost. The warehouse staff provides group advertising, merchandising, and other services. Also known as retailer-owned wholesaler.

Cooperative State Research, Education, and Extension Service (CSREES) A *U.S. Department of Agriculture (USDA)* agency created in October 1994 by merging the *Extension Service,* the *Cooperative State Research Service*, and the *National Agricultural Library.*

Cooperative State Research Service (CSRS) The name of the former agency of the *U.S. Department of Agriculture (USDA)* that administers certain federal funds appropriated for agricultural and forestry research at state agricultural experiment stations, forestry schools, 1890 universities and Tuskegee University, and selected veterinary schools. The agency coordinates research and research planning on a regional and national basis, conducts competitive and special research grants programs, and maintains the Current Research Information System (CRIS) of state and federal research records for publicly supported agricultural and forestry research. Under USDA's 1994 reorganization, CSRS was merged with the *Extension Service* and the *National Agricultural Library* to form the *Cooperative State Research, Education, and Extension Service (CSREES).*

Cooperator program A longstanding *market* development program administered by the *U.S. Department of Agriculture*'s (USDA) *Foreign Agricultural Service* that coordinates the *export* promotion efforts of more than 50 nonprofit *commodity* trade associations, including the U.S. *Wheat* Associates, *Cotton Council International*, and the National Potato Promotion Board.

Cooperators Trade and agricultural groups cooperating with the *U.S. Department of Agriculture's* (USDA) *Foreign Agricultural Service (FAS)* in developing commercial *markets* overseas for U.S. *farm* products such as *wheat, feed grains, soybeans, rice, livestock,* livestock products, *poultry,* fruit, *cotton, tobacco,* and others. Cooperating groups, from contributions of their memberships, provide personnel, services, and cash. FAS provides funds, as well as administrative guidance and assistance.

Copra Dried *coconut* meat, from which is extracted *coconut* oil (see *fats and oils).* Copra is produced primarily in the Philippines, Indonesia, India, and Malaysia.

Corduroy *Cloth* with ridges of pile running lengthwise, creating a ribbed surface.

Coriander The dried *seeds* of the *cilantro* plant. Coriander is often used as a pickling spice.

Corn The *seed* of a *cereal* grass and the only important cereal plant indigenous to America. Corn is used mainly for animal *feed*, but it is also used for oils (see *fats and oils), starches*, and *sugars* for human consumption and in some industrial products, such as *ethanol*. Major producers include the United States, China, Brazil, Argentina, South Africa, France, Romania, and Thailand. Also called *maize*.

Corn Belt The states of Iowa, Indiana, most of Illinois, and parts of Kansas, Missouri, Nebraska, Minnesota, Ohio, South Dakota, and Wisconsin, where *corn* is a principal *cash crop*.

Corn crib See *crib*.

Corn gluten A *byproduct* of corn *wet milling* that is used as a medium-*protein* (20–22 percent) and medium-*fiber* (10 percent) *feed*.

Corn harvester A *combine* with a special *header* designed to cut *corn* stalks. Corn headers may cut several rows at one time and convey the stalks to the beater where the ears are shelled. The *grain* is saved in a tank or *bin* on the combine, and the remainder of the plant is shredded and spread on the ground.

Corn-hog ratio The number of *bushels* of *corn* that are equal in value to 100 pounds of live hogs (see *swine*); that is, the *price* of hogs per *hundredweight* divided by the price of corn per bushel. The corn-hog ratio can be calculated in terms of U.S. average prices received by farmers, prices received by farmers in a given area, or on the basis of central *market* prices rather than *farm* price. This ratio has exhibited both seasonal and cyclical movements.

Corn planter A seeder designed to sow *corn* in rows. The machinery may also be used to apply *fertilizers* and *pesticides*.

Corn sweeteners Alternative sweeteners derived from *corn*. In the United States, the most important corn sweeteners are *high fructose corn syrup (HFCS), glucose* corn syrup (see *corn syrup*), *dextrose*, and *crystalline fructose*.

Corn syrup A purified concentrated solution of nutritive *saccharides* obtained from corn *starch* by partial *hydrolysis*, clarification, decolorization, and evaporation to syrup density. *Glucose* is often used synonymously with corn syrup.

Corporate farm A *farm* created and organized under the incorporation laws of the state(s) in which it operates. The farm can be of any size, including *family farms*. See *corporation*.

Corporation A joint-stock company, chartered by the state, that can conduct business as a legal entity, apart from the stockholders who own it. The corporation operates under the guidance of a board of directors.

Corral See *lot.*

Corrugation irrigation See *irrigation method.*

Cost-benefit analysis Economic appraisal of a project or action. It consists of adding up all benefits and costs of the project to society, adjusting them by using a *discount rate* to reflect the *opportunity cost* of the invested funds, and computing the value received from the project. The discount rate reflects either (1) the assumptions of the researcher about the preferences society has for consumption today over the future; (2) the amount that could have been earned if the funds had been invested elsewhere (opportunity cost); or (3) some combination of both.

Cost, insurance, and freight (c.i.f.) The seller's *price*, usually by ocean shipping, which includes the cost of goods, marine insurance, and transportation (freight) charges to the point of destination.

Cost of production The sum, measured in dollars, of all purchased *farm inputs* and other expenses necessary to produce *farm* products. The three components of the cost of production are total cash expenses, cash expenses plus capital replacement costs, and total economic costs. The latter includes an *opportunity cost* for owned inputs. Cost-of-production statistics may be expressed as an average per animal, per *acre*, or per unit of *production* (*pound*, *bushel*, or *hundredweight*) for all farms in an area or in the country.

Cost-saving input A *production* input (*see farm input*) that reduces the amount of at least one input needed to produce a unit of output. The relationship between inputs and output can be defined by the production technology.

Cost-sharing program A program whereby a recipient of government aid shares in the cost of the service or good. The *Emergency Feed Program* and *Emergency Feed Assistance Program*, for example, are operated as cost-sharing programs.

Cotton A soft, white vegetable (cellulosic [see *cellulose*]) *fiber* obtained from the *seed* pod of the cotton plant. Cotton is a member of the mallow family (*Gossypium*) and is produced in about 75 countries. The two principal types of cotton grown in the United States are *Upland Cotton* (*Gossypium hirsutum*) and *American Pima cotton* (*Gossypium barbadense*). *Upland cotton* is grown throughout the Cotton Belt, accounting for about 98 percent of U.S. cotton *production*. Pima cotton is grown primarily in the Southwest. (See also *extra long staple cotton* and *Sea Island cotton.*) Major cotton producers include China, the United States, India, Pakistan, Uzbekistan, Brazil, Turkey, and Egypt. *Yarn* and fabric (see *cloth*) are the principal products from cotton.

Cotton Board A quasi-governmental organization whose members are appointed by the secretary of agriculture from nominees of cotton *producer* and cotton *textile* organizations. Established in 1967 by the Cotton Research and Promotion Act, the board receives assessments and disburses funds primarily to *Cotton Incorporated* for use in *cotton* research and promotion activities.

Cotton compress The equipment that forms the ginned (see *gin*) raw *cotton* into a *bale*. The first compression, primarily to modified flat or universal bale dimensions, is performed at the gin. Further compression of flat or modified flat bales is performed at cotton warehouse locations.

Cotton Council See *National Cotton Council of America.*

Cotton Council International (CCI) The overseas operations service of the *National Cotton Council of America (NCAA)*. Established in 1956, CCI's primary objective is to develop *markets* for U.S. *exports*. CCI programs are operated in close cooperation with the *U.S. Department of Agriculture*'s

(USDA) *Foreign Agricultural Service* and trade groups in the United States and abroad. CCI's headquarters are in Washington, D.C.

Cotton count A measure of cotton density that varies for *yarn*, woven (see *weave*) *cloth*, and knitted (see *knitting*) fabric. (1) For yarn, cotton count is a numbering system based on the number of 840-yard lengths in a *pound*. Higher numbers indicate finer yarn. A single strand of #10 yarn is expressed as 10s or 10/1. A 10s yarn has 8,400 yards to the pound; a pound of 20s yarn is 16,800 yards long. (2) For woven cloth, the cotton count is the number of *warp* ends and *filling* picks per inch. If a cloth is 68 by 72, there are 68 *ends* and 72 *picks* per inch in the fabric. (3) For knitted fabric, the count indicates the number of *wales* and *courses* per inch.

Cotton exchange A membership organization that provides facilities where *cotton* futures and options contracts (see *futures contract* and *options*) are bought and sold. The New York Cotton Exchange is the major U.S. exchange. The base *grade* for the New York *contract* is Strict Low Middling 1 1/16-inch cotton. Each contract represents 50,000 pounds.

Cotton Incorporated (CI) A private organization established in 1971 as the sales-oriented marketing (see *agricultural marketing)* and research organization representing U.S. *cotton* growers. CI's objectives are to increase *producers'* profits and to expand the sale of products containing cotton. *CI*'s headquarters are in New York City.

Cotton lint The *fibers* separated from the *cottonseed* during ginning (see *gin*).

Cotton picker A large machine designed to separate *cotton* bolls (see *boll*) from the plant and place cotton in a bin or hopper.

Cotton quality Those characteristics of the *cotton* fiber that affect processing performance and/or the quality of the various end products. While there are numerous factors that affect quality, the seven most important are *fiber* length, length uniformity, strength, fineness, maturity, color, and trash content. Three major components of cotton quality—grade, staple, and micronaire—are included in official USDA cotton quality classifications. Added fiber properties, including length uniformity and strength, are also recognized as important and are increasingly being measured by instrument testing.

Cottonseed The *seed* of *cotton* from which the *lint* has been removed. See also *cottonseed oil*, *cottonseed meal*, and *cottonseed hulls*.

Cottonseed hulls One coproduct of the *cottonseed* crushing (see *crushing*) operation. The hulls are used as *livestock* feed (see *feed*). See also *cottonseed meal*.

Cottonseed meal One coproduct of the *cottonseed* crushing (see *crushing*) operation. The meal is used as *livestock* feed (see *feed*). See also *cottonseed hulls*.

Cottonseed oil An oil (see *fats and oils*) extracted (see *extraction, mechanical* and *extraction, solvent*) from *cottonseed* through a crushing process.

Cottonseed Oil Assistance Program (COAP) One of two programs under which the *Commodity Credit Corporation (CCC)* provides bonuses in vegetable *oils* to *exporters* to assist in *exports* to targeted *markets*. Funds for the programs were first authorized in fiscal year 1988 from *Section 32* of P.L. 74-320.

Cotton system A process originally used to manufacture *cotton* fiber (see *fiber*) into *yarn* and now used extensively to produce spun yarns of *man-made fibers*, including blends (see *blending*). The major manufacturing steps in the cotton system include opening of the fiber *bales*, picking, *carding*, *drawing*, *roving*, and *spinning*. The *combing* step is included after carding when combing yarns are made.

Cotyledons The first leaves of a plant as found in the *embryo*. The term is also applied to the two halves of a *legume* seed (see *seed*).

Coulter A circular *disk* used on *tillage equipment* to cut through stubble and *crop residue* and to prevent residue from clogging the machine. Disks come in several sizes and may be fluted, rippled, bubbled, or waved to add stability and to efficiently cut through crop residue.

Coulter chisel plow *Tillage equipment* designed to leave *crop residue* on the surface to prevent *erosion* of *soil*. *Coulters* cut through residue to make paths for *shanks* with narrow, pointed cultivation (see *cultivate*) *shovels*. Also known as mulch tillage plow.

Council of Economic Advisors An advisory group that engages in the analysis and appraisal of the national economy for the purpose of providing policy recommendations to the president of the United States.

Count See *cotton count*.

Counterpurchase A *countertrade* where an *exporter* agrees to make reciprocal purchases of an *importer's* products within a specified period of time.

Countertrade Any trade transaction where the seller is required to accept goods or other services from the buyer as either full or partial payment. Countertrade is common in East-West trade and is increasingly being practiced worldwide. Forms of countertrade include *barter*, *buy-back*, and *counterpurchase*.

Countervailing duty (CVD) An additional *levy* imposed on imported goods to offset subsidies (see *subsidy*) provided to *producers* or *exporters* by the government of the exporting country. A wide range of practices are recognized as constituting subsidies that may be offset. Under U.S. law, countervailing duties can only be imposed after the *U.S. International Trade Commission* has determined that the *imports* are causing or threatening to cause material injury to a U.S. industry.

Country commission firm A firm that represents the *producer* for a fee in negotiating sales of *livestock* on the *farm* to a *packer*.

Country elevator An *elevator* operated by a country *grain* dealer who buys grain from *producers*.

County committee See *Agricultural Stabilization Committee (ASC)*.

County extension agent A professional worker, employed by the county, state *Cooperative Extension Service*, and/or the *U.S. Department of Agriculture (USDA)*, to bring agricultural and homemaking information to local people and to help them meet *farm*, home, and community problems. The position is also called extension agent, farm and home advisor, agricultural agent, and extension home economist.

County loan rate The *nonrecourse loan rate* that varies between counties based on transportation costs to the nearest *terminal elevator*. The weighted average for all county loan rates (the actual loan levels received by farmers) in the United States must equal the national average loan rate established by the *U.S. Department of Agriculture (USDA)* according to limits set by Congress.

County office A county-level office of the *Consolidated Farm Service Agency (CFSA)*, where farmers go to conduct business associated with federal *farm* programs. The office may be co-located with other *U.S. Department of Agriculture (USDA)* agencies, such as the *Natural Resources Conservation Service.*

County type classification A *U.S. Department of Agriculture (USDA)* classification of nonmetro counties by principal economic activity or demographic base. The classification for farming-dependent, manufacturing-dependent, mining-dependent, and specialized government counties was updated in 1986. Categories include:

Farming-dependent: Counties in which farming contributed a weighted annual average of 20 percent or more of total labor and proprietor income in 1981, 1982, 1984, 1985, and 1986. (The year 1983 is not included because it was an extremely aberrant year for farm income.)
Federal lands counties: Counties in which federal *land* made up 33 percent or more of the land area in the county in 1979.
Manufacturing-dependent: Counties where manufacturing contributed 30 percent or more of total labor and proprietor income in 1986.
Mining-dependent: Counties where mining contributed 20 percent or more to total labor and proprietor income in 1986.
Specialized government counties: Counties where federal, state, and local government contributed 25 percent or more of total labor and proprietor income in 1986.
Persistent poverty counties: Counties where per capita income was in the lowest quintile in each of the years 1950, 1959, 1969, and 1979.
Retirement-destination: For the 1970–1980 period, counties where the net immigration rates of people aged 60 and over were 15 percent or more of the expected 1980 population aged 60 and over.

Course A predominantly horizontal (crosswise) row of loops produced by adjacent needles during the same cycle of the *knitting* machine.

Cover crop A close-growing crop grown to protect and improve *soil* between periods of regular crops, or between trees and vines in *orchards* and *vineyards.*

Cow A mature female *bovine*, usually at least two years old.

Cow and replacement heifer loss rate The number of *cows* and *replacement heifers* lost due to death, theft, and other causes during a calendar year per 100 cows.

Crambe A cool-season *annual* crop that yields erucic acid and a high-*protein* meal (see *meal*) used primarily for beef *cattle*. Much of the oil is used in the manufacture of plastic films and automotive and industrial lubricants. Other potential uses include cosmetics, paints and coatings, plasticizers, and pharmaceuticals.

Crawler tractor A *tractor* with continuous steel or rubber tracks instead of wheels. Tracks usually apply less pressure per square inch, causing less *soil* compaction than wheeled tractors. Track-type tractors are very stable, do not tip over easily, and are well adapted for hilly terrain.

Creamline milk *Milk* that was not *homogenized.* Creamline milk was sold in most *markets* at a lower *price* than homogenized milk but is no longer available.

Credit guarantees *U.S. Department of Agriculture* (USDA) programs that protect U.S. *exporters* or financial institutions against loss due to nonpayment by a *foreign* buyer. Maximum credit guarantee coverage period is three years under the *Export Credit Guarantee Program (GSM-102)* and up to 10 years under the *Intermediate Export Credit Guarantee Program (GSM-103)*. The programs are operated by USDA's *Commodity Credit Corporation*. The amount of coverage, including the interest rate and the guarantee fee, is established in the Office of the *General Sales Manager* and varies by country.

Creditor nation A country with a *balance of payments* surplus (the total amount owed to its government, private business, and banking interests from *foreign* sources exceeds the sum owed to foreign creditors).

Credit, supervised A technique of providing loans and intensive supervision by a management supervisor to help family farmers (see *family farm*) achieve successful commercial *farm* operation.

Creep feeder A feeding pen (see *lot*) built with a small entrance through which only young, small animals can enter. This prevents larger animals from crowding the smaller ones from the *feed*.

Creep, soil The downward mass movement of sloping (see *slope*) *soil*. The movement is usually slow and irregular and occurs most commonly when the lower soil is nearly saturated with water.

Crib A storage *bin* or granary exposed to the air and used for storing and drying *corn*. Also called a corn crib.

Crimped A process by which *grain* is rolled using corrugated rollers. The grain may be tempered or conditioned before crimping, and may be cooled afterward.

Crop acreage base A *farm's* average acreage (see *acre*) of *wheat*, *feed grains*, *cotton*, or *rice* planted for *harvest* and considered planted during each of the preceding five *crop years* or three crop years for cotton and rice. Acreage considered planted includes acreage reduced under an *acreage reduction program*, diverted, prevented from planting by natural disaster, and placed in *conserving uses* under the *0–85* provision or other provisions. Crop acreage bases are reduced during the life of the *contract* by the portion of base acreage placed in the *Conservation Reserve Program*.

Crop dryer A system of bins, *elevators*, or conveyers used to remove excess moisture from *corn*, *wheat*, and other crops. A crop dryer may provide heat to speed the *drying* process.

Crop duster A generic term for a number of different machines used to apply *pesticides*. Pesticides are agitated into a fine mist or dust and applied by plane, by a unit mounted on a truck (see *farm truck*) or specially designed *applicator* rig, or by a tank and applicator pulled by a *tractor*.

Crop failure Lack of or loss of harvested (see *harvest*) crops on acreage (see *acre*) due to poor *weather*, insect damage, and *diseases*. Also includes some crops not harvested due to lack of labor, low *market* prices (see *price*), or other factors. Excludes *cover crops* and *soil* improvement crops on acreage not intended for harvest.

Crop forecast Estimates of the probable *production* of a crop prior to *harvest*.

Crop insurance See *Federal Crop Insurance Program*.

Crop insurance indemnity The payment that a policyholder receives as settlement for a loss claim covered by the *Federal Crop Insurance Program*. The indemnity is the product of the *price* election for the crop and the number of *bushels* (or other unit) of loss that qualify for payment.

Crop insurance premium The amount that a farmer is charged for the purchase of "additional" crop insurance coverage (see *Federal Crop Insurance Program*). Among other factors, a farmer's per-*acre* premium depends on that farmer's *production* history and selection of coverage. The *premium* is subsidized by the federal government.

Cropland *Land* used for the *production* of adapted crops, like *corn*, *soybeans*, *wheat*, *hay*, and horticultural crops.

Cropland harvested Acreage (see *acre*) on which intertilled (see *till*) and closely sown crops, tree fruits, small fruits, planted tree nuts, and *hay* are harvested (see *harvest*).

Cropland pasture *Land* suitable for cultivation (see *cultivate*) and harvesting (see *harvest*) of crops used exclusively for grazing (see *graze*) during one or more consecutive growing periods; also, plant materials grazed from such land.

Cropland set-aside See *set-aside*.

Cropland used for crops *Cropland harvested*, *crop failure*, and *summer fallow*.

Crop report Forecasts or surveys of crop *production* from official or private sources. The *U.S. Department of Agriculture*'s (USDA) *National Agricultural Statistics Service* is the source of most U.S. crop reports. Similar agencies in Canada, Australia, and other *foreign* countries issue crop reports.

Crop residue pasture Plant materials grazed (see *graze*) on *land* from which a crop has been harvested (see *harvest*) during the *production* year.

Crop residues The remains of crop plants after *harvest* that supply organic matter to the *soil* and reduce soil *erosion*.

Crop rotation The practice of growing different crops in recurring succession on the same *land*. An example is *corn* followed by *soybeans*, followed by *oats*, and then *alfalfa*. Crop rotation plans are usually followed for the purpose of increasing *soil* fertility and to control weeds and insects.

Crop year The 12-month period over which a crop is harvested (see *harvest*) and marketed (see *agricultural marketing*). For *wheat*, *barley*, and *oats*, the crop year is from June 1 to May 31. For *corn*, *sorghum*, and *soybeans*, it is from September 1 to August 31. For *cotton*, *peanuts*, and *rice*, it is from August 1 to July 31.

Crop yield The amount measured in *bushels* or other units per *acre*, of a crop harvested (see *harvest*).

Cropping See *priming*.

Cross An offspring of two animals or plants of different genotypes, races, *breeds*, *varieties*, or *species*.

Crossbreeding The mating of two distinct *breeds* of animals.

Cross-compliance A requirement that a farmer participating in a program for one crop (*wheat, feed grains, cotton,* or *rice*) who meets the qualifications for *production* adjustment payments and loans for that crop must also meet the program provisions for other *program crops* that the farmer grows. This requirement is called full or strict cross-compliance. In a limited cross-compliance program, a *producer* participating in one *commodity* program must not plant in excess of the *crop acreage base* on that *farm* for any of the other program commodities for which an *acreage reduction program* is in effect. Strict cross-compliance provisions have not been enforced since the 1960s. Limited cross-compliance authority was implemented in the late 1970s and remained in effect under the *Food Security Act of 1985 (P.L. 99-198)* (see Appendix 3). The *Food, Agriculture, Conservation, and Trade Act of 1990 (P.L. 101-624)* (see Appendix 3) eliminated cross-compliance requirements.

Cross-contamination The *contamination* of one food by another. For example, vegetables cut on the same board as raw chicken can become contaminated with *salmonella* from the chicken.

Cross-fertilization *Fertilization* accomplished by the movement of *pollen* from one plant to another.

Crossflow dryer A type of on-*farm* commercial *grain* dryer (see *crop dryer*) in which grain and air flow in a perpendicular direction. This type of dryer tends to dry the grain nonuniformly, causing stress-cracking of the *kernels*.

Cross-pollinated crops A crop in which the flowers in individual plants may be fertilized (see *fertilization*) predominantly by *pollen* from other plants of the same *species*. Also known as open-pollinated.

Cross-pollination The process of transferring *pollen* from one plant to another plant of the same *species*. The process occurs naturally via wind or insects, or artificially as in the case of transferring pollen from a flower or plant to one of a different variety to produce a *hybrid*.

Crown Agency See *Canadian Wheat Board*.

Crude vegetable oil Extracted vegetable oils (see *fats and oils*) that have had no further *processing* or *refining* except possibly that of being degummed (see *degummed oil*) or filtered, settled, or both.

Crush See *crushing*.

Crusher See *crushing*.

Crushing The process that converts *oilseeds* into *meal* and oil (see *fats and oils*). Oilseed volume utilized in this process in the course of a year is called "crush." The *processors* who engage in this activity are called crushers.

Crust A thin, brittle layer of hard *soil* that forms on the surface of many soils when they are dry.

Crutching Removing *wool* from around the anus, vulva, and udder. Also known as tagging.

Crystalline fructose A form of *fructose* in a crystalline form similar to *sugar*. Crystalline fructose is usually made from *corn* and is sweeter than sugar, although it has different characteristics in baking and other uses. As of 1994, crystalline fructose is not competitive with sugar as a sweetener. It is, however, growing in use where some of its characteristics, such as attracting water, give it an advantage.

Cube cut (tobacco) See *smoking tobacco*.

Cucumber A fruit of a vine of the gourd family eaten fresh or used for pickling. The United States and Mexico are the major producers. The United States is a *net importer* of fresh cucumbers, with

Mexico being the single largest supplier in the U.S. *market*. Florida is the largest U.S. producer.

Cull The process of removing undesirable and/or unprofitable animals from a herd or flock. Culling is usually done to take unwanted breeding stock and send them to *slaughter*. See also *cull breeding animals*.

Cull breeding animals Male or female animals removed from the *breeding herd* generally for sale or *slaughter*.

Cultipacker A machine designed to first *till* the *soil* and then apply pressure to the soil surface, usually to prepare the ground for seeding (see *seed*).

Cultivar Certain cultivated (see *cultivate*) plants that are clearly distinguishable from others by one or more characteristics, which when reproduced retain those distinguishing characteristics. The term *variety* may be used synonymously with cultivar.

Cultivate 1: The process of preparing the *soil* for planting or loosening the soil between rows of growing crops to conserve moisture, uproot weeds, and aerate the soil.
2: Growing or tending plants or crops.

Cultivated summer fallow *Cropland* in subhumid regions of the West cultivated (see *cultivate*) for a season or more to control weeds and accumulate moisture before small *grains* are planted. Other types of *fallow* are excluded, such as cropland planted to *soil*-improvement crops but not harvested (see *harvest)* and cropland left idle all year.

Cultivator Any of several types of *tillage equipment* designed to work the *soil* to kill weeds, conserve moisture, and prepare the ground for seeding (see *seed*). A cultivator may have any number of different shaped *shovels* mounted on *shanks*. Shanks may be flexible or have spring tension devices to "give" when the shank hits a rock or other obstruction. Two main types include the *field cultivator* and *row crop cultivator* (designed to work soil between plant rows).

Cultivator disk hiller A *cultivator/disk* combination designed to build small mounds, or hills, to prepare the seedbed.

Cultivator fenders Flat, thin, vertical metal sheets mounted on *tillage equipment* to protect plants as the *cultivator* is pulled between crop rows.

Cultivator shank A steel bar, usually curved and flexible, that extends from the frame of a *cultivator* to the ground. *Shovels* or *shapes* are mounted on the end of the *shanks*.

Cultivator shape Any of different styles of *shovels* that are used to work the *soil*. Most have a single point with wings extending in two directions or may be double-pointed (reversible) with no wings. Shapes may be various angles, widths, and thicknesses.

Cultivator standards See *shank* and *cultivator shank*.

Cultural pest control Physical or mechanical farming methods to control pests, including *crop rotations*, removing *crop residue* soon after *harvest*, and clearing weeds from field borders.

Curd The coagulated or thickened part of *milk*. Curd from whole milk (see *milkfat content*) consists of *casein*, fat (see *fats and oils*), and *whey*. Curd from skim milk contains casein and whey but only traces of fat.

Cure 1: The process of preserving meat, fish, *tobacco*, *hides*, or lumber by salting or *drying*, for example. *Air-cured*, *flue-cured*, and *fire-cured* are the three basic methods of curing tobacco.

2: A treatment or procedure, such as a course of *drugs*, that eliminates the symptoms of a *disease* or illness.

Curing See *cure*.

Currency use payment A local currency payment under a *Public Law 480*, Title I agreement made to the U.S. government upon *commodity* delivery. See also *initial payment*.

Current account Part of a nation's *balance of payments* that includes the value of all goods and services imported and exported, as well as the payment and receipt of dividends and interest. A nation has a current account surplus if *exports* exceed *imports* plus net transfers to foreigners. The sum of the current and *capital accounts* is the overall *balance of payments*.

Current account surplus See *current account*.

Current dollars See *nominal dollars*.

Current Research Information System (CRIS) See *Cooperative State Research Service (CSRS)*.

Custom feeding The business of feeding *livestock* belonging to others, usually in a large-scale *feedlot*. The owner of the livestock pays for the *feed*, plus a fee for caring for the animals.

Custom grinding and mixing The service of grinding (see *grind*) customer-owned *feed* ingredients at a commercial feed mill (see *feedmilling establishment*) or by a mobile operator, usually for farmers feeding their own animals. However, the term applies to *toll milling*.

Customs A country's governmental agency authorized to collect *tariffs* or duties on imported (see *import*) and, less commonly, exported (see *export*) goods.

Customs classification A detailed classification, coding system, and description of goods that enter into international trade. The system is used by *customs* officials as a guide in determining which *tariff* rate applies to a particular *import* item. See *Harmonized Commodity Description and Coding System*.

Customs Cooperation Council Nomenclature (CCCN) A *customs classification* system used by most countries until 1988 when it was replaced by the *Harmonized Commodity Description and Coding System*.

Customs union A regional grouping of countries that levies common external duties (see *tariff*) on *imports* from nonmember countries but that eliminates tariffs, *quotas*, and many other governmental restrictions on trade among member countries. A customs union often is referred to as a *common market* or a tariff union.

Customs valuation The determination by *customs* officials of the value of imported (see *import*) goods for the purpose of collecting *ad valorem tariffs*. The customs valuation is based on the declared value of the goods.

Custom work Specific *farm* operations performed under *contract* between the farmer and the contractor. The contractor furnishes labor, equipment, and materials to perform the operation. Custom harvesting (see *harvest*) of *grain*, spraying and picking fruit, and *sheep* shearing are examples.

Cutability The expected *yield* of saleable meat from a *carcass.*

Cutback juice Fresh juice added to frozen concentrated *orange* juice to bring the soluble solids to 42 percent.

Cut Christmas tree Any evergreen conifer-bearing (see *softwood*) tree of the pine, spruce, or fir *species* cut for ornamental purposes, including such species as Douglas fir, noble fir, Fraser fir, grand fir, canceler fir, balsam fir, scotch pine, Virginia pine, eastern white pine, Austrian pine, red pine, Norway spruce, white spruce, and Colorado blue spruce.

Cut cultivated greens Plants grown and marketed (see *agricultural marketing*) for their showy foliage. Examples are cut leatherleaf ferns and chamaedorea.

Cut flowers Plants that are grown for their blossoms. Several crops of cut flowers can be produced from the same area each year. After the flowers are cut, the plants are either discarded or used for the production of more flowers, depending on the *variety* of the plant. Examples are roses, orchids, carnations, and chrysanthemums.

Cutout The *yield* of beef in percent of retail cuts that can be sold from a *carcass.* It is essentially the amount of meat versus fat (see *fats and oils*), bone, and waste.

Cutterbar A long, narrow, horizontal part of mowers, *combines*, *swathers*, and other harvesting (see *harvest*) equipment, mounted close to the ground, that cuts *hay*, *grains*, and other crops. Usually consists of stationary and moving sickle blades driven by a pitman.

Cutup A ready-to-cook *poultry* carcass (see *carcass*) cut into eight or nine pieces. Some are halved or quartered.

cwt See *hundredweight.*

Cyclamate An artificial sweetener banned in the United States in 1970 by the *Food and Drug Administration (FDA)* because it was believed to cause cancer. It is 30 times sweeter than *sugar.*

D

Dairy Diversion Program See *Milk Diversion Program.*

Dairy Export Incentive Program (DEIP) A program authorized by the *Food Security Act of 1985 (P.L. 99-198)*(see Appendix 3) that offered subsidies (see *subsidy*) to *exporters* of U.S. dairy products to help them compete with other subsidizing nations. Payments were made by the *Commodity Credit Corporation* on a *bid* basis either in cash, in-kind (see *payment-in-kind*), or through certificates redeemable for *commodities.* The payment rates reflected the type of dairy products exported, the domestic and *world prices* of dairy products, and other factors. Eligible sales had to be in addition to, and not in place of, those that would normally be made, and payments could not displace commercial *export* sales. The *Food, Agriculture, Conservation, and Trade Act of 1990 (P.L. 101-624)* (see Appendix 3) authorized the third DEIP. About 19,000 metric tons of dairy products were sold under the first DEIP and 5,000 metric tons of *butter* oils (see *fats and oils*) under the second.

Dairy Indemnity Payment Program A program that provides *indemnity* payments to dairy farmers whose *milk* has been removed from the commercial *market* because it contained residues of chemicals or toxic substances, including nuclear radiation or fallout.

Dairy Termination Program See *Milk Production Termination Program.*

Damping off A *disease* of planting *seeds* or very young seedlings caused by fungi (see *fungus*) that results in the death of the newly sprouted plants.

Danger zone The temperature zone (40 to 140 degrees F) in which bacterial (see *bacteria*) *disease* agents (see *agent*) and *molds* can grow in or on foods.

Dark air- and sun-cured tobacco Medium- to heavy-bodied *tobacco* that is used mostly for manufacturing *chewing tobacco* and *snuff,* although a substantial amount is also used in *cigarettes.* India, China, Burma, Indonesia, and Brazil are the major producers of dark air- and sun-cured tobacco. See also *burley tobacco*, *fire-cured tobacco*, *Oriental tobacco*, and *curing*.

Dark northern wheat See *hard red spring wheat.*

Dasheen See *taro.*

Day-haul operation A hiring process in which workers waiting at a pick-up point are hired, transported to the agricultural employment site, and returned to a drop-off point the same day.

Debt/asset ratio Total debt outstanding as of January 1, divided by the estimate of the current *market* value of assets (see *assets, current* and *assets, fixed*) owned by the *farm* business or sector.

Debtor nation A country with net *foreign* liabilities (the total amount that its government, private businesses, and banks owe to foreign creditors exceeds the sum due from foreign debtors). See also *balance of payments*.

Debt service The charges currently payable on a debt (usually for a given year), including interest and principle.

Deciduous fruit Fruit from trees that shed their leaves, such as *apple*, *pear*, and *cherry* trees. See also *deciduous trees*.

Deciduous trees Trees that lose their leaves at the end of the growing season. See also *deciduous fruit*.

Decitex (dtex) See *tex*.

Decontamination A food treatment, such as *pasteurization* and *irradiation*, that destroys a contaminant.

Decoupling Programs that separate government payments to farmers from the current or future quantity of a *commodity* produced or marketed (see *agricultural marketing*) and from the quantity of *farm inputs* used in *production*. Farmers make production decisions based on *market* prices (see *price*) but receive government payments independent of production and marketing decisions.

Defatted soy flakes *Flakes* formed when *soybeans* are cleaned, *dehulled*, flaked, oil-extracted (see *fats and oils*, *extraction, mechanical*, and *extraction, solvent*), heated, and cooled. The *protein* content ranges from 52 to 55 percent.

Defatted soy flour *Flour* produced by grinding (see *grind*) soy *flakes* after almost completely removing the oil (see *fats and oils*) from *soybeans* by solvent extraction (see *extraction, solvent*). Defatted soy flour is ground finer than soy *grits* and usually contains about 1 percent fat (see *fats and oils*).

Deferred delivery A *contract* provision calling for the seller to turn over ownership of a product to the buyer on one or more future dates.

Deferred payment A payment made under a negotiated *market* contract to farmers who do not wish to sell a *commodity* in the year it is produced. Under a deferred payment *contract*, the commodity may or may not be moved off the *farm* at the time of the contract signing.

Deferred pricing See *formula pricing*.

Deficiency payment A direct government payment (see *direct payments*) made to farmers who participate in *wheat*, *feed grain*, *rice*, or *cotton* programs. The payment rate is per *pound*, *bushel*, or *hundredweight*. It is based on the difference between the *price* level established by law (*target price*) and either the *market* price during a period specified by law or the *loan rate*, whichever is higher. The total payment is equal to the payment rate multiplied by the eligible payment acreage (see *crop acreage base*) planted for *harvest* and then multiplied by the *program yield* established for the particular *farm*.

For example, the 1990/91 target price for *corn* was $2.75 per bushel and the loan rate, $1.96.

If the market price during the first five months of the *marketing year* had been less than $1.96 per bushel, the deficiency payment would have been $0.79 per bushel. If the market price during this time had reached $2.00, the payment would have totaled $0.75 per bushel. When the market price during the specified time period exceeds the target price, no deficiency payment is made. The following actual case illustrates how to calculate the deficiency payment rate:

Calculating the Deficiency Payment Rate for Corn, 1990/91

Step 1	Target price	$2.75
	Market price	$2.28
	Difference	$0.47*
Step 2	Target price	$2.75
	Loan rate	$1.96
	Difference	$0.79

In this two-step process, the smaller difference (indicated with an asterisk) becomes the deficiency payment rate, which in this case is $0.47.

Additional deficiency (emergency compensation) payments for *wheat* and *feed grains* must be made whenever the *Findley loan rate* is in effect and season average market prices for wheat and feed grains are below the statutory loan rate. These payments are sometimes referred to as 12-month deficiency payments.

Defoliate 1: To strip off leaves to facilitate harvesting (see *harvest*). Defoliating is often done by spraying plants with a chemical.

2: To lose the leaves of a plant due to *disease* or insects.

Degermination The process whereby *corn* kernels (see *kernel*) are broken apart into *endosperm*, *germ*, and *pericarp*.

Degummed oil The product resulting from washing crude vegetable oil (see *fats and oils*) with water and/or steam for a specified period of time and then centrifuging the oil-and-water mixture to remove the phosphatides and other impurities.

Dehull To remove the outer covering of *grains* or other *seeds*.

Dehydrate To remove most or all of the moisture from a substance to preserve it, primarily through artificial *drying*.

Dehydrated onion Onion preserved by removing the water with a *processor*. See also *dehydrate*.

Dehydrating onion Dry *onions* intended for dehydration (see *dehydrate*). Dehydrating onions usually contain higher soluble solids than onions intended for fresh consumption.

Dehydration The removal or loss of water.

Delaney Clause The popular name of the 1958 Food Additive Amendment to the *Federal Food, Drug, and Cosmetic Act* (P.L. 75-717). The amendment prohibits the use of direct *food additives* that have been shown through appropriate tests to cause cancer in humans or laboratory animals. *Color additives* are also subject to the Delaney Clause. The clause implies a "zero cancer risk" standard for processed (see *processing*) foods.

Delayed pricing A farmer-first *handler* contract provision calling for *price* to be determined by formula some time after ownership has been transferred to the buyer. *Contracts* including this feature are sometimes called "price-later" contracts. See *formula pricing*.

Delinting Separating the very short *cotton* fibers (*linters*) remaining on the *seed* after the longer *fibers* have been removed in the *ginning* (see *gin*) process.

Delivered sale The terms of sale calling for the seller to deliver the product either to the buyer's place of business or another location agreed upon by both parties. The seller assumes the risk of damage and loss in transit.

Delivery month The specified month within which a *futures contract* matures and can be settled by delivery.

Delivery point A location where a *commodity* may be delivered to fulfill a *futures contract.*

Demand The desire to possess a *commodity* coupled with the willingness and ability to pay prevailing *market* prices.

Denier A metric measure of the linear density of *fiber,* defined as the weight in grams, of a 9,000 meter length of fiber or the fiber assembly (*yarn*, *roving*, etc.)

Denim A relatively heavy, *yarn*-dyed twill fabric (see *cloth*) traditionally made of *cotton* with colored *warp* yarns and undyed fill yarns.

Denitrification The bacterial (see *bacteria*) reduction of nitrate to *nitrogen* gas, nitrous oxide, and nitric oxide resulting in a loss of nitrogen available from the *soil.*

Deodorizing Producing bland edible oils (see *fats and oils*) by removing trace constituents that result in undesirable flavors and odors. Steam is passed through the hot oil, usually under vacuum, to deodorize it. This step is usually done after *refining* and *bleaching*.

Deoxyribonucleic acid See *DNA*.

Desert Land Act of 1877 See *Homestead Act of 1862.*

Designated nonbasic commodities *Commodities* other than *basic commodities* for which the secretary of agriculture is authorized and directed to provide price support (see *price-support program*). These commodities include *tung* nuts, *soybeans*, *honey*, *milk*, *sugar beets*, and *sugar-cane*.

Devaluation An official reduction of the *exchange rate* of a nation's currency that lowers the *price* of domestic currency to foreigners and raises the price of *foreign* currency. Prices of a nation's *imports* rise after devaluation and the cost of *exports* to foreigners declines. Devaluation is done to address *balance of payments* problems.

Developing country A country whose economy is mostly dependent on *agriculture* and primary *resources* and does not have a strong industrial base. In 1992, the *World Bank* defined developing countries as low income (less than $610 per capita) and middle income ($610 to $7,620 per capita). The phrase is often used synonymously with less developed country (LDC) and underdeveloped country.

Dewberry See *blueberry*.

Dextrose A *monosaccharide* produced commercially by the complete conversion of *starch*. Since dextrose has been produced largely from *corn* starch, it is commonly called refined corn syrup. The term dextrose is often used synonymously with *glucose*. Hydrate and anhydrous are the two principal types of dextrose. The hydrate type, containing approximately 8 percent moisture, comprises the largest share of dextrose. Anhydrous dextrose contains less than 0.5 percent moisture.

Dialdehyde starch A chemical derivative of *starch* derived from cereal (see *cereals*) *grains* used to improve wet strength of paper products and *tanning* leathers (see *leather*) and for other purposes.

Dietary Guidelines for Americans Seven guidelines developed by the *U.S. Department of Agriculture* and the *U.S. Department of Health and Human Services*, emphasizing variety, balance, and moderation in the total diet without making recommendations regarding specific foods to include or to eliminate.

Digestible protein The portion (usually 75 to 85 percent) of the *protein* in *feed* that an animal can digest.

Digger Equipment designed especially for harvesting (see *harvest*) root crops, such as *sugar beets*, *potatoes*, or *carrots*.

Diglyceride See *emulsifier*.

Dillon Round The fifth round of *multilateral trade negotiations* held under the *General Agreement on Tariffs and Trade (GATT)* during 1960–1961 in Geneva. See *GATT rounds*.

Dimethyl sulfoxide See *black liquor*.

Diosgenin A compound derived from wild yams that was used to produce steroid medications. Today, such medications are made from *byproducts* of *soybean* processing (see *processing*). See *sitosterol* and *stigmasterol*.

Direct-consumption sugar Any crystalline or liquid *sugars* that do not need to be further refined (see *refined sugar*) or otherwise improved in quality.

Direct (export) credit A federal *export promotion program* (designated as GSM-5) operated by the *Foreign Agricultural Service* of the *U.S. Department of Agriculture*. Loans were made directly by the *Commodity Credit Corporation (CCC)* at *market* interest rates to *foreign* buyers of agricultural *commodities*. The program was replaced by federally guaranteed commercial credit in fiscal year 1981. See *blended credit* and *guaranteed export credit*.

Direct government subsidies See *government payments*.

Direct marketing The practice of bypassing assembly points (middlemen). For example, a farmer may sell directly to a *processor* or a port terminal.

Direct payments Payments in the form of cash or *commodity* certificates made directly to *producers* to supplement their incomes in low *price* years, induce them to participate in certain programs, and help when natural disasters occur. These payments include *deficiency payments*, *paid land diversion* payments, *incentive payments*, storage payments, *Conservation Reserve Program* payments, and disaster payments. See also *generic commodity certificates*.

Direct sales 1: Sales by *processors* to retailers without passing through a wholesaler.
2: Sales of *livestock* by growers or feeders to *meat packers* without passing through a *terminal market.*

Disappearance See *utilization.*

Disaster payments See *Ad Hoc Disaster Assistance.*

Discounting A process used by economists to calculate the present value of costs or returns that are not expected to occur until some time in the future. *Market*-driven incentives are reflected in discounted net present values, calculated using market rates of interest. For example, at an interest rate of 7 percent, the present value of $1 million that will not be received for 200 years is only $1.32. The present value of a future cost of $1 billion is only $1.53 if those costs can be deferred for 300 years.

Discount rate The rate at which the *Federal Reserve* typically lends to banks. The rate is set by the Federal Reserve.

Discrimination The unequal treatment of internationally traded goods or services according to their source or destination.

Disease An abnormal condition in the body, the causes and symptoms of which can vary widely. See *acute* and *chronic.*

Disinfectant Any material that kills *microorganisms.*

Disk A *harrow* or *plow* composed of circular plates arranged at an angle with the line of pull. A disk is used to prepare *soil* for seeding. Disks come in a variety of styles, such as tandem (see *tandem disk harrow*), off-set, plow (see *disk plow*), and one-way (see *one-way disk*).

Disk blade A circular, concave, thin, steel wheel, sometimes with a serrated edge, used to lift and move *soil.* A *disk* blade is similar to a *coulter* or a disk opener (see *double-disk opener*), which is flat. Disk blades are found on implements, such as *tandem disk harrow* and *one-way disks.*

Disk harrow A *tillage tool* consisting of several groups of concave *disk blades* that throw the *soil* to the left and right. See also *disk plow.*

Disk hiller A machine with specially designed flat, circular steel plates to construct small ridges or mounds in fields to prepare a seedbed for planting or seeding (see *seed*).

Disk marker A flat, circular blade mounted at the end of an articulated boom for making a visible track in the field to let the operator know what part of the field has been seeded (see *seed*) or cultivated (see *cultivate*).

Disk plow A tool with a series of concave *disk blades* that throw the *soil* in the same direction. A *disk harrow*, in contrast, moves the soil in both directions.

Disk ripper *Tillage equipment* with *coulters* in front of *shanks.* A *disk* ripper is designed to cut through stubble or *mulch* and break up the *soil* and leave *crop residue* on the surface to reduce soil *erosion.*

Disposable personal income The income available to individuals after deducting income taxes.

Distant futures One or more of the latest *futures contract* delivery months, among those currently traded for a *commodity*.

Distressed sale Sale of merchandise that the seller is anxious to move either because of its location (rejected as below specification at delivery to the buyer), possible quality deterioration if held longer, or because of large unsold current *production* or *stocks*.

District Banks for Cooperatives See *Bank for Cooperatives (BC)*.

Ditcher A machine used to make ditches for irrigation (see *acres irrigated* and *irrigated farms*), laying pipe, or for drainage.

Diversification When a firm undertakes activities outside of its primary product line or service.

Diversion payments Government payments made to farmers in some years for not planting a specified portion of their *crop acreage base* or *permitted acreage*. A specified acreage is usually diverted to *soil* conserving uses.

Diversion terrace An earthen *terrace* designed to prevent *runoff* from precipitation crossing *feedlots* and becoming contaminated with feedlot wastes.

Divestiture The sale of a factory, division, or subsidiary of a firm, either to another firm, to management of the unit, or to independent investors.

DNA The acronym for deoxyribonucleic acid, the molecule that carries the genetic information for most living systems. The DNA molecule consists of four bases (adenine, cytosine, guanine, and thymine) and a *sugar* phosphate backbone arranged in two connected strands to form a double helix. See also *RNA*.

Dockage The *foreign* material in *market* grain, such as dust, *chaff*, *seeds*, other *grains*, stems, weeds, and sand, which is readily removed by ordinary cleaning devices. Dockage is a factor in the *grading* of grains and *oilseeds*.

Dockworker See *longshoreman*.

Dominant A characteristic possessed by one of the parents of a *hybrid* which is manifested in the hybrid to the apparent exclusion of the contrasted character from the other (recessive) parent.

Dormancy A condition of a *seed* or *bud* that prevents it from germinating (see *germination*) or sprouting under normal conditions. The term is also applied to the period between leaf fall and spring when there is no growth on *deciduous trees* or vines.

Downy mildew A type of fungal (see *fungus*) *disease* of plants caused by various parasitic (see *parasite*) lower fungi that adversely affects a wide variety of fruits, vegetables, and *grains*. The disease is generally characterized by whitish masses on the lower surface of the leaves of the infected *host* plant or any of the various parasitic lower fungi that cause these diseases.

Double cross A cross between two first-generation *hybrids*.

Double cropping The practice of growing successive crops on the same *land* during a single year. The crops may be the same or different. For example, after harvesting (see *harvest*) *wheat* in early summer, a farmer could plant *corn* or *soybeans* on the same acreage (see *acre*) for harvest in the fall.

Double-disk opener A set of two *disks* mounted at an angle so that the "V" part of the angle enters and spreads the *soil*, making a *furrow* for seeding (see *seed*). A *drill* has several openers, usually preceded by *coulters* to cut through stubble, and followed by *packers* to push the soil in place around the seed.

Double-zero rapeseed *Canola,* so-called because it has less than 2 percent erucic acid in the oil (see *fats and oils*) and low to zero levels of *glucosinolates* in the *meal.*

Drafting Attenuating a *fiber* assembly. Often this is achieved by passing the mass from one pair of weighted, rotating rollers to another pair moving faster, and set at a distance a little greater than the *staple length* of the fibers.

Draft horse See *horse.*

Drainage The removal of excess surface water or excess water from within the *soil* by means of surface or subsurface drains.

Drainage, soil The rapidity and extent of the removal of water from the *soil* by *runoff* and flow through the soil to underground spaces. As a condition of the soil, the frequency and duration of periods when the soil is free of saturation. For example, in well-drained soils, the water is removed readily, but not rapidly; in poorly drained soils, the root zone is waterlogged for long periods and the roots of ordinary crop plants cannot get enough oxygen; and in excessively drained soils, the water is removed so completely that most crop plants suffer from lack of water.

Drainage tile Clay, concrete, or elastic pipe used to provide *drainage.*

Drawback A practice authorized by the *U.S. Customs Service* that allows an *exporter* to obtain a refund for up to 99 percent of any duties (see *tariff*) and fees paid on an imported good that is *reexport*ed or used in the manufacture of exported products. An exporter is eligible for drawback if (1) the product was made within three years of the date the components were imported; (2) the product was exported within two years of manufacture; and (3) documents were received by the U.S. Customs Service within three years of the date the product was exported.

Drawbar A hitch mounted on the rear of a *tractor*, consisting of a length of flat steel, varying in width and thickness depending on the size of the tractor. A drawbar is used to pull *farm* implements. Most are designed to move either from side to side or to be held rigid by drawbar pins.

Drawbar horsepower The amount of power developed by a *tractor* at the *drawbar* (hitch point). It is usually equivalent to about 80 percent of the engine's *horsepower.*

Drawframe A machine that performs the process of drawing sliver, to parallelize and blend *fibers* in addition to minimizing the mass variation along the length of the sliver. It is used after *carding* and before and after *combing*. Generally, sliver is "drawn" twice before it is supplied to the roving frame or *spinning* machine.

Drawing The process in which a number of *slivers*, typically six or eight, are simultaneously drafted to produce a sliver of similar size to one of those supplied.

Dressed animal The body of an animal without *hide*/feathers, head, feet, tail, and most internal organs. The parts removed when an animal is dressed vary slightly among *species.*

Dressed weight The weight of a chilled animal *carcass.* See *dressed animal.*

Dressing percentage A measure of the percentage *yield* from slaughtered (see *slaughter*) animals derived by comparing the weight of a chilled *carcass* with its *liveweight*. For example, a live hog (see *swine*) that weighed 200 pounds on hoof and yielded a carcass weighing 140 pounds would have a dressing percentage of 70. Dressing percentage is important in determining the *market* price (see *price*) of meat animals sold for slaughter.

Range of Avg. Dressing Percentages

U.S. Choice Beef	59–65 percent
U.S. Choice Lamb	48–52 percent
U.S. No. 1 Hog	67–74 percent

Dried onion See *dehydrated onion.*

Drill A *farm* machine for seeding (see *seed*) *grain* in narrow rows. A drill usually consists of a series of *double-disk openers*, usually preceded by *coulters* and followed by *packers, harrows*, and other devices to spread *soil* over the seed. Bins or seed buckets are mounted on a frame above the openers to hold grain or grass seed. There are geared assemblies below the bins to meter the number of seeds entering the openers. See *grass seeder, no-till drill*, and *range drill.*

Drilling Opening the *soil* to receive the *seed*, planting the seed, and covering it in a single operation.

Drinking places See *foodservice.*

Drip irrigation See *irrigation methods.*

Drone A male *honeybee.*

Drone egg The unfertilized *egg* of the *honeybee.*

Drought A prolonged period of dryness. *Soil* moisture is generally depleted to such an extent that plant growth is seriously retarded.

Drug A substance (1) intended for use in the diagnosis, *cure*, mitigation, treatment, or prevention of *disease* in man or other animals or (2) a substance other than food intended to affect the structure or any function of the body of man or other animals.

Drug residue Trace amounts of a veterinary *drug* in meat and poultry products resulting from treating a sick animal or from low-level drug use in feed to promote *feed* efficiency. Sensitivity of testing methods means residues may be detected below the level harmful to human health. See also *insecticide residue* and *pesticide residue.*

Dry-bulk carrier See *bulk carrier.*

Dry cow A *cow* that is not producing *milk*. Generally, the period before the next *calving* and lactation (see *lactation period*).

Dry edible bean The edible *seed* of a number of *varieties* of leguminous (see *legume)* plants. Important varieties in the United States include *Navy, Great Northern, Pinto,* Pink, Small Red, Blackeye, *Red Kidney*, *Black,* Small White, Lima, and *Garbanzo*. The United States, Brazil, Mexico, Canada, Argentina, and Chile are the major producers.

Dry edible peas (beans) The round *seeds* of a number of leguminous (see *legume)* plants that are harvested (see *harvest*) in dry form. Primarily cool season crops, dry edible peas are grown in the

Pacific Northwest portion of the United States. See also *dry edible beans.*

Dryer See *crop dryer.*

Dry farming Producing crops that require tillage (see *till*) in subhumid or semiarid regions without irrigation (see *acres irrigated* and *irrigated farms*). The system usually involves strips of *fallow* between crops from which water from precipitation is absorbed and retained.

Drying Removing moisture from foods, such as *grain,* by various methods. In the case of grain, air temperature, grain velocity, and air-flow rate during the drying process have a greater influence on grain quality than all other grain handling operations combined.

Drying oils Oils (see *fats and oils*) that have the ability to polymerize or "dry" after they have been applied to a surface to form tough, adherent, impervious, and abrasive-resistant film. Their film-forming properties are closely related to their degree of saturation (see *fatty acids, classification of*).

Dryland farming A system of producing crops without irrigation (see *acres irrigated* and *irrigated farms*) in semiarid regions that usually receive less than 20 inches of annual rainfall. Frequently, part of the *land* will be *fallow* in alternate years to conserve moisture.

Dry lot A relatively small enclosure without vegetation, either with a shelter or an open yard, in which animals may be confined indefinitely.

Dry lot cattle feeding Fattening *cattle by* confining them in a small area and feeding them carefully mixed, high-*concentrate* feed (see *feed*).

Dry-milling The basic process used to mill (see *milling*) *wheat* and *corn*, involving the cleaning, conditioning (see *tempering*), grinding (see *grind*), and sifting of *grain.* The process uses a small amount of water or steam to facilitate the separation of the various component parts of the *kernel.*

Dry onions Dry bulb *onions. Bulb* onions are usually dug and left to dry in the field several weeks before *harvest.*

Dry rendering Residues of animal tissues cooked in open steam-jacketed vessels until the water has evaporated. *Fat* (see *fats and oils*) is removed by draining and pressing the solid residue.

Dry weight The weight of *tobacco* as it is usually traded internationally.

Dual exchange rate system See *multi-tier system.*

Dual pricing The selling or buying of an identical product for different *prices* in different *markets.* Under a dual pricing arrangement, for example, a *commodity* might be sold for a lower price in the international market than in the domestic market. This can result from *export subsidies* and *dumping.*

Duck See *poultry.*

Dumping The sale of products on the world *market* below the *cost of production* to dispose of *surpluses* or gain access to a market. Dumping is generally recognized as an *unfair trade practice* because it can disrupt markets and injure *producers* of competitive products in an importing country.

Dump rake An implement designed to collect *hay* and periodically release (dump) it into rows. A dump rake consists of large, curved, flexible tines mounted on a movable frame that lifts the tines when a trip mechanism is activated. Now technologically obsolete.

Duopoly A *market* characterized by two sellers.

Durable press A finishing treatment for *textile* materials designed to assure the retention of specific contours, including creases and pleats, during normal use, washing, and/or dry cleaning. Used synonymously with "permanent press" and "wash-and-wear."

Durum wheat (*Triticum durum*) A spring-*seed* very hard, high-*protein* wheat. The *wheat* is milled (see *milling*) for *semolina*, which is used in *pasta* products.

Dust Fine, dry, pulverized particles of matter usually resulting from cleaning or grinding (see *grind*) *grain*.

Duster See *crop duster*.

Duty See *tariff*.

E

Earth pea See *bambara groundnut.*

Eating places See *foodservice.*

EC **92** The *European Union (EU)* plan to eliminate impediments to the free movement of goods, services, capital, and people among member countries. EC 92 went into effect on January 1, 1993.

E. coli See *Escherichia coli.*

***E. coli* O157:H7** See *Escherichia coli* and *virulence.*

Ecology The branch of biology that deals with the mutual relations among organisms and between organisms and their *environment.*

Economic Community of Central African States An organization of 10 nations established in 1983 to promote regional economic cooperation and establish a Central African Common Market (see *Common Market*). Members are Burundi, Cameroon, Central African Republic, Chad, Congo, Equatorial Guinea, Gabon, Rwanda, Sao Tome, and Principe, Zaire. Also called Communaute Economique des Etats de l'Afrique Centrale (CEEAC).

Economic Community of West African States (ECOWAS) An organization of 16 West African nations established in May 1975 to promote regional economic cooperation. The members are Benin, Burkina, Cape Verde, The Gambia, Ghana, Guinea, Guinea-Bissau, Cote d'Ivoire, Liberia, Mali, Mauritania, Niger, Nigeria, Senegal, Sierra Leone, and Togo.

Economic costs A measurement of *farm* costs and returns. It includes cash and noncash costs, calculated as if the *producer* fully owned all *production* assets (see *assets, current* and *assets, fixed*). Cash expenses (except interest payments), capital replacement, and imputed returns to capital, *land,* and unpaid labor are included. The measure is typically used to adjust for inflation to account for quantities so as to maintain the production asset base.

Economic indicators Statistics used by the governments to describe the state of the economy. Examples include the unemployment rate and the *Gross National Product (GNP).*

Economic Research Service (ERS) A *U.S. Department of Agriculture* (USDA) agency responsible for economic data and analyses and social science information needed to develop, administer, and evaluate agricultural and *rural* policies and programs.

Economic sanctions Restrictive trade or fiscal actions by one or more countries to try to force another country to change a certain course of action.

Economies of scale See *economies of size*.

Economies of scope The lower unit costs resulting from combining two or more functions or stages, such as assembly and *processing*. Economies of scope may also result from joint processing of products made from the same raw materials or joint distribution of products using common facilities.

Economies of size The increase of returns as factor use is expanded in least-cost combinations. Economies of size permit large *producers*, therefore, to manufacture and *market* their products at lower average costs per unit than small producers. Once an operation reaches a certain size, the marginal cost of producing additional output begins to decline.

The different types of economies include: (1) product-specific that are associated with the volume of any single product made and sold; (2) plant-specific that are related to the total output of an entire plant (see *plant or establishment*) or plant complex; and (3) multiplant that are associated with an individual firm's operation of multiple plants. Also referred to as economies of scale.

Economies of speed Economies that result from running the *production* unit at higher speed because of integration and coordination of the flow of materials through the production line.

Ecosystem An ecological community involving plants, animals, *microorganisms*, moisture, heat, wind, solar radiation, and surface materials in a landscape, which, considered together with the nonliving factors of its *environment*, form a unit.

Efficient production The least-cost manufacture of goods.

Effluent The outflow of water from a subterranean storage space.

Egg The reproductive unit produced by a female. *Chicken*, duck, *turkey*, or *goose* are the principal types for hatching or table use.

Egg dealer A *farm* or firm that assembles, *grades*, packs, and wholesales (see *wholesaling*) *eggs*.

Egg grader A machine that sorts *eggs* by weight into size categories and is typically combined with a *candler* to determine *grade* or quality (e.g., AA, A, B).

Egg layers The type of *poultry* used for *production* of *eggs*.

Eggplant The fruit of a *perennial* herb (see *herb*). Major producers include the United States (Florida) and Mexico.

Egg Products Inspection Act A law to ensure that *eggs* and egg products are wholesome and unadulterated. Under the act, the *U.S. Department of Agriculture*'s (USDA) *Food Safety and Inspection Service (FSIS)* provides continuous mandatory inspection (see *inspection, federal*) in all plants (see *plant or establishment*) *processing* liquid, dried, or frozen egg products. The act also controls the disposition of restricted shell eggs that might contain harmful *bacteria* that could cause foodborne illness. Under the act, egg products from a *foreign* country can be imported into the United States only if the country's inspection system is equivalent to that of the United States. Canada and the Netherlands are the only countries eligible to *export* egg products to the United States.

Egg size Categories of *eggs,* depending on the size by weight. *Market* sizes are: jumbo (30-ounce dozen); extra large (27-ounce dozen); large (24-ounce dozen); medium (21-ounce dozen); small (18-ounce dozen); and pewee (15-ounce dozen).

Egg washer A machine used to wash *eggs* with hot water and a cleanser.

Elasticity of demand The responsiveness of quantity demanded to changes in the *price* of a product. The elasticity of demand is measured by the percentage change in quantity demanded relative to percentage change in price.

Electronic Benefits Transfer (EBT) An electronic system for transferring money from a food stamp (see *Food Stamp Program)* or other public assistance account to a retailer's account. Food stamp customers make purchases by using a plastic debit card at store terminals to access their available benefits.

Electronic control Any number of devices used to regulate the function and rate of various *farm*-related operations. For example, electronic control may be used to change the temperature of heated buildings at different times during the day; to apportion *feed* mixes for *livestock*; to regulate ignition timing in *tractors* and trucks (see *farm truck*); and to change settings for *pesticide* and *fertilizer* applicators (see *applicator*), depending on requirements in specific parts of the field.

Electronic GPS (Global Positioning System) Satellite navigation, sometimes supplemented by ground-based location towers and used to accurately locate ground position. The system is increasingly being used for precise application of *fertilizers* or *pesticides* based on particular *soil* and plant requirements in different parts of the field.

Electronic market Selling *commodities* by *auction* conducted over an electronic communication mechanism (computer, teletype, or telephone conference call). All buyers having access to the system can participate. A rigorous product description understood by all potential traders is necessary for conducting such an auction without the physical presence of buyer, sellers, or product being traded.

Electronic milking monitor A system that records and stores data on *milk* production (see *production*) for each animal. The information can be tied in with *feed* consumption types, rates, and costs to determine the economic efficiency of each animal in a dairy herd.

Electronic monitor Any number of devices used to observe various *farm*-related operations and found mounted in *tractor* cabs, farm buildings, and dwellings. The monitoring devices may show a variety of observations, including position in the field, fuel consumption rates at various speeds, loads and gears, on-the-go crop *yields* by position in the field, field *ration* mixes, and planting and seeding (see *seed*) rates. The devices are often computer-controlled and may be used in conjunction with data bases.

Electronic yield measurer A device mounted in the cab of a *combine* that monitors the speed, width of cut, and amount of *grain* conveyed to the bin to determine crop *yield*, depending on the position in the field. The device may have the capability to record yield and field position into a data base.

Electrophoresis A technique used to identify and categorize *grain* varieties (see *variety*). Molecules are separated from a mixture of similar molecules by passing an electric current through a medium containing the mixture. Each type of molecule travels through the medium at a rate corresponding to its electric charge and size. Separation is based on differences in net electrical charge and in size and arrangement of the molecules.

Elevator 1: A machine for moving *grain, forage,* and other *farm* crops from the ground to a bin (see *bin, grain*), from one bin to another, and so forth. The machine uses buckets, chains, belts, augers, or other means of conveyance and can be powered by *power-take-off (PTO)* or gas or electric motors.
2: A commercial building where grain is brought by farmers to be stored, sold, and loaded on trains or large trucks (see *farm truck*) for delivery to central terminals.

Elevator leg Part of the belt-bucket system used in commercial *grain* facilities. It consists of an endless vertical belt with buckets spaced evenly along it. The buckets scoop up the grain at the bottom (boot) of the leg and discharge it at the top.

Embargo A government-ordered prohibition of trade with another country restricting all trade or that of selected goods and services.

Embryo A fertilized (see *fertilization*) mammal *egg*. See also *embryo transfer*.

Embryo transfer The transplantation of a fertilized (see *fertilization*) *embryo* into a recipient female animal for the duration of gestation (see *gestation period*). Most widely practiced with *cattle*, it is used to improve herd characteristics by selecting genetic material, both male and female, superior to existing stock.

Emergency Compensation Payments See *Findley loan rates*.

Emergency Conservation Program (ECP) A *U.S. Department of Agriculture* program that provides emergency funds for sharing with farmers and ranchers (see *ranch*) the cost of (1) restoring farmland (see *farm*) to productive use that has been seriously damaged by natural disasters or (2) carrying out emergency water conservation measures during periods of severe *drought*. ECP assistance is available only to help solve new conservation problems caused by natural disasters that impair and endanger the *land* or its *productive capacity*. The damage must be unusual (with the exception of wind *erosion*) and not likely to recur frequently in the same area. Assistance is based on the type and extent of the damage.

Emergency Feed Assistance Program (EFAP) A program that provides for the sale of *Commodity Credit Corporation*-owned *grain* at 75 percent of the basic county *loan rate* to *livestock* producers whose *harvest* of *feed* has suffered because of *drought* or excess moisture. Eligible livestock *producers* must have insufficient feed available to preserve and maintain their breeding stock. The secretary of agriculture must declare a county a natural disaster area before this program can be implemented in that county. See also *cost-sharing program*.

Emergency Feed Program (EFP) A program that provides disaster assistance to eligible *livestock* owners by sharing the cost of *feed* purchased to replace the *farm's* normal *production* and feed purchased in quantities larger than normal because of an emergency. This program requires the secretary of agriculture to declare the county a natural disaster area before implementation. The program is also called the Feed Cost-Sharing Program (see *cost-sharing program*).

Emulsifier An agent that markedly lowers the interfacial tension between oil (see *fats and oils*) and water or other liquid, permitting them to mix. *Lecithin* and mono- and diglycerides are emulsifiers derived from fats and oils and widely used in food products, such as *margarine*, *shortening*, salad dressing, frozen desserts, peanut butter, and candy.

End A *warp* thread or *yarn* that runs lengthwise or vertically in the fabric (see *cloth*). Ends interlace at right angles with *filling* yarn (picks) to make woven (see *weave*) fabric.

Endangered species Plants and animals whose numbers are so small that the *species* is threatened with extinction. The list, compiled by the U.S. Fish and Wildlife Service, includes approximately 70 mammals, 100 birds, 50 reptiles and amphibians, 250 plants, 50 clams and snails, 25 insects, and 90 *varieties* of fish. Federal laws prohibit hunting, killing, capturing, harming, or harassing a protected species.

End-gate seeder A small machine with rotating *disks* fastened behind a *wagon* or truck (see *farm truck*) used to broadcast *seed.*

Ending stocks Inventories of a product in storage and ready for sale at the end of a reporting period, or end of a *marketing year.*

Endosperm The nutritive tissue formed within the *seed* of plants and on which the *embryo* feeds while germinating (see *germination*). In *wheat*, it is the starchy (see *starch*) portion of the *grain* that is ground (see *grind*) into *flour*. It contains *protein*, riboflavin, and B vitamins. In *corn*, the quantity of vitreous or horny endosperm relative to floury endosperm in the *kernel* determines the hardness of the grain.

End-use The final form in which a product is consumed. In the case of *cotton*, for example, end uses include apparel, household products, and industrial items.

Enrollment (signup) period A period prescribed by the *U.S. Department of Agriculture* (USDA) during which farmers can enroll in a crop *price support* or *production adjustment program* for the coming year. An enrollment period can last up to several months.

Ensiled The process of blowing finely cut plant material into an air-tight chamber such as a *silo*. The material is pressed to remove the air and undergoes an acid *fermentation* that retards spoilage.

Enterprise for the Americas Initiative (EAI) An effort announced by President George Bush on June 27, 1990, to encourage political and economic reform in Latin America and the Caribbean by promoting *free trade*, entrepreneurship, and economic growth.

Entrepot An intermediary center of trade where goods are stored temporarily for distribution within a country or for *re-export*. See also *foreign trade zone*, *re-export,* and *sufferance*.

Entry (customs) A statement of the kinds, quantities, and values of goods imported together with duties (see *tariff*) due, if any, and declared before a *customs* officer or other designated officer.

Entry price The calculated *price* in the *European Union (EU)* of certain imported fruits and vegetables and certain fish at the border.

Environment The complex of climatic (see *climate*), edaphic, and biotic factors that act upon an organism or an ecological community and determine its form and survival, including sunlight, temperature, moisture, and other organisms.

Environmental Conservation Acreage Reserve Program (ECARP) A program authorized by the *Food, Agriculture, Conservation, and Trade Act of 1990 (P.L. 101-624)*(see Appendix 3) that includes the previous *Conservation Reserve Program (CRP)*. The ECARP places greater emphasis on preserving and upgrading water quality, identifying environmentally sensitive areas for special conservation treatment, and planting trees. The ECARP also includes a *Wetland Reserve Program* that has a goal of enrolling 1 million acres, including farmed and converted *wetlands*.

Environmental Defense Fund A national organization whose membership of scientists, lawyers, economists, and others work to find scientifically sound solutions to environmental problems.

Environmental Easement Program (EEP) A program authorized by the *Food, Agriculture, Conservation, and Trade Act of 1990 (P.L. 101-624)* (see Appendix 3) and designed to obtain easements for long-term protection of environmentally sensitive *lands*. Participating *producers* would be required to implement approved natural *resource* management plans in exchange for a maximum of $250,000 for the easement or, if less, the value of the land without an easement. Annual payments, not to exceed $50,000 a year, may be paid for up to 10 years. The EEP also allows for up to 100 percent cost sharing (see *cost-sharing program*) with the federal government to implement the management plan. The program had not been implemented as of mid-1994.

Environmental horticulture The *production* of live plants for water and soil conservation (see *conservation, soil*), wildlife preservation, air quality enhancement, and to provide shade and an aesthetic space. Most environmental horticulture plants are *perennial* and last several years. Examples of environmental horticulture plants include landscape trees, shrubs, bulbs, sod, groundcovers, and *nursery plants*.

Environmental Protection Agency (EPA) A federal agency created in 1970 to permit coordination of effective governmental action on behalf of the *environment*. The agency works to control pollution by proper integration of a variety of research, monitoring, standard setting, and enforcement activities.

Enzyme A substance produced by living *cells* that can bring about or speed up chemical reactions without undergoing change themselves.

Enzyme-linked immunosorbent assay (ELISA) A test to identify *proteins* and plant *pathogens* by using antibodies. A protein-antibody complex is incubated with an *enzyme*-coupled antibody that recognizes and binds to the protein. The reaction is measured spectrophotometrically (see *spectrophotometer*) to identify the presence of the specific protein that is attached to the antibodies.

Ephemeral erosion See *erosion*.

Epidermis The outermost layer of *cells* of a leaf or other plant part.

Equilibrium A state of balance between opposing forces or actions.

Equine Pertaining to *horses*.

Equity level A measurement of net worth. It is the balance that would remain if assets (see *assets, current* and *assets, fixed*) were sold and existing debt repaid. It is calculated as total operator assets minus debt outstanding.

Eradication The deliberate extinction of a *pathogen*. *Hog cholera*, for example, was eradicated from the United States in 1978 after systematic diagnosis and destruction of infected herds. See *sanitary and phytosanitary regulations*.

Erodibility index A value that combines a *soil's* inherent erodibility with its susceptibility to damage by *erosion*. Soil type, amount of rainfall and *runoff*, and *slope* length and steepness determine a soil's inherent erodibility. The susceptibility of a soil to erosion damage is inversely related to the soil's natural rate of formation. *Cropland* with an erodibility index of 8 or more is subject to the *conservation compliance provisions* of the *Food Security Act of 1985 (P.L. 99-198)* (see Appendix 3).

Erodible (soil) *Soil* susceptible to *erosion.*

Erosion The process by which water or wind moves *soil* from one location to another. Types of erosion are: (1) sheet and rill—a general washing away of a thin, uniform sheet of soil, or removal of soil in many small channels or incisions, caused by rainfall or irrigation (see *acres irrigated* and *irrigated farms*) *runoff*; (2) gully—channels or incisions cut by concentrated water runoff after heavy rains; (3) ephemeral—a water-worn, short-lived, or seasonal incision, wider, deeper, and longer than a rill, but shallower and smaller than a gully; and (4) wind—the carrying away of dust and sediments by wind in areas of high prevailing winds or low annual rainfall.

Escherichia coli A bacterium (see *bacteria*) found in the lower intestinal tract of most vertebrates. A particularly virulent strain, *E. coli* O157:H7, caused a severe foodborne *disease* outbreak in 1993 associated with undercooked hamburger.

Essential amino acid An *amino acid* that is needed for an animal's *metabolism* but cannot be synthesized by the animal in the amount needed and thus must be present in the diet of that animal.

Essential oils Oils (see *fats and oils*) extracted (see *extraction, mechanical* and *extraction, solvent*) from a variety of fruit, *herbs*, spices, flowers, and other plants to obtain concentrated flavorings and fragrances for food and beverage manufacturing and the cosmetics and toiletries industries. For fruit and vegetables, the oils are often *byproducts* of *processing* activities and are extracted from pits, *seeds*, *husks*, skins, or rinds. Conversely, oils extracted from nuts, herbs, flowers, and other plants are often the primary product of the processing function.

Ethanol An alcohol fuel that may be produced from an agricultural feedstock, such as *corn*, *sugarcane*, or wood. Ethanol may be blended with gasoline to enhance octane, reduce automotive exhaust pollution, and reduce reliance on petroleum-based fuels. Also known as ethyl alcohol.

Ethyl alcohol See *ethanol.*

EU Commission The executive body of the *European Union (EU)* that proposes legislation, implements EU policy, enforces EU *treaties*, manages the EU budget, and represents the EU in *trade negotiations.* The commission has investigative powers and can take legal action against member states and companies. There are 20 commissioners, two each from France, Germany, Italy, Spain, and the United Kingdom and one each from the remaining member states. The commissioners are appointed for four years by unanimous agreement of the member states and are charged to act in the EU's interest independent of national interest.

EU Council The *European Union's (EU)* legislative body composed of the 15 ministers of the member states. When the agricultural agenda is under discussion, the council is made up of the 15 agricultural ministers. The council acts on *EU Commission* proposals and is the final decisionmaking body in the EU. The presidency of the council rotates among member states every six months.

EU Court of Justice The *European Union's (EU)* Supreme Court. It interprets EU law for national courts and rules on matters pertaining to EU *treaties* raised by EU institutions, member states, or individuals. Its rulings are binding, and it has consistently ruled in favor of EU law in matters of intra-EU trade.

EU Parliament The only directly elected body in the *European Union (EU).* The Parliament's members are elected every five years. Its members debate issues, question the *EU Commission* and

EU Council, review the budget, propose amendments, and have final budget approval.

Eurodollar market The name given to the operations outside U.S. boundaries in which *foreign* banks and foreign branches of U.S. banks make loan and security transactions denominated in U.S. dollars rather than foreign currency.

European Agricultural Guidance and Guarantee Fund (EAGGF) A fund used to finance the *European Union's (EU) Common Agricultural Policy*. All members of the EC contribute to the fund. Also known by its French abbreviation FEOGA.

European Community (EC) The former name of the *European Union (EU)*. The EC represented a fusing of the European Atomic Energy Community (Euratom), the European Coal and Steel Community (ESC), and the European Economic Community (EEC or *Common Market)*.

European currency unit (ECU) A weighted average of all *European Union (EU)* currencies. The ECU fluctuates against third country currencies and is used for internal EU accounting purposes. In agriculture, common *farm* prices (see *price*), *subsidies*, and import *levies* are established in the ECU. Similar to the previously used European unit of account.

European Economic Community (EEC) See *European Community (EC)* and *European Union (EU)*.

European Free Trade Association (EFTA) A regional *free trade area* established in 1960, concerned with eliminating *tariffs* on manufactured goods and some agricultural products that originate in and are traded among member countries. Most agricultural products are not subject to EFTA schedule tariff reductions. Members include Austria, Finland, Iceland, Norway, Sweden, Liechtenstein, and Switzerland.

European Monetary Cooperation Fund See *European monetary system*.

European monetary system (EMS) A monetary system established in 1979 to move Europe toward closer economic integration and avoid the disruption in trade that can result from fluctuations in currency *exchange rates*. The EMS member countries deposit gold and dollar reserves with the European Monetary Cooperation Fund in exchange for *European currency units*.

European Union (EU) An organization established by the Treaty of Rome in 1957; also known as the *Common Market*. The EU was formerly called the *European Community (EC)* and the European Economic Community (EEC). Originally composed of the six European nations of Belgium, the Federal Republic of Germany, France, Italy, Luxembourg, and the Netherlands, the EU has expanded to 15 nations. The EU attempts to unify and integrate member economies by establishing a *customs union* and common economic policies, including the *Common Agricultural Policy (CAP)*. Member nations include the original six nations, plus Austria, Denmark, Finland, Greece, Ireland, Portugal, Spain, Sweden, and the United Kingdom.

European unit of account See *European currency unit (ECU)*.

Eutrophication The process by which a body of water becomes rich in *nutrients*. This can occur naturally or can result from agricultural *runoff* and industrial or municipal wastewater flowing into streams, rivers, and so on.

Eviscerated Having had the organs in the great cavity of the body removed.

Ewe A female *sheep* of any age.

Excess capacity The productive capacity of an industry that is not currently utilized but that could be if needed.

Excess foreign currencies *Foreign* currencies received by the U.S. government that exceed U.S. government currency needs in that country for such expenses as U.S. embassy operating costs. The U.S. government receives foreign currencies from sales of *farm* products under *Public Law 480* or other programs.

Exchange rate The number of units of one currency that can be exchanged for one unit of another currency at a given time. A decline in the value of the U.S. dollar drops the "*price*" of U.S. *farm* products in terms of the currency of many *importers*. Conversely, an appreciation in the value of the dollar means that *foreign* importers must spend more of their currency to buy U.S. farm products.

Exchange rate management Efforts by some governments to either directly or indirectly influence the level of *imports* and *exports* in their countries through *exchange rate* policies. In an exporting country, overvaluation of the currency will curtail exports as export *prices* rise relative to import prices valued in the currencies of importing countries. Conversely, an undervalued currency will help increase exports as importers perceive imports as being less expensive when valued in their own domestic currencies.

Exchange restrictions Limits imposed by governments on the buying and selling of their nation's currency.

Exercise To elect to buy or sell, taking advantage of the right, without the obligation, conferred by an *option* contract (see *contract*).

Eximbank See *Export-Import Bank.*

Ex (point of origin) The point at which a *price* of a good is quoted. Examples include ex factor, ex mill, and ex warehouse.

Export The process of legally selling and shipping goods to *foreign* nations.

Export allocation or quota Controls applied by an exporting country to limit the amount of goods leaving that country. Such controls usually are applied in time of war or during some other emergency requiring conservation of domestic supplies (see *supply*).

Export assistance programs See *export programs.*

Export Credit Guarantee Program (GSM-102) The largest U.S. agricultural *export promotion program*, functioning since 1982. It guarantees repayment of private, short-term credit for up to three years. See also *Intermediate Export Credit Guarantee Program (GSM-103).*

Export credit insurance Insurance coverage against nonpayment risks obtained by a seller who is granting credit terms to a *foreign* buyer.

Export Credit Revolving Fund A fund authorized by the Agriculture and Food Act of 1981 (P.L. 97-98) (see Appendix 3) to provide short- and intermediate-term direct credit for *export* sales of agricultural *commodities*, breeding animals, and handling facilities in developing *markets*.

Export declaration A form filed by an *exporter* or an exporter's *agent* at the port of departure that states the content of goods being exported. *Export* declarations are used to compile export statistics and as an export control document.

Export deficit See *balance of trade*.

Export Enhancement Program (EEP) A program initiated in May 1985 under a *Commodity Credit Corporation (CCC)* charter to help U.S. *exporters* meet competitors' *prices* in subsidized (see *subsidy*) *markets*. The program was formally authorized by the *Food Security Act of 1985 (P.L. 99-198)* (see Appendix 3). Under the EEP, exporters are awarded *generic commodity certificates* or cash payments. The certificates are redeemable for CCC-owned *commodities*. The certificates or payments enable an exporter to sell certain commodities to specified countries at prices below those of the U.S. market.

Exporter A person, firm, or other entity that legally sells and ships goods to *foreign* nations.

Export-Import Bank of the United States A U.S. government institution that administers programs to assist the U.S. exporting community, including direct lending and the issuance of guarantees or insurance to minimize risk for private banks and *exporters*. Also known as the Eximbank.

Export Incentive Program (EIP) A program administered by *U.S. Department of Agriculture's* (USDA) *Foreign Agricultural Service* that assists private firms to promote their branded products overseas. An EIP is developed for a specific *commodity* or product based, in part, on a determination that *export* markets (see *market*) for the product can be developed most effectively by brand promotion and that there is sufficient U.S. industry interest to support such a program.

Exportkhleb The official entity in the former Soviet Union responsible for *imports* and *exports* of food and *feed grains*, *pulses*, *flour*, *oilseeds* and products, and *seeds* and seedlings.

Export license A government document authorizing *exports* of specific goods in specific quantities to a particular destination. Some countries require an export license only under special circumstances, while others require it for most or all exports.

Export limitation Any restriction on the quantity of a country's *exports*. A *Public Law 480*, Title I, agreement, for example, limits exports from the recipient country to *commodities* that are similar to those being supplied by the United States.

Export Marketing Certificate Program See *Wheat and Feed Grain Export Certificate Programs*.

Export payment A *subsidy* paid to exporting firms so that higher-priced U.S. *commodities* can be sold on the world *market* at a competing *price*.

Export programs Government programs that support the *export* of U.S. agricultural *commodities* or products. Government export programs include overseas food aid and concessional loans, credit and credit guarantees, export price *subsidies*, and generic and branded nonprice promotion programs (see *export promotion programs*).

Export promotion programs Activities carried out by the *U.S. Department of Agriculture's* (USDA) *Foreign Agricultural Service* to encourage the *export* of U.S. agricultural products. Export promotion activities include *market* development, information and exhibit assistance, data services, market intelligence, and technical assistance.

Export quota agreement A mechanism for regulating trade in a *commodity*, whereby *export* quantities are allocated among member exporting countries. To prevent nonparticipating *exporters* from undermining such an agreement, importing countries may agree to limit *imports* from nonmember exporting countries. See also *international commodity agreement.*

Export restitutions Direct *export subsidy* payments used to promote *exports* of agricultural goods by the *European Union (EU).* The "restitution" refunds the difference between the domestic *market* price and the lower *price* needed to export.

Export restraints Restrictions by an exporting country on the quantity of *exports* to a specified importing country. The restrictions usually follow a formal or informal request by the importing nation. See *voluntary export agreement.*

Exports Domestically produced goods and services that are sold abroad.

Export subsidies Special incentives, such as cash payments, tax exemptions, preferential *exchange rates*, and special *contracts*, extended by governments to encourage increased *foreign* sales. These *subsidies* are most often used when a nation's internal *prices* exceed *export* prices.

Export surplus See *balance of trade.*

Export tax A fee paid on *exports* to the government of the originating country. An export tax may be a constant value per unit, or it may be assessed as a percentage of the *commodity* price (see *price*).

Ex (tariff number) A notation often used in discussions of *tariff* reductions, *binding*, and so forth. It is used to identify or excerpt specified tariff items within the tariff classification number and text for particular products or lists of products. For example, tariff classification number 16.01-000 includes sausage of all meats, whereas the specification "ex 16.01-000, sausage of bovine origin" would include only bovine sausage. See also *Customs Cooperation Council Nomenclature.*

Extension agent See *county extension agent.*

Extension home economist See *county extension agent.*

Extension Service The *Cooperative Extension Service*, the *U.S. Department of Agriculture*'s educational agency and one of the three partners in the *Cooperative Extension System* with state and local governments. All three share in financing, planning, and conducting educational programs. Under a 1994 USDA reorganization, the *Extension Service* was merged with the *Cooperative State Research Service* and the *National Agricultural Library* to form the *Cooperative State Research, Education, and Extension Service.*

Externality A situation where one person's or group's actions benefits or costs another person or group. The benefits or costs from the externality-generating activity are not priced (see *price*) in the *market* place and are often outside the control of those affected. The typical example of a negative externality is a manufacturing plant (see *plant or establishment*) that does not incur financial costs for polluting the air or nearby waters.

Extraction, mechanical The process of removing fat or oil (see *fats and oils*) from materials using heat and mechanical pressure.

Extraction rate The proportion by weight of a processed (see *processing*) product to its raw material. For example, an extraction rate of 72 for *wheat* means that 100 pounds of that wheat will produce 72 pounds of *flour* and 28 pounds of wheat *byproducts.* The extraction rate varies according to the *commodity* and the *crushing* or *milling* facilities.

Extraction, solvent A process that uses an organic solvent to extract oil (see *fats and oils*) from oil-bearing materials. In the United States, nearly all *soybeans* are processed (see *processing*) by solvent extraction, as well as a substantial part of the *cottonseed*, flaxseed (see *linseed*), and other *oilseeds.*

Extractor A machine that rotates uncapped *honeycombs* at a speed sufficient to remove the *honey* but without destroying the combs.

Extra-long staple cotton *Cotton* having a *staple length* of 1 3/8 inches or more, according to the classification used by the *International Cotton Advisory Committee.* This cotton is also characterized by fineness and high *fiber* strength, contributing to finer and stronger *yarns* needed for thread and higher-valued fabrics (see *cloth*). American types include *American Pima* and *Sea Island cotton.*

Extrusion A process by which a product, such as *feed*, is pressed, pushed, or protruded through orifices under pressure.

F

Fabric See *cloth.*

Fabrication 1: The *production* of meat cuts (*primals, subprimals,* or retail cuts) from a *carcass.* 2: The process of manufacturing prepared and semiprepared foods that are sold to restaurants or institutions to serve with minimal labor input. For example, meats are cut, wrapped, and boxed at the packing plant and delivered oven- or grill-ready. Other suppliers prepare main courses or complete meals in a fashion analogous to TV dinners.

Fabricator A firm or enterprise that fabricates (see *fabrication)* meat.

Face The side of a fabric (see *cloth*) that, because of weave, finish, or other characteristics, presents a better appearance than the other side, or back.

Failed acreage Crops that were planted properly but did not grow or were destroyed by a natural disaster, such as a *drought.*

Fair share A requirement of the *Public Law 480* program that the United States should benefit equitably from any increase in commercial purchases of agricultural *commodities* by recipients of U.S. food aid.

Fair value The reference against which U.S. purchase *prices* of imported (see *import*) merchandise are compared during antidumping (see *dumping*) investigations. Fair value is generally expressed as the weighted average of the *exporter*'s domestic *market* prices or prices to third countries during the period of investigation. However, it may be a constructed value if there are no, or virtually no, home-market or third-country sales or if there are too few sales made at prices above the *cost of production* to provide an adequate basis for comparison.

Fallow *Cropland* left idle in order to restore *productivity*, mainly through accumulation of water, *nutrients*, or both. It is usually tilled (see *till*) to control weeds and conserve moisture in the *soil.* See also *summer fallow* and *bush or forest fallow.*

Falling object protective structure (FOPS) A *tractor* cab constructed of heavy gauge steel to prevent falling objects from causing injury to the operator. FOPS are more commonly found on forest or construction machines than on *farm* tractors, which use *roll-over protective structures (ROPS).*

Family farm An often-used term for which there is no official, single definition. One description of a family farm is an agricultural business that (1) produces agricultural *commodities* for sale in such quantities as to be recognized as a *farm* rather than a *rural* residence; (2) produces enough income (including off-farm employment) to pay family and farm operating expenses, to pay debts, and to maintain the property; (3) is managed by the *farm operator*; (4) has a substantial amount of labor provided by the operator and family; and (5) may use seasonal labor during peak periods and a reasonable amount of full-time hired labor.

Family flour Retail *flour,* specialized or all-purpose, used in the home for a variety of purposes including baking breads, cakes, biscuits, and pastries or in gravies and batters. This flour is typically packaged in 5-pound bags for supermarket (see *foodstore*) sale.

Fancy clear See *first clear.*

Fancy patent flour A high-quality *flour* that is the finest *grind* of all flour.

Fanning mill A machine, usually stationary, used to clean *grain* and other crops by removing unwanted *seeds,* dirt, *chaff,* and *foreign materials.* A fanning mill uses a large power-driven fan and sieves with different size holes that allow some objects to pass and not others.

Farina Pure *wheat* endosperm (see *endosperm*) that is ground (see *grind*) medium to fine and has a variety of uses.

Farm A tract or tracts of *land,* improvements, and other appurtenances available to produce crops or *livestock,* including fish. Since 1974, the *Bureau of the Census* has defined a farm as any place that has or would have had $1,000 or more in gross sales of farm products. In 1993, there were 2.1 million farms in the United States, accounting for about 978 million acres.

Farm acreage base The annual total of the *crop acreage bases* (*wheat, feed grains, upland cotton,* and *rice*) on a *farm,* the average acreage (see *acre*) planted to *soybeans, peanuts,* and other approved *nonprogram crops,* and the average acreage devoted to *conserving uses.* Conserving uses include all uses of *cropland* except crop acreage bases, acreage devoted to nonprogram crops, acreage enrolled in annual *acreage reduction programs* or acreage limitation programs, and acreage in the *Conservation Reserve Program.* Farm acreage bases are not currently used to administer *commodity* programs.

Farm and home advisor See *county extension agent.*

Farm balance sheet A statement of financial position that provides information about the *farm* business entities' assets (see *assets, current* and *assets, fixed*), debt, and equity, and their relationships to each other at a given point in time, usually the end of the year.

Farm bill Major omnibus agricultural legislation, usually enacted every four or five years. The bill usually includes provisions on *commodity* programs, trade, conservation, credit, agricultural research, food stamps (see *Food Stamp Program*), and marketing. See *Food, Agriculture, Conservation, and Trade Act of 1990 (P.L. 101-624)* and Appendix 3.

Farm Bureau The largest *farm* organization in the United States. The Farm Bureau is a nonpartisan, voluntary organization comprised of more than 2,700 county Farm Bureaus, which offer programs to meet farm families' needs.

Farm Costs and Returns Survey (FCRS) A nationwide survey that includes, for a given year, data on *costs of production,* capital expenditures, debts and assets (see *assets, current* and *assets, fixed*), earnings, *production* practices, and other *farm* business and *farm operator* characteristics. These data are used to update cost-of-production estimates. They are also used to measure annual *farm* *income* and to assess changes in the financial conditions of farms. The survey is managed by the *U.S. Department of Agriculture*'s *Economic Research Service* and conducted by the *National Agricultural Statistics Service* and its 44 state offices. The survey covers every state except Hawaii and Alaska.

Farm Credit Administration (FCA) The independent federal financial regulator of the *Farm Credit System*. The FCA is responsible for examining and ensuring the safety and soundness of all System institutions. The FCA is an independent agency and is financed by direct assessment of the System institutions that the FCA regulates. The FCA is directed by a three-member board nominated by the U.S. president and confirmed by the Senate. It is located in McLean, Virginia.

Farm Credit Bank (FCB) Part of the *Farm Credit System* authorized to provide funding for the *Production Credit Associations*, *Federal Land Bank Associations*, *Federal Land Credit Associations,* and other financial institutions serving the *rural* United States. The FCBs make long-term mortgage loans to farmers and ranchers for *land* and other purchases through the Federal Land Bank Associations. The FCBs were created in July 1988 by the mergers of the *Federal Intermediate Credit Banks* and the Federal Land Banks in 11 of the 12 Farm Credit districts. Plans have been approved for mergers in 1994 that would reduce the number of FCBs from 10 to eight.

Farm Credit Council (FCC) A national trade association representing the interests of the *Farm Credit System (FCS)*. There are two FCC offices: Washington, D.C., where the Council represents the FCS's legislative and regulatory concerns; and Denver, Colorado, where the council provides business-related services to its FCS membership.

Membership in the FCC is made up of Farm Credit District Councils and CoBank (see *Bank for Cooperatives*). These member councils, in turn, draw their membership from *Farm Credit Banks* and associations, as well as other cooperative *farm* lenders and *cooperatives*.

Farm Credit Leasing Services Corporation An entity that provides leasing and other services to eligible *Farm Credit System* borrowers, including agricultural *producers*, *cooperatives*, and *rural* utilities. It has a nine-member board, five elected from its owner-banks and four representing its customers.

Farm Credit System (FCS) A network of borrower-owned lending institutions and related service organizations serving all 50 states and the Commonwealth of Puerto Rico that specializes in providing credit and related services to farmers, ranchers, and *producers* or harvesters (see *harvest*) of aquatic products. Loans may also be made to finance the *processing* and marketing (see *agricultural marketing*) activities of these borrowers. In addition, loans may be made to *rural* homeowners, certain *farm*-related businesses, and agricultural, aquatic, and public utility *cooperatives*. The system provides about one third of the total credit used by the farmers, ranchers, and cooperatives of the United States.

All system banks and associations are governed by boards of directors elected by the stockholders who are farmer-borrowers of each institution. Additionally, federal law requires that at least one member of the board be elected from outside the system by the other directors. System institutions, unlike commercial banks or thrifts, do not solicit funds for deposit.

On January 1, 1993, the system was comprised of the following lending institutions:

Farm Credit Banks (ten) that make direct long-term real estate loans through 77 *Federal Land Bank Associations (FLBAs)* and/or provide loan funds to 71 *Production Credit Associations (PCAs)*, 69 *Agricultural Credit Associations (ACAs)*, and 27 *Federal Land Credit Associations (FLCAs)*. PCAs make short- and intermediate-term loans, ACAs make short-, intermediate-, and long-term loans, and FLCAs make long-term loans.

Banks for Cooperatives (three) that make loans of all kinds to agricultural, aquatic, and public utility cooperatives.

Federal Intermediate Credit Bank (FICB) of Jackson, Mississippi, provides short- and intermediate-term loan funds to two PCAS, which serve Alabama, Mississippi, and Louisiana.

See also *Farm Credit System Financial Assistance Corporation* and *Farm Credit Leasing Services Corporation.*

Farm Credit System Assistance Board A board that was created by the Agricultural Credit Act of 1987 (P.L. 100-233) (see Appendix 3) to approve *Farm Credit System* lender requests for federal financial assistance. The three members of the Board were the secretary of agriculture, the secretary of treasury (or their appointees), and an agricultural *producer* with financial experience. The Assistance Board terminated operations December 31, 1992, as specified in the Agricultural Credit Act of 1987. See *Farm Credit System Financial Assistance Corporation* and the *Farm Credit System Insurance Corporation.*

Farm Credit System Financial Assistance Corporation An institution created by the Agricultural Credit Act of 1987 (P.L. 100-233) (see Appendix 3) to provide funds under the direction of the *Farm Credit System Assistance Board* to assist financially stressed system institutions. Approximately $1.26 billion in funds was provided by the Assistance Corporation, and its authority to raise additional funds expired on December 31, 1992. The corporation is managed by the same board of directors as the *Federal Farm Credit Banks Funding Corporation* and will continue to operate until all funds used to provide the assistance are repaid. As of January 1, 1993, the Farm Credit System Insurance Corporation is required to provide funds to retire equities of the *Farm Credit System* institutions placed in receivership.

Farm Credit System Insurance Corporation An institution established by the Agricultural Credit Act of 1987 (P.L. 100-233) (see Appendix 3) primarily to insure the timely payment of principal and interest on consolidated or *Farm Credit System* debt securities. It is managed by a board of directors consisting of the *Farm Credit Administration* board.

Farmer cooperative See *agricultural cooperative.*

Farmer feedlot A *feedlot* with capacity for less than 1,000 head of *cattle.*

Farmer Mac See *Federal Agricultural Mortgage Corporation.*

Farmer-Owned Reserve (FOR) Program A program designed to provide storage when *wheat* and *feed grain* production are in abundant *supply* to provide a buffer against unusually sharp *price* movements. Farmers can place eligible *grain* in storage and receive extended loans for up to three years. The loans are *nonrecourse* in that farmers can forfeit the *commodity* held as collateral to the government in full settlement of the loan without penalty and without paying accumulated interest. Interest on the loan also may be waived by the secretary of agriculture, and farmers may receive annual storage payments from the government. Under the *Food, Agriculture, Conservation, and Trade Act of 1990 (P.L. 101-624)* (see Appendix 3), farmers may sell reserve grain at any time. Formerly, "release prices" or "trigger price levels" governed release of grain from the FOR.

Farmers Educational and Cooperative Union of America See *Farmers Union.*

Farmers Home Administration (FMHA) A former *U.S. Department of Agriculture (USDA)* agency that provided credit at reasonable rates and terms to farmers unable to get credit from other sources.The agency was abolished in October 1994 and its functions transferred to the *Consolidated Farm Service Agency* and the *Rural Housing and Community Development Service.*

Farm financial condition A measure of the financial position of a *farm* based on income and the ratio of debts to assets (see *assets, current* and *assets, fixed*). The catagories are as follows:

Favorable: A farm operation with positive net income and a low *debt/asset ratio* (0.40 or lower);
Vulnerable: A farm with negative net income and a high debt/asset ratio (above 0.40);
Marginal Solvency: A farm with a positive net income but a high debt/asset ratio (above 0.40);
Marginal Income: A farm with a low debt/asset ratio (0.40 or lower) but negative net income.

Farm income Measures of the earnings of a farming (see *farm*) operation over a given calendar year. See *gross cash income, gross farm income, net cash income,* and *net farm income.*

Farming-dependent county See *county-type classification.*

Farm inputs Resources used in *farm* production, such as labor, *feed, seed, fertilizer,* and equipment.

Farm labor contractor A person other than an agricultural employer, an *agricultural association,* or an employee of an agricultural employer or association who receives a fee for recruiting, soliciting, hiring, employing, furnishing, or transporting *migrant farmworkers* or *seasonal agricultural workers.*

Farmland protection State and local programs designed to limit conversion of agricultural *land* to other higher-valued uses.

Farm lot feeding A *farm* operation where *cattle* feeding is complementary with other farming enterprises. See also *feedlot.*

Farm management The coordination and supervision of a *farm* business in order to increase long-term profits or to achieve other specified goals.

Farm operator A person who operates a *farm*, either by working or supervising and by making the daily operating decisions. *U.S. Department of Agriculture (USDA)* data sources count one operator per farm; in the case of equal partners, the senior partner is considered the operator.

Farm program payment yield The *farm* commodity (see *commodity*) *yield* of record determined by a procedure outlined in legislation. The farm program payment yield applied to eligible acreage (see *acre*) determines the level of *production* eligible for *direct payments* to *producers.*

Farm-related income Cash income earned from agricultural sources other than *commodity* sales or government payments. This category of income includes the sale of forest products, such as timber; income from machine hire and *custom-work* on other *farms;* recreational uses of farms; and other miscellaneous sources, such as dividends from *cooperatives.*

Farm sale weight (FSW) The weight of *tobacco* as it is sold by the grower after *curing.* Further *processing* or conditioning is usually needed before it can be used.

Farm size Commonly measured by the value of gross *farm* sales or by acreage (see *acre*).

Farm Storage Facility Loan Program A *U.S. Department of Agriculture (USDA)* program that helps qualifying *producers* obtain needed on-*farm* storage for their crops. Applications for these loans are accepted by county *Consolidated Farm Service Agency* offices only during periods announced by the secretary of agriculture. No farm storage facility loans have been made since 1982.

Farm-to-retail price spread A measure of all *processing*, transportation, *wholesaling*, and retailing charges incurred after products leave the *farm*. This figure accounts for about 70 percent of the retail cost of a *market basket* of U.S. farm foods sold in *foodstores*. The remainder is *farm value*, or the payment for the raw product.

Farm tractor A *tractor* designed for use on a *farm*. The *U.S. Department of Agriculture* definition includes only tractors having 40 or more *horsepower*. Tractors under 40 horsepower include residential garden tractors and lawn tractors that are generally used for nonfarm purposes. Industrial, nonfarm tractors may be used for road and building log construction and other commercial purposes and sometimes appear identical to tractors used on farms. However, most industrial tractors are designed for specific purposes, such as contruction, and carry unique equipment, such as protective cages for operators, industrial backhoes, and so forth.

Farm truck Trucks used on *farms* and often indistinguishable from trucks used for nonfarm purposes. Farm trucks used to haul *grain* have tight metal or wooden boxes, usually equipped with hydraulic dumps. *Cattle* trucks have tall, slatted sides that allow ventilation. Many other farm trucks are uniquely outfitted for specific purposes, including *beet* trucks, utility trucks, and flatbed trucks. Most farm trucks are rated from 3/4 ton to 5 tons. See also *farm tractor*.

Farm value A measure of the *prices* farmers receive for the raw *commodities* equivalent to foods in the *U.S. Department of Agriculture*'s (USDA) *market basket*. The *farm* value is calculated by multiplying farm prices by the quantities of farm products equivalent to food sold at retail. For example, it takes 2.4 pounds of live animal to equal 1 pound of Choice *grade* beef. In 1988, therefore, the farm value of Choice grade beef selling for $2.55 was $1.47.

Farm value share A measure of the proportion farmers get from the amount *consumers* spend on the *market basket* of food purchased in retail foodstores. The farm value share reflects relative changes in *farm* and retail food *prices*. If farm prices increase more than retail, for instance, the farm value share will increase. Farm value shares vary greatly among foods. *Farm value* is a much larger percentage of foods with lower *processing* and marketing (see *agricultural marketing*) costs, such as meats, *eggs*, *poultry*, and dairy products.

Farmworker An individual who is employed on a *farm* either for cash wages or salary (hired farmworker), as a *farm operator*, or unpaid (not receiving a cash wage or salary, or receiving only a "token" cash allowance, room and board, or *payment-in-kind*).

Farrow To give birth to a *litter* of *pigs*.

Farrowing crate Equipment used to protect piglets by restricting the movement of a *sow* during farrowing (see *farrow*) and lactation (see *lactation period*).

Farrowing house Shelter for the *farrowing* of *pigs*.

Farrow-to-finish operation One of three types of *hog* enterprises where *pigs* are produced and fed to *slaughter* weight on the same *farm*. See also *feeder-pig finishing operation*.

Fast food restaurant An operation that sells prepared, ready-to-eat food for consumption on or off the premises and that does not provide table service. These firms may be either part of a *chain* or independently owned. See also *table-service restaurant.*

Fast-track negotiating authority Presidential authority granted by Congress to negotiate trade agreements with the understanding that the negotiated agreement will go before Congress for an "up" or "down" vote without possibility of amendment and within a specified period of time.

Fat A concentrated source of food energy. See also *fats and oils.*

Fats and oils Substances of plant or animal origin that are predominantly the trifatty acid (see *fatty acids*) esters of *glycerine*, commonly called triglycerides. This excludes *essential oils* and mineral (petroleum) oils. Animal fats come from the tissue of *hogs*, *cattle*, *sheep*, *poultry*, or from marine animals, such as whales. *Butter* is a special type of animal *fat* obtained from *milk.* Vegetable oils are obtained from a variety of plant *seeds*, such as *soybeans*, *cottonseed*, *peanuts*, *corn* germ (see *germ*), *olives*, *coconuts*, *rapeseed*, sesame, *sunflowerseed*, cocoa *(cacao)* beans, and various oil palms.

The term "Fat" usually refers to materials that are solid or plastic at ordinary temperatures, while oils are liquid. No rigid distinction can be made between the two, and it is common practice to use the terms interchangeably, except where a distinction between solid and liquid materials is important. Fats and oils make up one of three major classes of food materials, along with *carbohydrates* and *proteins.* Industrial products utilizing fats and oils include soap, cosmetics, and lubrication.

Fat splitting (hydrolysis of fat) A process to separate *fatty acids* and *glycerine* from vegetable oils (see *fats and oils*) or animal fats (see *fats and oils*) by use of heat, water, and a catalyst or *enzyme.*

Fat substitute A product designed to simulate the taste and feel of *fat* but with fewer calories. Examples include Olestra, a calorie-free fat substitute made from *sucrose* and edible *fats* and *oils.* See *fatty-acid-based fat substitute* and *protein-based fat substitute.*

Fatty-acid-based fat substitute A *fat substitute* that uses *fatty acids* that have been chemically altered to provide no calories or fewer calories than *fat.* Polyglycerol esters, for example, have about one-third less than a gram of fat and are used in low-calorie versions of ice cream, other frozen desserts, *margarine*, *shortening*, *peanut* butter, whipped topping, and bakery items.

Fatty acids The product obtained when *glycerine* is separated from the *triglycerides* of natural *fats and oils* by any method of *hydrolysis.* After hydrolysis, 100 grams of fat *yield* approximately 95 grams of fatty acids.

Fatty acids (classification of) *Fats and oils* can be classified according to their degree of saturation. Saturated fatty acids contain only single-bond *carbon* linkages and are the least active chemically. Saturated acids (such as stearic, palmitic, and lauric) are normally solid at room temperature.

Unsaturated fatty acids contain one or more double-bond carbon linkages and are normally liquid at room temperature. Fats and oils contain both saturated and unsaturated fatty acids but in differing proportions.

Monounsaturated fatty acids contain one double bond (such as oleic acid). If it contains more than one double bond, it is called polyunsaturated. Linoleic acid has two double bonds, and linolenic acid has three double bonds.

Fed cattle *Cattle* that have been fed a high *grain* ration (see *ration*) diet.

Federal Agricultural Mortgage Corporation (Farmer Mac) An organization authorized by the Agricultural Credit Act of 1987 (P.L. 100-233) (see Appendix 3) that creates a *secondary (resale) market for agricultural loans*, enabling lenders to obtain cash for further lending. Mortgages from lenders are pooled into securities and sold on the capital *market*. Farmer Mac is fashioned after similar home mortgage secondary markets such as the Federal National Mortgage Association ("Fannie Mae"), the Government National Mortgage Association ("Ginnie Mae"), and the Federal Home Loan Mortgage Corporation ("Freddie Mac").

Federal Crop Insurance Corporation (FCIC) A federal corporation within the *U.S. Department of Agriculture* that administers the *Federal Crop Insurance Program.*

Federal Crop Insurance Corporation Reinsurance A system of risk-sharing between the *Federal Crop Insurance Corporation (FCIC)* and reinsured private companies, under which FCIC protects the company from extraordinarily large actuarial losses. This risk-sharing relationship is governed by a "Standard Reinsurance Agreement," which is revised annually. Reinsurance is common in all types of insurance industries.

Federal Crop Insurance Program A subsidized insurance program providing farmers with a means to manage the risk of crop losses resulting from natural disasters. Federal crop insurance is available for about 50 different crops, although not all crops are insurable in every county. With the Federal Crop Insurance Reform Act of 1994, coverage is classified as "catastrophic" (CAT) or "additional." CAT Coverage guarantees 50 percent of a farmer's average *yield*, at 60 percent of the price election, for a nominal processing fee. Farmers who participate in the annual commodity programs or receive certain *Consolidated Farm Service Agency* loans must purchase at least CAT coverage for crops they produce that are of economic significance. Farmers may buy additional coverage at up to 75 percent of their average yield. Both CAT and additional coverage are subsidized by the government. Reform provisions are in effect beginning with 1995 crops. The *Group Risk Plan* is a federal crop insurance option.

Federal Farm Credit Banks Funding Corporation The *Farm Credit System (FCS)* institution responsible for managing the sale of FCS securities in the United States' money and capital *markets* through a network of investment dealers and dealer banks in New York City. Farm Credit Consolidated Systemwide Debt Obligations are the primary source of loan funds for the FCS and are the joint and several liability of System banks.

The Funding Corporation also provides financial advisory services and supports the FCS banks in the management of interest-rate risk. The Funding Corporation is owned by FCS banks and is governed by a 10-member board. Seven elected board members appoint two members from outside the System. The Funding Corporation president serves as a nonvoting board member.

Federal Food, Drug, and Cosmetic Act (FFDCA) (P.L. 75-717) An act, passed in 1938, that is the basic food and *drug* law of the United States. The law is intended to ensure that foods are pure and wholesome, safe to eat, and produced under sanitary conditions; that drugs and devices are safe and effective for their intended uses; that cosmetics are safe and made from appropriate ingredients; and that all labeling and packaging is truthful, informative, and not deceptive. The *Food and Drug Administration* is responsible for enforcing the FFDCA.

Federal Grain Inspection Service (FGIS) The name of the former *U.S. Department of Agriculture (USDA)* agency that established official U.S. standards for 11 *grains* and other assigned *commodities* and administered a nationwide inspection (see *inspection, federal*) system to certify those *grades*. FGIS was abolished in October 1994 and its functions transferred to the new *Grain Inspection, Packers, and Stockyards Administration.*

Federal Insecticide, Fungicide, and Rodenticide Act of 1947 (FIFRA) Legislation administered by the *Environmental Protection Agency (EPA)* requiring registration of *pesticide* products to ensure that they meet stated health, safety, and environmental criteria. Amendments to the law require previously registered pesticides to be reregistered by 1997 to meet updated standards. Under the law, EPA can cancel registration of pesticides that do not meet the required criteria, require label changes, or order immediate termination of use.

Federal Intermediate Credit Banks (FICB) A group of 12 banks that made discount loans to *Production Credit Associations* and to other financial institutions, both cooperative and otherwise. They obtained funds by selling debentures on the open *market*. In July 1988, the FICBs were merged with the *Federal Land Banks Associations* in 11 of the 12 Farm Credit districts. See *Farm Credit System*.

Federal Land Bank Associations (FLBAs) *Farm Credit System* organizations through which farmers obtain long-term (up to 40 years) loans on *land*. The FLBAS serve in an *agent*-like capacity for *Farm Credit Banks* in making long-term real estate mortgage loans to eligible borrowers.

Federal Land Credit Associations (FLCA) *Farm Credit System* institutions that make long-term real estate mortgage loans to eligible borrowers. Most FLCAS are former *Federal Land Bank Associations* that have received direct lending authority from the *Farm Credit Banks*.

Federal lands counties See *county type classification*.

Federally inspected slaughter 1: The U.S. requirement that all meat crossing state lines must be federally inspected (see *inspection, federal* and *meat and poultry inspection*), or it cannot be sold. State inspection must meet federal standards.
2: The number of animals slaughtered (see *slaughter*) under federal inspection.

Federal marketing orders and agreements A means authorized by legislation for agricultural *producers* to promote orderly marketing (see *agricultural marketing*) and to collectively influence the *supply*, *demand*, *price*, or quality of particular *commodities*. A marketing order may be requested by a group of producers and must be approved by the secretary of agriculture and a required number of the commodity's eligible producers (usually two-thirds) in specified areas in a referendum. Conformance with the order's provisions is mandatory for all producers and *handlers* covered by the order. It may limit total marketings, prorate the movement of a commodity to *market*, or impose size and *grade* standards.

Federal Meat Inspection Act The law requiring inspection (see *inspection, federal*) of meat processed (see *processing*) in privately owned plants (see *plants or establishments*) in the United States to ensure that domestically produced meat sold in interstate and *foreign* commerce is safe, wholesome, and accurately labeled. The *U.S. Department of Agriculture*'s (USDA) *Food Safety and Inspection Service (FSIS)* is responsible for carrying out federal inspection laws.

Federal milk marketing orders A regulation issued by the secretary of agriculture specifying minimum *prices* and conditions under which *milk* can be bought and sold within a specified area. See also *federal marketing orders and agreements* and *classes of milk*.

Federal Reserve See *Federal Reserve System*.

Federal Reserve System The independent central banking authority in the United States. It consists of 12 district banks and a centralized decisionmaking body, the board of governors. The Federal Reserve provides currency upon demand by member banks, provides check-clearing services, and regulates the money *supply*.

Federal Seed Act A law prohibiting false labeling and advertising of *seed* in interstate commerce and seed *imports* contaminated with noxious weeds. The act also complements seed laws of the 50 states by prohibiting the shipment of seed containing excessive noxious weed seeds. The law is administered and enforced by the *U.S. Department of Agriculture*'s (USDA) *Agricultural Marketing Service.*

Federal-State Marketing Improvement Program A program designed to improve marketing (see *agricultural marketing*) of food and agricultural products through projects in a number of states. The projects cover domestic *market* development, development of alternative crops, and studies of new marketing concepts. Under the program, states must contribute an amount equal to the federal grants they receive. The program is administered by the *U.S. Department of Agriculture*'s (USDA) *Agricultural Marketing Service (AMS).*

Federal-state market news service A service operated by the *U.S. Department of Agriculture*'s (USDA) *Agricultural Marketing Service (AMS),* in cooperation with 44 State agencies, the District of Columbia, and three territories, that provides information on *prices*, *supply*, and *demand* for most agricultural *commodities*, usually daily or weekly.

Feed Edible materials consumed by animals that contribute energy and *nutrients* to their diet.

Feed bunk A trough or slotted enclosure used to provide *livestock* access to *hay* or *grain.*

Feed conversion ratio (feed efficiency) The amount of *feed* consumed by animals or *poultry* compared to the amount of meat or *eggs* produced by them. A beef animal, for example, requires approximately 7.5 pounds of feed for each pound gained.

Feed Cost-sharing Program See *Emergency Feed Assistance Program.*

Feeder calf An animal mature enough to be placed on *feed* but less than one year old.

Feeder cattle Young stock eventually destined to finish gaining weight in commercial or on-farm (see *farm*) *feedlots*.

Feeder pig finishing operation One of three types of hog (see *swine*) enterprises where feeder pigs are purchased and fed to *slaughter* weight. See also *farrow-to-finish operation.*

Feeders Animals of various weights ready to be finished for *market.*

Feeder yearling An animal suitable for *feedlot* placement that is older than a year but less than two years old.

Feed grain Any of several *grains* most commonly used for *livestock* or *poultry* feed (see *feed*), including *corn*, grain *sorghum*, *oats*, *rye*, and *barley*. The term is used interchangeably with "*coarse grains.*" See also *feed.*

Feed grinder A machine used to crush, crunch, pulverize, or mill (see *milling*) *oats*, *corn*, *barley*, and other *grains* to prepare grain for *livestock* consumption. See also *feed.*

Feeding margin The value of an animal when sold for *slaughter* minus the difference between the cost of the feeder animal (see *feeder calf*, *feeder cattle*, and *feeder pig finishing operation*) and the cost of gain.

Feedlot A *farm* or commercial operation that places 600–800 pound *steers* or *heifers* in pens (see *lot*) and feeds them a high *grain* ration (see *ration*) diet composed primarily of a concentrate (see *concentrate*) *feed* (often *corn*), a *protein* supplement (see *supplement* and *supplementing*) such as *soybean meal* or *cottonseed meal*, and some roughage from *silage* or *hay*. See *dry lot feeding, farm lot feeding,* and *commercial feedlot.*

Feedlot placements The inventory of *cattle* in *feedlots* at a reported time.

Feedmilling establishment An operation consisting of a stationary mill (see *milling*) operating at a single location together with any mobile mills based at that location. It may also consist of one or more mobile mills not associated with a stationary mill but based at one location. An establishment (see *plant or establishment*) is not necessarily the same as a business firm, which may include a number of establishments. The feed-milling activities of establishments are classified as (1) *primary feed manufacturing* and (2) *secondary feed manufacturing*.

Feed mixer A system of bins (see *bin grain*), *elevators*, conveyors, and chutes that combine different *grains* and *concentrates* into a prescribed combination for *livestock* consumption.

Feedstuff See *feed.*

Feijoa A gray-green, *egg*-shaped fruit native to South America. Also known as pineapple guava.

Fence charger A device used to provide high voltage electricity to wire that encloses a *pasture*. A fence charger is used to control *livestock* by giving them a harmless electric shock if they touch the wire.

Fence line bunk feeder A trough designed to hold *silages* or *concentrates*, constructed as part of the *lot* enclosure and serviced from outside the lot.

FEOGA French abbreviation, see *European Agricultural Guidance and Guarantee Fund (EAGGF)*.

Fermentable protein *Protein* that is directly available for digestion by a *ruminant*'s intestines.

Fermentation The enzymatic (see *enzyme*) decomposition of materials, especially *carbohydrates*, by various *microorganisms*. Fermentation is used in the production of alcohol, bread, vinegar, and other food and industrial materials.

Fertility, soil The quality that enables a *soil* to provide plant *nutrients* in the proper amounts and in the proper balance for the growth of specified plants when other factors such as light, temperature, and the physical condition of the soil are favorable.

Fertilization The union of the male nucleus with the female *egg*.

Fertilizer Material used to supply *nutrients* for plants. *Nitrogen* (N), *phosphate* (P), potassium (K), and other minor ingredients, such as sulphur, are combined in various concentrations to make commercial fertilizers. Livestock (see *livestock*) *manure* and *green manure* can also be used as fertilizers.

Fertilizer analysis The chemical composition of the *fertilizer* as determined by laboratory analysis.

Fertilizer formula The quantity and *grade* of crude stock materials used in making a *fertilizer* mixture.

Fertilizer grade The percentage of plant *nutrients* in a *fertilizer*. For example, a 10-20-10 *grade* contains 10 percent *nitrogen*, 20 percent phosphoric oxide, and 10 percent potash.

Fertilizer Institute An organization representing plant food producers, *fertilizer* distributors and retail dealerships, equipment suppliers and engineering construction firms, brokers and traders, rail and barge transportation firms, and a wide variety of other companies and individuals involved in *agriculture*.

Fertilizer material or carrier Any substance that contains one or more plant *nutrients*.

Fertilizer side dresser An attachment to a *drill* or *cultivator* used to place commercial *fertilizer* beside a crop row.

Fertilizer spreader A machine that applies *fertilizer* by broadcasting it over the surface of the ground. *Livestock* fertilizers are generally broadcast with a *manure spreader*, but some are broadcast with high pressure pumps through pipes and spray nozzles. Commercial fertilizers are broadcast with machines, which may be towed or mounted on trucks (see *farm truck*), *tractors*, or other power units.

Fiber 1: A slender strand of natural (see *natural fiber*) or man-made (see *man-made fiber*) material usually having a length at least 100 times its diameter and characterized by flexibility, cohesiveness, and strength. Several strands may be combined for *spinning*, *weaving*, and *knitting* purposes.
2: The thick-walled *cells* giving strength to plant tissue. The fiber in edible plant foods is important for human nutrition and health. Plant fiber is resistant to human digestive *enzymes*, is critical for proper bowel function, and helps reduce symptoms of *chronic* constipation, diverticular disease, and hemorrhoids. A high-fiber diet may also help lower *cholesterol* levels.
Dietary fiber is divided into two basic groups, soluble and insoluble. Insoluble fiber, found mainly in whole *grains* and on the outside of *seeds*, fruit, *legumes*, and other foods, absorbs many times its weight in water, expanding in the intestine. This type of fiber may alleviate some digestive disorders and play a role in colon cancer prevention. Soluble dietary fiber is found in fruit, vegetables, seeds, brown *rice*, *barley*, *oats*, and oat *bran* and works to increase cholesterol excretion in the bowel by binding bile acids and preventing their reabsorption.

Fiberboard A material, used for furniture and sheathing, made by compressing wood *fibers* into stiff sheets. Bonding agents or other materials are usually added to increase strength or to achieve other properties. The United States and Canada are major producers of fiberboard.

Fiber crop A crop, such as *cotton*, *hemp*, or *flax*, grown for *fibers* for *cloth*, rope, or twine.

Fiber maturity A measure of the thickness of the *cotton* fiber (see *fiber*) walls. Fiber maturity is one characteristic of cotton used to measure quality.

Fiber strength A measure of *cotton* quality that shows how much weight the *fiber* can handle before breaking. Fiber strength is an important factor in *spinning*.

Fiddlehead fern A tightly coiled, bright green, baby fern used as a cooked vegetable or decorative garnish.

Field bees Worker bees (see *honeybee*) two and a half to three weeks old that collect *nectar*, *pollen*, water, and *propolis* for the *hive.*

Field capacity The moisture content of *soil* in a field measured two or three days after a thorough wetting of a well-drained soil by rain or irrigation (see *acres irrigated* and *irrigated farms*) water. Field capacity is usually expressed as a percentage of the ovendry weight of soil.

Field cultivator A machine used to till (see *till*) *soil* to kill weeds, conserve moisture, and prepare ground for seeding (see *seed*) close-grown crops. A field *cultivator* is different from a *row cultivator*, which is used to *cultivate* rows of crops.

Field work Work related to planting, cultivating (see *cultivate*), or harvesting (see *harvest*) operations that occurs in the field rather than in a *processing* plant (see *plant or establishment*) or packing shed. Also includes work in nursery and *mushroom*-growing operations and similar planting, cultivating, and/or harvesting.

Field worker See *hired worker.*

50/85/92 A provision, originally known as 50/92, that allows *cotton* and *rice* growers who plant at least 50 percent of their *maximum payment acreage* (MPA) to receive *deficiency payments* on 92 percent of the *farm*'s MPA under certain conditions. The Farm Disaster Assistance Act of 1987 (P.L. 100-45) (see Appendix 3) also authorized 50/92 for *wheat, feed grain*, cotton, and rice *producers* who were affected by a natural disaster in 1987 and met certain criteria stated in the law. The Omnibus Budget Reconciliaton Act of 1993 (see Appendix 3) reduced the percentage of deficiency payments producers may receive from 92 to 85 percent of MPA. However, producers who plant approved industrial or experimental crops, or who suffer prevented planting or failed acres, may earn deficiency payments on up to 92 percent of the farms' MPA.

Fig One of the oldest cultivated (see *cultivate*) fruits. Fresh *varieties* include Black Mission, Brown Turkey, Brunswick, Celeste, and Kudota. Smyrna is the most common imported dried fig, while Calimyrna is a variety for *drying* grown in California.

Filament An individual strand of *fiber* of almost indefinite length. *Man-made fibers* are initially in this form. Silk is the only *natural fiber* available in filament form and may run several hundred yards in length.

Filbert A shrub or tree found in Europe and Asia that yields edible nuts known as filberts or hazelnuts. Major producers include the United States, Turkey, Italy, and Spain.

Filler tobacco A class of cigar *tobacco* used mainly in the core or body of a *cigar.* Filler tobaccos are medium to heavy in body. Also cheaper types of *cigarette* tobacco blended with higher quality types to reduce overall costs.

Filling An individual *yarn* that interlaces with *warp* yarn at right angles in woven (see *weave*) fabric (see *cloth*). Also known as pick or filling pick. Usually has less *twist* than warp yarn, which runs lengthwise in the fabric.

Filling pick See *filling.*

Filly A female *foal* until it reaches maturity. See also *colt.*

Filter strip See *buffer strip.*

Findley loan rates An option available to the secretary of agriculture to improve U.S. competitiveness by lowering the statutory *loan rate* for *wheat* and *feed grains* up to 20 percent. The first 10 percent of the reduction is determined by legislative formulas depending on the *stocks-to-use ratio.* Originally proposed by former Representative Paul Findley (R-IL), this provision was adopted in the *Food Security Act of 1985*. If the loan rate is reduced under the Findley provision, USDA is required to make additional *deficiency payments* to *producers* to provide the same total return as if there had been no reduction. The Findley payment rate equals the statutory loan rate minus (1) the national weighted season average *farm* price for the *marketing year*; or (2) the announced loan level, whichever is higher. If the season average *price* is above the statutory loan rate, no Findley payments are required. Findley payments are also called emergency compensation payments.

Findley payments See *Findley loan rates*. Also known as emergency compensation payments.

Fine breaks The rolls that *grind* the smaller particles of *grain* in a classified *break system* (see *coarse breaks*).

Finishing 1: The last stage of *production* before *livestock* emerge from the *feedlot* and are sent to packing (see *packer*) plants. See also *overfinished.*
2: The processes to which a fabric (see *cloth*) is subjected after its formation, including desizing, scouring, *bleaching*, dying, waterproofing, applying a *durable press* finish, and removing defects.

Finishing house (barn) A building for growing and *finishing* animals until they are ready for *market.*

Fire-cured tobacco A medium- to heavy-bodied *tobacco*, light to brown in color, and strong in flavor. It acquired the name because of the smoky flavor and aroma acquired from "firing" over open fires in the *curing* barns. It is used to make *snuff*, roll and plug *chewing tobacco*, strong *cigars*, and heavy smoking tobacco.

First clear flour The portion of *flour* that remains after the *patent flour* has been removed. First clear, which may be further divided into fancy clear and second clear (see *second clear flour*), is higher in *protein* and ash content than patent flour but poorer in color and lower in commercial value.

First litter gilt A female *pig* producing her first *litter*.

Fiscal policy See *macroeconomics.*

Fiscal year The official U.S. government operating year, which begins October 1. The fiscal year is used by program agencies in reporting much of their data on federal programs.

Fish meal A commercial *feed* for *poultry* and other *farm* animals. It consists of the clean, dried, ground tissue of decomposed fish and fish cuttings, with or without the oil (see *fats and oils*) extracted. In the United States, fish meal is manufactured mainly from menhaden.

Fixation (in soil) The conversion of a soluble material, such as a plant *nutrient* like phosphorus, from a soluble or exchangeable form to a relatively insoluble form.

Fixed costs *Costs of production* that do not change as a result of the type or volume of crop produced. Examples of fixed costs include interest, rent or *land* mortgage payments, and insurance. See also *variable costs.*

Fixed exchange rates See *controlled exchange rates.*

Fixed-price contract A *forward contract* that establishes the specific *price* to be paid by the buyer to the seller.

Flail cutter A machine that cuts *hay* and *forage* crops using rotating knives or flails.

Flakes An ingredient rolled or cut into flat pieces with or without prior steam conditioning (see *tempering*).

Flaking A process of "super*hydrogenation*" of oil (see *fats and oils*) that produces a much harder product than normal hydrogenation. Flaked vegetable oil (flaked *shortening*) is used primarily by the baking and confectionery industries because of its stability, baking qualities and other desired qualities affecting the finished product.

Flaking grits A product of *dry-milled* hard *corn*. Low-fat (see *fats and oils*), large flaking *grits* are used primarily in the manufacture of breakfast food, and coarse and regular grits are used in the brewing industry.

Flame cultivator Equipment mounted on a *tractor*, truck (see *farm truck*), or other power unit which includes fuel tank (usually propane), hoses, nozzles, valves, and shields (to protect the operator and to protect commercial plants), used to destroy weeds or insects.

Flash candling Mechanized *candling* of *eggs* that enables the operator to check up to 35 cases of eggs an hour.

Flax A plant whose *fiber* is used to make *linen*. Argentina, Canada, India, the former Soviet Union, and the United States are major flax producers.

Flaxseed See *linseed.*

Fleece Shorn *wool* from one *sheep.*

Flex acres See *normal flex acres.*

Flexible exchange rate The *market*-determined rate of exchange of a nation's currency. The value of a country's currency is determined by the *supply* and *demand* for the currencies, which are based on the supply and demand for goods and services produced by the trading nations.

Float A device pulled by a *tractor* and used to level or smooth *soil.*

Floating elevator A mechanical device placed on a floating platform used for transferring *grain* between ships and barges.

Flood irrigation See *irrigation methods.*

Flood plain A lowland, relatively flat area adjoining inland and coastal waters, including floodprone areas of islands. This *land* includes, at a minimum, those areas that are subject to a 1 percent or greater chance of flooding in any given year.

Floriculture The *production* of cut flowers, decorative greens, potted flowering and foliage plants, and bedding and garden plants. Although usually grown in *greenhouses* or under other protective cover, some floriculture crops, such as cut flowers, may be grown outdoors.

Flour The soft, finely ground *endosperm* and *meal* of *grains* or other *seeds*. Wheat flour consists mainly of the *starch* and *gluten* of the endosperm of the wheat *kernel*. At retail, there are five major wheat flour types: hard wheat flour, whole *wheat* flour, *soft wheat* flour, all purpose flour, and *semolina*. However, an infinite number of flours may be produced from blends for commercial food use. Flour is classified according to strength. Strong flours, derived from hard wheat and used mainly for baking bread, are high in *protein* and elastic gluten. These include semolina, which is made from *durum wheat* and used to manufacture *pasta*. Weak flours, derived from soft wheat, are used for biscuits and pastries and are low in protein and gluten. All purpose flour is a blend of hard and soft wheat flour. See also *patent flour*, *fancy patent flour*, *first clear flour*, and *second clear flour*.

Flour extraction rate The pounds (percent) of *flour* extracted from 100 pounds of *wheat*.

Flour stream The *flour* resulting from each separate *dry milling* process. Flour from each point of the process has different characteristics and baking properties. In large flour mills, 30 or more separate flour streams of varying composition and purity may be collected, grouped, and merchandised.

Flow The movement of a *commodity* through channels of trade.

Flower and vegetable seed Any *seed* used for growing flowers or vegetable plants.

Flow-to-market A quantity provision in a *federal marketing order* that does not change the total quantity that can be marketed during a season, but controls the rate that quantities can be shipped to *market*. The provisions include shipping holidays and prorates.

Flue-cured tobacco One of the principal classes of *tobacco* grown in the United States. Its name comes from the metal flues of the heating apparatus originally used in *curing* barns. The tobacco is cured by a process of regulating the heat and ventilation without allowing smoke or fumes from the fuel to come in contact with the tobacco. Flue-cured tobacco is yellow to reddish orange in color, thin to medium in body, and mild in flavor. It is used mainly in *cigarettes*. China, the United States, Brazil, Zimbabwe, Canada, and India are the major producers of flue-cured tobacco. See also *burley tobacco*.

Fluid (Class I) differential The amount added to the basic manufactured class price to determine a *federal milk marketing order's* minimum price for *milk* used for fluid purposes (see *fluid milk products*).

Fluid (Class I) utilization The proportion, or share, of the total *producer* milk (see *milk*) receipts in a market order (see *federal milk marketing orders*) used to produce *fluid milk products*.

Fluid milk products Dairy products packaged in fluid form for beverage use.

Fluid utilization The proportion of *Grade A milk* pooled in a *market* and used to produce *fluid milk products (Class I)*.

Flush gutter A gutter in the floor of a hog (see *swine*) house through which *manure* is flushed with water periodically into a storage or disposal facility. Gutters may be open or under a slotted cover.

Foal A newborn *horse* of either sex until it is of weaning age.

Foam marker A device used to leave a visible white application of bubbles in the field at the end of a *sprayer* boom to let the operator know what part of the field has been covered.

Fobbing The physical handling of bulk *grain* from alongside an *elevator*. The bulk grain passes from a barge, railroad car, or truck (see *farm truck*), through an elevator to a vessel tied up at the elevator's loading berth.

F.o.b. (free-on-board) acceptance final A term of sale calling for the seller to load a product of specified quality on the buyer's conveyance at the seller's loading dock. This term of sale is essentially the same as *f.o.b. shipping point* except that the buyer explicitly relinquishes any right to a price adjustment if the quality on arrival is below specifications.

F.o.b. (free-on-board) shipping point A term of sale calling for the seller to load a product of specified quality on the buyer's conveyance at the seller's loading dock. The buyer pays hauling costs and assumes all risk of damage and delay in transit. The buyer has the right of inspection at the shipping point or arrival to verify that the product shipped met *contract* terms at shipment time. The buyer is not required to accept delivery if it can be determined that the seller did not ship the product satisfying the contract terms at shipment time. In practice, the seller and the buyer may negotiate a *price* adjustment if the product is below specifications on arrival.

F.o.b. (free-on-board) shipping point price The average *price* (by shipping carton) of any *commodity* shipped from a major producing area. F.o.b. represents the buyer's cost before freight, insurance, and other transportation costs are added.

Foliage plants Plants produced for their decorative shape, size, color, and stem and leaf characteristics and usually sold by the pot or hanging basket. Includes house plants for indoor or patio use and large specimens for the interiors of hotels, restaurants, and offices. Examples are ficus, dieffenbachia, schefflera, spathiphyllum, and philodendrons. See also *potted flowering plants.*

Food, Agriculture, Conservation and Trade Act of 1990 (P.L. 101-624) The omnibus food and agriculture legislation (see *farm bill*) signed into law on November 28, 1990, that provides a five-year framework for the secretary of agriculture to administer various *agriculture*, food, conservation, and trade legislation. The law, with its 25 titles, amends the *permanent legislation*—the Agricultural Adjustment Act of 1938 (P.L. 75-430) and the Agricultural Act of 1949 (P.L. 89-439) (see Appendix 3)—and supersedes the *Food Security Act of 1985 (P.L. 99-168).*

Food Aid Convention See *International Wheat Agreement.*

Food and Agriculture Organization (FAO) An agency of the *United Nations* concerned with the distribution and *production* of food and agricultural products around the world. Founded in 1945, FAO is responsible for collecting, analyzing, and disseminating country data on food, *agriculture*, and rural affairs. The agency also offers technical assistance and operates training projects in many *developing countries.* It is officially known as the United Nations' Food and Agriculture Organization.

Food and Consumer Service (FCS) A *U.S. Department of Agriculture (USDA)* agency created in October 1994 by merging the *Food and Nutrition Service (FNS)* and the *Office of the Consumer Advisor.* The FCS is responsible for administering the federal food assistance programs, such as the *Food Stamp Program,* the *National School Lunch Program,* and the *Special Supplimental Food Program for Women, Infants, and Children.*

Food and Drug Administration (FDA) An agency of the *U.S. Department of Health and Human*

Services that has the primary responsibility for assuring the safety and wholesomeness of the processed food *supply* (except meat, which is the responsibility of the *U.S. Department of Agriculture [USDA]*). The agency also regulates *drugs* for humans or animals, cosmetics, and medical devices. See *Federal Food, Drug, and Cosmetic Act (FFDCA)*.

Food and fiber system Farming and the industries linked to it, including *farm* machinery and equipment, food and fiber (see *fiber*) *processing* and manufacturing, wholesale and retail *foodstores*, as well as the labor, energy, and transportation used to produce and market (see *agricultural marketing*) *agriculture* and food products. The food and *fiber* system is one of the largest sectors in the U.S. economy.

Food and Nutrition Service (FNS) The name of the former *U.S. Department of Agriculture (USDA)* agency responsible for administering federal food assistance programs, such as the *Food Stamp Program*, the *National School Lunch Program*, and the *Special Supplemental Food Program for Women, Infants, and Children*. In October 1994, FNS merged with the *Office of the Consumer Advisor* to form the *Food and Consumer Service Agency*.

Food balance The *supply* of food in relation to the population to be fed.

Food broker A marketing specialist who serves as an *agent* of the producer (canner, *processor*, *packer*, manufacturer, etc.) and assists in sales negotiations of the seller's food and grocery products to buyers (wholesalers, *chains*, supermarkets [see *foodstore*], industrial users, etc.) in their trade area.

Food chamber A bee (see *honeybee*) *hive* area where *honeycombs* are filled with *honey* for winter food storage.

Food disappearance data A measure of the amount of the major food *commodities* entering the marketing (see *agricultural marketing*) channels, regardless of their final use. The food disappearance data estimate the total amount available for consumption as the residual after *exports*, industrial uses (see *industrial users*), *seed* and *feed* use, and year-end inventories are subtracted from the sum of *production*, beginning inventories, and *imports*. The use of *conversion factors* accounts for subsequent *processing*, trimming, shrinkage, or loss in the distribution system. However, the estimates also include residual uses for which data are not available, such as miscellaneous food uses and changes in *consumer* and retail *stocks*. Because the food disappearance data come from *market* channels, the data are available only on a per capita basis and cannot be used to estimate consumption by sex, age, or demographic group. See also *Continuing Survey of Food Intakes by Individuals (CSFII)* and *Consumer Expenditure Survey*.

Food, farm-produced Food products originating on U.S. *farms*. These included processed (see *processing*) products made mainly from farm-produced ingredients, as well as *eggs*, fresh fruits and vegetables, and other products sold to *consumers* without processing. Nonfarm foods are those not originating on U.S. farms, such as imported products.

Food for Development Program See *Public Law 480*.

Food for Peace Program See *Public Law 480*.

Food for Progress A food aid program created by the *Food Security Act of 1985 (P.L. 99-198)* (see Appendix 3) to provide agricultural *commodities* under *Section 416* or *Public Law 480*. Food is provided to countries committed to encouraging free enterprise in their agricultural sector.

Food grain *Cereal* seeds (see *seeds*) used for human food, chiefly *wheat* and *rice.*

Food manufacturing The process of mechanically or chemically transforming raw materials into foods and beverages for human consumption. Power-driven machines and materials-handling equipment are typically used. Food manufacturing may also include certain related industrial products, such as *feeds*, and vegetable and animal *fats and oils.*

Food Marketing Institute A national trade association of firms engaged in retail and wholesale (see *wholesaling*) distribution of food and grocery products.

Food Research and Action Center (FRAC) A nonprofit, nonpartisan research, public policy, and legal center working to eradicate hunger and undernutrition in the United States.

Food reserves Supplies of food stored against contingencies either by food exporting (see *export*) or food importing (see *import*) countries.

Food Safety and Inspection Service (FSIS) A *U.S. Department of Agriculture (USDA)* agency that oversees food labeling and inspection (see *inspection, federal*). FSIS inspects all meat and *poultry* sold in interstate and *foreign* commerce and tests for *chemical residues* and *drug residues.* The agency is also responsible for all functions under the Egg Products Inspection Act and USDA's *salmonella enteritidis* reduction program and pathogen reduction activities.

Food Security Act of 1985 (P.L. 99-198) The omnibus food and *agriculture* legislation (see *farm bill*) signed into law on December 23, 1985, that provided a five-year framework for the secretary of agriculture to administer various agriculture and food programs. The act amended *permanent legislation*—the Agricultural Adjustment Act of 1938 (P.L. 75-430) and the Agricultural Act of 1949 (P.L. 89-439) (see Appendix 3)—for the 1986 through 1990 crop years.

Food Security Wheat Reserve A special *wheat* reserve of up to 4 million metric tons (see Appendix 6) to be used for humanitarian purposes. The reserve, created by the Agriculture Act of 1980 (P.L. 96-494) (see Appendix 3), is generally used to provide famine and other emergency relief when commodities are not available under *Public Law 480.*

Foodservice The dispensing of prepared meals and snacks for on-premise or immediate consumption. Vended foods qualify as foodservice only when tables or counters are available in the immediate area. Foodservice operations include commercial and noncommercial establishments:

Commercial establishment: A public establishment (free standing or part of a host establishment) that prepares, serves, and sells meals and snacks for profit to the general public.

- *Drinking place.* An establishment with foodservice that does not operate as part of a different and separately identifiable business and whose primary function is to sell alcoholic beverages for consumption on the premises. Drinking places include bars, beer gardens, taverns, night clubs, and saloons.
- *Eating place.* An establishment that does not operate as part of a different or separately identifiable business and whose primary function is to sell prepared meals and snacks for on-premise or immediate consumption. Eating places include restaurants, lunchrooms, fast-food outlets, and cafeterias.
- *Lodging place.* An establishment that provides both lodging and foodservice to the general public. Includes hotels, motels, and tourist courts but not rooming and boarding houses and private residences.
- *Recreation/entertainment.* Foodservice operations in theaters; bowling, billiard, or pool halls; commercial sports establishments (racetracks and stadiums); membership golf or country clubs; public golf courses; and miscellaneous commercial amusement and recrea-

tional establishments (tennis clubs, athletic clubs, and amusement parks).

- *Retail host.* A foodservice establishment that operates in conjunction with or as part of a retail establishment, such as a department store, limited-price variety store, drugstore, or miscellaneous retailer.

Noncommercial establishments: An establishment where meals and snacks are prepared and served as an adjunct, supportive service to the primary purpose of the establishment. Noncommercial establishments include schools, colleges, hospitals and extended care facilities, vending areas, plants and offices, correctional facilities, military kitchens, and transportation (trains, cruise ships, and airplanes).

Food Stamp Program (FSP) A U.S. program that helps low-income households improve their diets by providing them with coupons, based on their income levels, to purchase food at any authorized retail *foodstore.* The program began as a pilot operation in 1961 and was made part of *permanent legislation* in the Food Stamp Act of 1964 (P.L. 88-525) (see Appendix 3).

Foodstore A retail outlet with at least 50 percent of sales in food products intended for off-premise consumption.

Grocery store: A *foodstore* that sells a variety of food products, including fresh meat, *produce,* packaged and canned foods, frozen foods, other processed (see *processing*) foods, and nonfood products.

- *Supermarket.* A *grocery store,* primarily self-service in operation, providing a full range of departments, and having at least $2.5 million in annual sales in 1985 dollars.
    - Combination food and drug store—A supermarket containing a pharmacy, a nonprescription drug department, and a greater variety of health and beauty aids than that carried by conventional supermarkets.
    - Superstore—A supermarket distinguished by its greater variety of products from conventional supermarkets, including specialty and service departments and considerable nonfood (general merchandise) products.
    - Warehouse store—A supermarket with limited product variety and fewer services provided, incorporating case lot stocking and shelving practices. Superwarehouse stores are larger and offer expanded product variety and often offer meat, delicatessen, or seafood departments.
- *Convenience store.* A small *grocery store* selling a limited variety of food and nonfood products, typically open extended hours.
- *Superette.* A primarily self-service *grocery store* selling a wide variety of food and nonfood products with annual sales below $2.5 million in 1985 dollars (see *real dollars*).

Specialized foodstore: A *foodstore* primarily engaged in the retail sale of a single food category, such as meat and seafood stores, dairy stores, candy and nut stores, and retail bakeries. See also *wholesale club store.*

Foot-and-mouth disease (FMD) A viral *disease* of all cloven-footed animals. The disease is spread by direct or indirect contact with an infected animal or by *milk,* meat, or slaughter (see *slaughter*) *byproducts* of an infected animal but is rarely transmitted to humans. Because the disease is highly contagious, infected animals are usually destroyed. Trade with the United States in animals and meat products from areas with FMD is restricted. Also called hoof-and-mouth disease and aftosa.

Foot rot A common ailment of *sheep* and *cattle* characterized by inflammation between the toes and in the hoofs that causes limping and swelling. Foot rot is caused by a combination of *fungus* and *bacteria.*

Foots See *soapstock.*

Forage Plant materials grazed (see *graze*) or harvested (see *harvest*) for livestock (see *livestock*) *feed*. Harvested materials can be fresh, dried (see *drying*), or *ensiled*. Forage includes *hay*, *oats*, *corn*, *wheat*, and *barley*.

Forage blower See *blower*.

Forage harvester A machine used to cut, chop, and blow *forage* into a large high-sided *wagon*.

Foreign 1: A country or geographical region other than the home or native country.
2: An object in a place where it does not normally belong, such as nongrain materials in a shipment of *grain*. See *foreign material*.

Foreign Agricultural Service (FAS) The *export* promotion and service agency for the *U.S. Department of Agriculture* (USDA). FAS coordinates and directs the USDA's activities related to international trade agreement programs and *trade negotiations* to improve access for U.S. *farm* products abroad. The agency is also responsible for gathering and disseminating information on worldwide *production*, *supply*, and *demand* for agricultural *commodities*; operating statutory programs to facilitate the *export* of U.S. agricultural products; and representing U.S. agricultural interests abroad.

Foreign Credit Insurance Association (FCIA) An association of private insurance companies and the *Export-Import Bank* that provides *export credit insurance* for short- and medium-term transactions. FCIA insurance covers political and commercial risks.

Foreign exchange controls Government limitations or restrictions on the use of certain types of currency, bank drafts, or other means of payment in order to regulate *imports*, *exports*, and the *balance of payments*.

Foreign exchange rate See *exchange rate*.

Foreign investment The ownership of domestic assets (see *assets, current* and *assets, fixed*) by *foreign* persons or firms.

Foreign material A factor in *grading* commodities, the exact definition of which varies by *commodity*. In the case of *oilseeds* and most *grains* (except *wheat* and *sorghum*), it is all nongrain material in a shipment. For wheat and sorghum, nongrain material is divided into *dockage* and foreign material, defined as nongrain, nonmillable (see *milling*) material that cannot be easily removed from the grain.

Foreign trade zone Special commercial and industrial areas in or near ports of entry where *foreign* and domestic merchandise, including raw materials, components, and finished goods, may be imported (see *import*), stored, or processed (see *processing*) without being subject to duties (see *tariff*) and taxes. Merchandise can be brought into these zones before being reexported (see *reexport*) or transferred into the national customs territory. Examples include Gibraltar, the Canary Islands, Aden, and Manaus, Brazil. Also called a free trade zone, free zone, and free port. See also *bonded warehouse*.

Forestry Incentives Program (FIP) A program administered by the *U.S. Department of Agriculture*'s (USDA) *Forest Service* in cooperation with state forestry agencies. The FIP authorizes the federal government to share with private landowners the cost of planting trees and improving timber stands. The federal share of these costs can be up to 65 percent. See also *cost-sharing*.

Forest Service (FS) The largest *U.S. Department of Agriculture (USDA)* agency, responsible for managing and protecting the national forests and grasslands, cooperating in managing and protecting certain nonfederal *lands*, and conducting research in forestry and forest products utilization.

Forklift A self-propelled unit with two hydraulically operated forks used to lift heavy loads. A squeeze-type forklift is used to lift baled (see *bale*) *hay* on and off trucks (see *farm truck*).

Formed steaks Steaks formed by a machine into a specific shape. The raw material fed into the machine is usually chunked or flaked.

Formula approach A method of negotiating *tariff* reductions using an agreed upon formula applied to tariff rates (with limited exceptions being granted for very sensitive items) by all *contracting parties*.

Formula feeds *Feeds* containing two or more ingredients that are processed (see *processing*) or mixed according to specifications.

Formula pricing Agreeing on a rule to be used on a later date to determine the *price* to be paid for a *commodity*. The rule usually specifies a differential to be added to a specified *market* price quotation on a date selected by the seller or buyer to determine the price to be paid. When the rule sets the price relative to a futures (see *futures contract*) price it is called '*booking the basis*' in the *grain* trade and *call pricing* in the *cotton* trade. Also called deferred pricing.

Fortified foods Foods improved by the addition of *nutrients*, such as *vitamins*, minerals, and *amino acids* or other *protein* supplements, without any detectable change in appearance, flavor, and technological properties. These nutrients may or may not have been present in the food at *harvest*. For example, certain B vitamins are added to *flour* in the United States to replace the quantities lost during *milling*. Iodine is added to salt and vitamin D to *milk* although both nutrients are not present in the unprocessed (see *processing*) food.

Forward buying or selling Contracting for *deferred delivery* in which the *price* is set at the time the *contract* is entered.

Forward contract See *contract*.

Forwarder An *agent* who receives goods for transportation, delivers them to the carrier, and performs various services for the shipper. Forwarding agents provide information to the shipper regarding inland and ocean freight rates, insurance, and *customs* requirements. Also called a freight forwarder or forwarding agent.

Forwarding agent See *forwarder*.

Forward market An institutional arrangement for trading in *contracts* calling for *deferred delivery*.

Forward pricing An agreement between seller and buyer on *price*, or a minimum or maximum price, for a delivery to occur in the future.

Foundation herd Breeding stock retained for replacement.

Founder An inflammation of the foot and lower leg of *ruminant* animals caused by overeating *grain* or green grass. Also known as laminitis.

4-H Youth Programs Organized groups of young people (ages 9 to 19), through which the *Cooperative Extension Service*, the *U.S. Department of Agriculture* (*USDA*), and state *land-grant universities* carry on educational work in farming and homemaking projects, career development, citizenship, leadership, and other youth development activities. The H's stand for head, hand, heart, and health.

Frame A rectangular wood case for holding *honeycomb.*

Free choice feeding *Grains* and *protein* feeds (see *feed*) offered separately without constraint as to the amount consumed.

Free fatty acids *Fatty acids* easily removed from *fats and oils* by deacidification to prevent rancidity. Deacidizing edible oils is usually done by an alkali *refining* process while industrial oils are washed with acid or water or both.

Free in (FI) A *contract* term that specifies that the buyer of cargo, and not the charterer, pays loading costs.

Free in and out (FIO) A *contract* term that specifies that the buyer of cargo, and not the charterer, pays loading and unloading costs.

Free list A list of goods designated as free from import *duties* or *import licensing requirements* in a given country.

Free market A system in which the *market* forces of *supply* and *demand* determine *prices* and allocate available supplies. A free market approach in *agriculture* would eliminate barriers to international trade (see *trade barriers*) and programs that distort the market forces that determine supply and demand.

Free of particular average (FPA) The minimum marine insurance coverage in general use for shipping *commodities*, covering (1) a total loss of goods shipped or (2) a partial loss of goods occurring after the vessel is stranded, sunk, burnt, or involved in a collision.

Free-on-board See *f.o.b. (free-on-board) shipping point price.*

Free out (FO) A *contract* term that specifies that the buyer of cargo, not the charterer, pays unloading costs.

Free port See *foreign trade zone.*

Free range Animals, such as chickens and turkeys (see *poultry*), that are not raised in confined conditions. Used because the animals have free run of their barnyards or other area during at least part of the day.

Free rider A firm or person who benefits from a collectively funded activity without contributing to its costs. A producer or manufacturer, for example, who does not contribute to a *generic advertising* campaign for their *commodity*, may still benefit if the promotion effort results in greater *demand* for the product.

Free stocks *Stocks* of *commodities* available for purchase in commercial *markets*. Excludes any stocks held under government or private reserve programs.

Free trade Exchange of goods between countries with no *trade barriers* or restrictions, such as *tariffs* or *import quotas*.

Free trade area A cooperative arrangement by a group of nations to eliminate *trade barriers* among the members. Each member may maintain its own trade regime with nonmember nations. The *European Free Trade Association* is the best known example.

Free trade zone See *foreign trade zone*.

Free zone See *foreign trade zone*.

Freight corrections-base point pricing The process used to determine *prices* in locations other than those of the quoted price. Freight corrections-base point pricing is determined by adding or subtracting from the quoted price the normal transportation charge from the location of the quoted price to the location where a price is being determined.

Freight forwarder See *forwarder*.

Freight tariff Published rates for transporting *commodities* or classes of commodities between specific points. In addition to rates, the *tariff* may include available routings, transit privileges, and special services.

Freight ton See *measurement ton*.

French plow A *plow* for the cultivation (see *cultivate*) and care of *vineyards*.

Freshen When a *cow* comes into *milk* production after *calving*.

Fresh weight equivalent The weight of *processed vegetables and fruits* converted to an equivalent weight of the fresh *produce*.

Fresno Equipment used to move dirt by dragging a bucket that can be rotated forward for dumping or spreading. Also known as a Fresno scraper.

Fresno scraper See *fresno*.

Friable soil *Soil* that crumbles easily.

Front-end loader A hydraulically operated implement mounted on the front of a *tractor* used for collecting and lifting materials.

Front of the mill particles *Wheat* particles removed near the beginning of the *milling* process that are not finely ground. These are larger than *tail of the mill* particles.

Fructose A *monosaccharide* that has the same elements in the same proportions as *glucose* but whose molecular structure is different. As a result, fructose has a sweeter taste than glucose. Fructose is usually a liquid.

Fryer See *broiler*.

Fryer-roaster turkey A young, immature bird, usually under 16 weeks of age.

Full bloom A stage of maturity when two-thirds or more of the plants are in bloom.

Full or strict cross-compliance See *cross-compliance.*

Fumigant A chemical that, at a required temperature and pressure, can exist in the gaseous state in sufficient strength and quantities to be lethal to a given pest organism. Fumigants are some of the most toxic and unique *pesticides*. They are usually used in *soils* or in closed structures, such as warehouses. Methyl bromide and hydrogen phosphide, for example, are the fumigants most commonly used on *grain.*

Fumigation The application of *fumigants* to destroy pests.

Functioning of the GATT System (FOGS) A *Uruguay Round* negotiating group whose function was to strengthen the *General Agreement on Tariffs and Trade (GATT)* process. The major stated goals of the group were to improve GATT *surveillance* of trade policies and practices, to encourage greater involvement of trade ministers in the GATT, and to strengthen the GATT's relationship with other international organizations, such as the *International Monetary Fund.*

Fungicide Any substance used to kill *fungi.*

Fungus A simple form of plant life that, lacking *chlorophyll* and being incapable of manufacturing its own food, lives off dead or living plant or animal matter. Fungi include *molds*, *rusts*, *mildews*, and *mushrooms*. *Aspergillus flavus*, for example, is a fungus that grows on *corn*, *peanuts*, and *soybeans* and produces *aflatoxin.*

Furrow A trench in the *soil* made by a *plow* or *cultivator.*

Furrow irrigation See *irrigation methods.*

Further processed A product prepared by cutting-up, deboning, cooking or otherwise *processing* beyond the initial stage to change the form, appearance, or texture or to maintain quality.

Further processor A plant (see *plant or establishment*) that prepares *further processed* products from raw *commodities*. In the case of *poultry*, for example, it is usually a specialized plant function following *slaughter*, and may be in the same or a separate facility.

Futures See *futures contract.*

Futures contract A standardized agreement calling for *deferred delivery* of a *commodity,* or its equivalent, entered through an organized exchange, and guaranteed by *margin* deposits made by the trading parties. Most agricultural futures contracts call for physical delivery, but *feeder cattle* futures contracts call for cash settlement at *contract* maturity.

Futures exchange See *commodity futures exchange.*

Futures market See *commodity futures exchange.*

Futures trading The buying and selling of standardized fixed-price (see *price*) *forward contracts* under the rules of an organized *commodity futures exchange.*

G

Gaits The forward movements of a *horse*, including walking, trotting, cantering, or galloping. There are also gaits acquired through training, such as the pace, stepping pace, foxtrot, running walk, rack, and amble.

Galled spots Areas in fields that are practically sterile from *erosion* or removal of top *soils*.

Garbanzo bean The dry, threshed *seed* of an Asiatic leguminous (see *legume*) *herb*. Also known as chick pea. See also *beans*.

Garlic An *herb* of the lily family, the *bulb* of which is used as flavoring. The United States, Mexico, Guatemala, Chile, and Argentina are major producers.

Gasohol A mixture of *ethanol* and gasoline.

Gate price See *minimum import price* and *variable levy*.

GATT See *General Agreement on Tariffs and Trade*. The designation also refers to the organization headquartered in Geneva that carries out the *General Agreement on Tariffs and Trade*. The organization provides a framework within which international *trade negotiations* are conducted and trade disputes resolved.

GATT **Rounds** Cycles of *multilateral trade negotiations* conducted under the *General Agreement on Tariffs and Trade (GATT)*. Eight rounds have been completed since the GATT was established in 1947.

1947: Geneva, Switzerland. The GATT was created during this round.
1949: Annecy, France. This round involved negotiations with nations that desired GATT membership. Principal emphasis was on *tariff* reduction.
1951: Torquay, England. This round continued *accession* and tariff reduction negotiations.
1956: Geneva, Switzerland. This round proceeded along the same track as earlier rounds.
1960-1962: Geneva, Switzerland. This round, referred to as the *Dillon Round*, involved further revision of the *GATT* and the addition of more countries.
1964-1967: Geneva, Switzerland. Known as the *Kennedy Round*, this round was a hybrid of the earlier product-by-product approach to *trade negotiations* and the new formula tariff reduction approach with across-the-board tariff reductions (see *across-the-board [linear] tariff negotiations*).
1973-1979: Geneva, Switzerland. This round, also called the *Tokyo Round*, centered on the negotiation of additional tariff cuts and developed a series of agreements governing the use of a number of nontariff measures.

1986-1993: Geneva, Switzerland. This round, launched in Punta del Este, Uruguay, focused on strengthening the GATT and expanding its disciplines to new areas, including *agriculture*.

GATT Secretariat The administrative body of the GATT, headed by the director-general and headquartered in Geneva, Switzerland.

Gelding A male *horse* castrated before reaching sexual maturity.

Gene The portion of a *DNA* molecule that constitutes the basic functional unit of heredity.

General Agreement on Tariffs and Trade (GATT) An agreement, originally negotiated in Geneva, Switzerland, in 1947, among 23 countries including the United States, to increase international trade by reducing *tariffs* and other *trade barriers*. This *multilateral agreement* provides *codes of conduct* for international commerce. GATT also provides a framework for periodic *multilateral trade negotiations* on *trade liberalization* and expansion. The eighth and most recent round of negotiations began in Punta del Este, Uruguay, in 1986 and concluded in December 1993. See *GATT* Rounds.

General Sales Manager (GSM) General Sales Manager of the *U.S. Department of Agriculture*'s (USDA) *Foreign Agricultural Service*. The GSM office administers the *Export Credit Guarantee Program* (GSM-102) and the *Intermediate Export Credit Guarantee Program* (GSM-103), the *Export Enhancement Program*, *Public Law 480*, and other USDA *export assistance* programs.

Generalized System of Preferences (GSP) A policy that permits *tariff* reductions or possibly duty-free entry of certain *imports* from designated *developing countries*. Among other things, the GSP may increase economic growth in developing countries, help maintain favorable *foreign* relations with free world developing countries, and may serve as a low-cost means of providing aid to these nations. It is part of a coordinated effort of the industrial trading nations to bring developing countries more fully into the international trading system. Under the GSP, the United States provides nonreciprocal tariff *preferences* for designated developing nations.

Generally Recognized as Safe (GRAS) A category of food ingredients *grandfathered* as safe in the 1958 Food Additive Amendment to the *Federal Food, Drug, and Cosmetic Act* (P.L. 75-717). These substances have been part of the food *supply* for many years. According to qualified experts with evidence based on scientific procedures, they pose no threat to human safety. Few GRAS substances have been subjected to the extensive testing applied to new products. Such items as *sucrose*, salt, *corn syrup*, and *dextrose* are included on the GRAS list.

General partnership A business organized as a partnership in which partners share in profits and losses.

Generic advertising Promotion of a *commodity* without reference to the specific farmer, brand name, or manufacturer. Generic advertising has been used to overcome competition from other products, to increase awareness of lesser known products, and to alter negative opinions about a product. Dairy and beef promotion campaigns are examples of generic advertising. Overseas *market* development is also an application of generic advertising.

Generic commodity certificates Negotiable certificates, which do not specify a certain *commodity*, issued by the *U.S. Department of Agriculture (USDA)* in lieu of cash payments to commodity program participants and sellers of agricultural products. The certificates, frequently referred to as *payment-in-kind (PIK)* certificates, can be used to acquire *stocks* held as collateral on government loans or owned by the *Commodity Credit Corporation* (CCC). Farmers have received generic certificates as payment for participation in numerous government programs including *acreage reduction, paid land diversion*, the *Conservation Reserve Program*, rice (see *rice*) *marketing loans*, disaster, and *emergency feed programs*. *Grain* merchants and commodity groups also have been issued certificates through the *Export Enhancement Program* and the *Targeted Export Assistance Program*.

Genetic engineering Technologies (including *recombinant DNA* methods [see *DNA*]) used by scientists to isolate *genes* from one organism, manipulate them in the laboratory, and then insert them stably in another organism. See *biotechnology*.

Genetics The science of heredity, variation, sex determination, and similar factors determining physical characteristics and conditions.

Genome All the genetic material carried by a single *germ* cell (see *cell*).

Genotype The hereditary makeup of an individual plant or animal, which, with the *environment*, controls the individual's characteristics.

Genotypic variability The range of expression for a specific trait in an individual plant or animal. For example, the *protein* percentage of a plant or animal can range from 7 to 30 percent.

Germ 1: The *embryo* (rudimentary plant) within the *seed* and frequently separated from the *bran* and *starch endosperm* during the *milling*.
2: A *cell* capable of or sharing in reproduction.
3: A *disease*-causing microorganism.

Germicide A substance that kills *microorganisms*. See also *bactericide*.

Germination 1: The beginning of growth in a plant.
2: The percentage of *seeds* capable of producing normal seedlings under ordinarily favorable conditions.

Germplasm The hereditary material transmitted to offspring through *germ* cells. In the case of plants, it refers to *pollen*, the female germinal *cells*, *seeds*, vegetative parts used for reproduction, or a total collection of plant material. For animals, it includes sperm cells, *egg* cells, and entire animals. Also refers in a broad sense to the total hereditary makeup of organisms.

Gestation period The length of a pregnancy, which varies by *species*. In *cows*, for example, the average period is 281 days; *sheep*, 147 days; *swine*, 114 days.

Ghee A product made from *butter*, similar to *butteroil* but with a granular texture. It is popular in some countries of the Middle East and Central Asia.

Gilt An immature female *pig* who has not yet had a *litter* of pigs. See also *farrow*.

Gin A machine that separates *cotton* lint from *seed* and removes most of the trash and *foreign material* from the *lint*. The lint is cleaned, dried, and compressed into *bales* weighing approximately 500 pounds, including wrapping and ties.

Ginger A small, brown, spicy *tuber* frequently used in Asian cooking. It is used to flavor cakes, ice cream, and sweet beverages, as well as sauces, condiments, and stir-fry dishes.

Ginseng An aromatic root used as a medicine in Asia. Major producers include Korea, China, Japan, Canada, and the United States.

Give-up charge The amount charged to cover the increase in cost per unit in a reserve *milk* supply plant (see *plant or establishment*) when milk is withdrawn from the plant to make supplemental shipments to fluid milk *processors* (see *fluid milk products*). The costs per unit of production for the reserve *supply* plant increase because of less volume to spread the *fixed costs* over.

Gleaning The gathering of unharvested (see *harvest*) crops from the fields, or obtaining agricultural products from farmers, *processors*, or retailers without charge.

Gliadin Simple *proteins* obtained by alcoholic extraction of *gluten* from *wheat* or *rye*.

Global quota Limits established by a country on the value or quantity of goods that may be imported (see *import*) or exported (see *export*) through its borders during a given period.

Glucose The molecule used in the human body as an energy source. All other *carbohydrates* that are consumed as *starch* are broken down by *enzymes* into glucose, which is then oxidized for energy.

Glucose isomerase An *enzyme* capable of converting *dextrose* to *fructose*.

Glucose syrup See *dextrose*.

Glucosinolates Sulfur-containing *glucose* compounds found in the *seeds* from all plants of the mustard family, such as industrial *rapeseed*, *crambe*, *broccoli*, *turnips*, and *cabbage*.

Glume A small, dry, light leaf next to the flower of a grass.

Glut A *supply* of a *commodity* in excess of the effective *demand* for it at the *price* being quoted.

Gluten The rubberlike proteinaceous (see *protein*) material remaining after water solubles and *starch* are washed out of dough. The quantity of gluten in *flour* is a measure of flour quality. One-third of the weight of wet gluten approximates the protein content of the flour.

Glycerine A *byproduct* of producing soaps, fatty acids, and fatty alcohols from vegetable oils and animal *fats*. Synthetic glycerine is manufactured from petroleum raw materials. Glycerine has over 1,500 commercial applications, including *drugs*, cosmetics, resins, polymers, and explosives.

Goat A horned, *ruminant* animal closely related to *sheep* and used for *milk* (Alpine and Anglo-Nubian varieties), meat (Nubian *variety*), and *fiber* (Angora). North Africa, the Middle East, and India are the major producers.

Gobbler A mature male turkey (see *poultry*).

Goose See *poultry*.

Gossypol A phenobic, yellow pigment in *cottonseed* that is toxic to some animals.

Government-dependent county See *county type classification*.

Government payments Cash or *payment-in-kind* payments made to farmers directly from the Treasury for complying with provisions of *farm* legislation (such as removing *land* temporarily from *production* in a given year). These *subsidies* include *deficiency payments*, as well as payments for *paid land diversion* programs, *commodity* storage programs, and conservation programs. See also *direct payments*. Not included are indirect payments (such as the dairy *price support program*) and commodity loan programs (see *nonrecourse loans*). These indirect payments show up as higher cash receipts through *price* enhancements. Government payments are also referred to as *direct payments* or direct government subsidies.

Grade A dairy A *farm* producing *Grade A milk* according to federal and/or state approved sanitary conditions.

Grade A milk *Milk*, also referred to as fluid *grade*, produced under sanitary conditions that qualify it for use in fluid (beverage) milk (see *fluid milk products*). Only Grade A milk is regulated under *federal milk marketing orders*.

Grade and size (produce) *U.S. Department of Agriculture* (USDA) regulations establishing produce *grade* and size standards. Grades categorize *produce* by the maturity, taste, color, and shape attributes. Grades set the minimum requirements for produce to be eligible for shipment to regulated *markets* (often fresh domestic). Sizes set the ranges of produce sizes eligible for shipment to regulated markets.

Grade and yield (weight) A method of determining value of a *slaughter* animal based on quality and weight of the dressed (see *dressed animal*) *carcass* rather than *liveweight* of the animal.

Grade B dairy A *farm* that produces *Grade B milk*. Because the sanitation requirements are not as strict as for *Grade A milk production*, the milk cannot be sold or used for fresh fluid (beverage) consumption (see *fluid milk products*).

Grade B milk *Milk*, also referred to as *manufacturing grade milk*, not meeting *Grade A milk* standards. Less stringent sanitary production standards generally apply.

Grade-determining factors Indicators of quality and value that help determine the numerical *grade* of *grain*. See *grading*.

Gradeout The percentage of dressed *poultry* graded as *grade* A. Eighty percent, for example, is usually considered standard for *turkeys*.

Grades Quality and *yield* designations that provide convenience in trading and pricing (see *price*) without both buyer and seller having to see the product. See *grading*.

Grading 1: An alphabetical or numerical system for indicating particular characteristics of a *commodity*. The commodity or products are assessed against particular standards or rules to determine a *grade* designation to assign to each item or lot. For example, *wheat* is graded according to established quality and value factors (see *grade-determining factors*). Meat, *poultry*, and *eggs* can also be assigned official quality and *yield* grades for a fee by *U.S. Department of Agriculture* (USDA) graders.

2: Using machinery or other means to smooth or level a field or other *soil* area.

Graduation The process of removing *developing countries* that have increased their economic independence and competitiveness through increased *production* or *export* earnings, for example, from the list of *Generalized System of Preferences* countries. Requiring that "graduated" countries undertake the obligations of developed countries allows the developing nations to be fully integrated into the trading system.

Grafting Creating a permanent union between two plants by inserting an offspring of one specific *variety* into a stem, root, or branch of another.

Graham plow An implement with a series of *shanks* and wide sweeps. Graham plows are used to cut weed-plant growth under the ground with a minimum of surface disturbance. It is used in *summer fallow* to conserve ground moisture.

Grain The *seeds* or fruits of *cereal* grasses (*wheat, corn, barley, oats,* and *rye*) and other plants.

Grain bin unloader An *elevator* or conveyor used to move *grain* or other *seed* crops from storage to a truck (see *farm truck*), *wagon*, or other storage structure.

Grain blower A machine equipped with a large fan and power source, either as part of the unit or provided by tractor (see *tractor*) *power-take-off* or other source, used to force *grain* under air pressure through tubes from one area to another.

Grain breakage Mechanical damage to *grain* that results in broken grain and fine material. This can result from harvesting (see *harvest*) grain that is too dry. It reflects the cumulative damage inflicted on grain during repeated handling. Grain breakage causes decreased quality, greater storage problems, and increased rates of *mold* and insect infestation.

Grain cart A *wagon* or trailer with a tightly constructed box to haul *wheat, barley, oats,* and other *grains.*

Grain cleaner Any of a number of different machines used to separate weed *seeds*, *chaff*, and other unwanted debris from *wheat, oats, barley,* and other *grains*. See also *fanning mill* and *carter mill.*

Grain-consuming animal unit (GCAU) An aggregate *feed* consumption index that converts the *corn*-equivalent *concentrate* requirements for all animals to a common unit. *Livestock* and *poultry* numbers are weighted by the concentrate consumption factors developed for high-energy, low-*fiber* feeds over a base period.

Grain conveyor See *elevator* and *grain bin unloader.*

Grain crop silage The whole plant of *corn*, grain *sorghum*, or other *grain* crops harvested (see *harvest*) at a moisture content high enough to allow *fermentation* and preservation in a *silo.*

Grain drill Equipment used to *seed* or sow *wheat, barley, oats,* and other *grain* crops, as opposed to a *grass drill* that is used to sow fescues or other grass seeds.

Grain elevator See *elevator.*

Grain-fed animal Slaughtered (see *slaughter*) animals fed *rations* that were largely *grain* for an extended period before slaughter.

Grain harvester See *combine.*

Grain Inspection, Packers, and Stockyards Administration (GIPSA) A *U.S. Department of Agriculture (USDA)* agency created in October 1994 by merging the *Federal Grain Inspection Service* and the *Packers and Stockyards Administration.* The GIPSA is responsible for establishing official U.S. standards for 11 *grains* and other assigned commodities and administering a nation-wide inspection system to certify those *grades.* The GIPSA is also responsible for enforcing provisions of the *Packers and Stockyards Act* (PSA), the Truth in Lending and Fair Credit Billing Acts, and the Equal Credit Opportunity Act with respect to firms subject of the PSA.

Grain lupins A tropical *legume* grown in many countries as a *fertilizer* or as animal *feed.* Grain lupins contain a high proportion of *protein* (34–42 percent) and 4–25 percent oil (see *fats and oils*). The oil has a high proportion of polyunsaturated fats (see *fatty acids, classification of*).

Grain quality In general, the physical, sanitary, and intrinsic characteristics of *grain.* See *intrinsic quality, physical quality,* and *sanitary quality.*

Grain reserve A planned inventory of *grain.* See *Food Security Wheat Reserve* and *Farmer-owned Reserve Program.*

Grain Reserve Program See *Farmer-owned Reserve Program.*

Grain screenings The small, imperfect *grains, wheat* seeds (see *seed*), and other *foreign material* that have feeding value and are separated when grain is cleaned.

Grain sizes In the United States, three government-established *grain* sizes and shape types for *rice*—long, medium, and short. The length/width ratio is 3.0 or more for *long-grain rice,* 2.0–2.9 for *medium-grain rice,* and 1.9 and below for *short-grain rice.*

Grain storage Any of three methods for storing *grain*: vertically in upright metal *bins* or concrete *silos*; horizontally, in flat warehouses or other facilities; and on-ground in piles.

Grain truck See *farm truck.*

Gramm-Rudman-Hollings Deficit Reduction Act The common name for the Balanced Budget and Emergency Deficit Control Act of 1985 as amended in 1987 (P.L. 100-119) (see Appendix 3). The law mandated annual reductions in the federal budget deficit to eliminate it by 1993. Under the law, if Congress and the U.S. president could not agree on a targeted budget package for any specific fiscal year, automatic cuts could have occured for almost all federal programs. Social Security, interest on the federal debt, veterans' compensation, veterans' pensions, Medicaid, Aid to Families with Dependent Children, the *Special Supplemental Food Program for Women, Infants, and Children,* Supplemental Security Income, the *Food Stamp Program,* and the *child nutrition programs* were exempt from the cuts. The original Gramm-Rudman-Hollings law (P.L. 99-177) was declared unconstitutional in 1986.

Grandfather clause A clause providing exemptions to certain regulations, agreements, and so forth, because of conditions applying before they were enacted. See also *Generally Recognized as Safe (GRAS).*

Granger legislation Laws favoring agricultural interests enacted in many western states between 1870 and 1890. The legislation brought the rates and services of railroads, warehouse, grain (see *grain) elevators,* and others under public control to eliminate monopolistic (see *monopoly*) and discriminatory practices towards farmers.

Granular fertilizer A *fertilizer* composed of particles of roughly the same composition. This kind of fertilizer contrasts with the normally fine or powdery fertilizer. Particle sizes within the different types of fertilizer are fairly uniform. Between types, however, particle size can range from 0.8 millimeters (mm) to 4.8 mm.

Granulated tobacco See *smoking tobacco.*

Grape A fruit sold fresh and used for juices, wines, jellies, and jams. Grape *varieties*, including (European) Emperor, Flame Seedless, Ribier, Cardinal and (American and hybrids) Concord, Seneca, and Delaware, are seedless or seeded (see *seed*), and have a range of colors, tastes, and textures. The *European Union (EU)*, the United States, Canada, Mexico, Argentina, Chile, and South Africa are major producers.

Grapefruit A member of the citrus (see *citrus fruit*) family used as a fresh and processed (see *processing*) fruit, juice, *pectin*, *essential oil*, and animal *feed.* Major producers include the United States, Israel, Cuba, Argentina, and Cyprus. Florida produces over 75 percent of the U.S. crop.

Grass drill See *grass seeder* and *grain drill.*

Grass-fed animal slaughter Slaughtered (see *slaughter*) animals that were fed grass or *roughage* and little or no *grain* or *concentrate.* Grass feeding usually produces leaner animals with less *marbling.*

Grass seeder Equipment used to sow *pasture* or range (see *rangeland*) grasses. Range-type grass *seeders* or *drills* are specially constructed with heavy steel frames and openers to sow *seed* in sod or hard, packed *soil.* The term "grass seeder" also includes a wide variety of other devices, from hand-held, crank-operated lawn seeders to pulley-type broadcast seeders (see *end-gate seeder*).

Grass silage Any harvested (see *harvest*) nongrain (see *grain*) crop *forages* stored at a moisture content high enough to allow *fermentation* and preservation in a *silo.*

Gravitational water Water that is not available for use by plants because it runs off or percolates through *soil.*

Gravity-drain gutter A gutter under a slotted floor with capacity for holding *manure* accumulated for three to seven days after which it is drained by gravity into a manure storage tank.

Gravity wagon A towed unit with metal or wood sides to form a box or hopper, tapered to a spout underneath and fitted with an *auger* or conveyor to move *feed* or *grain* from the *wagon* troughs or bunks.

Gray or greige fabric Woven (see *weave*) or knitted (see *knitting*) goods direct from the *loom* or knitting machine, before they have been given any kind of finishing (see *finishing [textile]*) treatment.

Graze Feeding by *livestock* or wild animals on standing vegetation.

Grease 1: Animal fats (see *fats and oils* and also *tallow* and *lard*) with a melting point generally below 40 degrees centigrade. It may often contain some fat of vegetable origin if processed (see *processing*) from restaurant grease.

2: A petroleum lubricant used for *farm* machinery.

Grease mohair *Mohair* as it comes from Angora *goats* before applying any process (see *processing*) to remove the natural *fats and oils.*

Grease weight The weight of animal *fiber* (particularly *wool* and *mohair*) before scouring (washing).

Grease wool *Wool* as it comes from the *sheep* or *lamb* before applying any process (see *processing*) to remove the natural oils (see *fats and oils*).

Great Northern bean A *variety* of *dry edible bean* sold mainly in dry form, but a small amount is canned. Nebraska produces 85 percent of the U.S. Great Northern crop. France is the major U.S. *export* market for Great Northern beans.

Great Plains A level-to-gently-sloping region of the United States that lies between the Rocky Mountains and the 100th meridian, stretching from Canada to Mexico. The area is subject to recurring *droughts* and high winds. It consists of parts of North and South Dakota, Montana, Nebraska, Wyoming, Kansas, Colorado, Oklahoma, Texas, and New Mexico.

Great Plains Conservation Program (GPCP) A program established in 1956 to help stabilize *Great Plains* agriculture. The program helps land users change their *farm* and *ranch* operations to mitigate natural hazards of the area, such as those related to *climate, soil, topography*, floods, and salinity. The changes include measures for *erosion* control, water conservation, and *land* use adjustment.

Green chop See *soiling crops.*

Greenhouse A glass-enclosed structure with controlled temperature, light, and humidity used for cultivating (see *cultivate*) and growing plants, primarily horticultural and *nursery plants.* See also *floriculture* and *green industry.*

Green industry The *floriculture* and *environmental horticulture* industries. In 1994, the green industry ranked as one of the fastest growing sectors of U.S. agriculture.

Green manure The use of leguminous (see *legume*) crops as a source of *nitrogen* by plowing them into the *soil.* See also *fertlizer.*

Green rate of exchange An administratively determined *exchange rate* used by the *European Union (EU)* to convert agricultural *prices* from the *European currency unit* to the currencies of member countries. The green rate of exchange was established to ensure uniformity of *farm* prices throughout the EU despite the *devaluation* or revaluation of individual currencies. Usually referred to as green rates. See also *monetary compensatory account.*

Green rates See *green rate of exchange.*

Greens Dark, leafy vegetables, such as collards, kale, *mustard greens*, and turnip greens.

Green wood Freshly sawed or undried wood.

Grind To reduce *grains* and other *seeds* to small segments or particle size by impact, shearing, or attrition.

Grinder See *feed grinder.*

Grits Coarsely ground (see *grind*) *grain* from which the *bran* and *germ* have been removed, usually screened to uniform particle size.

Groats The berry from *oats* after removal of the *hull.*

Grocery Manufacturers of America (GMA) A trade association of the manufacturers and *proces-*

sors of food and nonfood products primarily sold in retail grocery stores (see *foodstore*) throughout the United States.

Grocery store See *foodstore.*

Gross cash income The sum of direct government payments (see *direct payments*) and all receipts from the sale of crops, *livestock*, and *farm*-related goods and services.

Gross domestic product (GDP) A measure of the *market* value of goods and services produced by the labor and property of a nation. Unlike *gross national product*, GDP excludes receipts from that nation's business operations in *foreign* countries and the share of reinvested earnings in foreign affiliates or domestic corporations.

Gross farm income Income that the *farm* sector realizes from farming. It includes *gross cash income*, plus the value of food and fuel produced and consumed on farms where grown, the rental value of farm dwellings, and an allowance for the value of the change in year-end *inventories* of crops and *livestock.*

Gross margin A retailer's markup (over the costs of goods sold) as a percentage of total sales.

Gross national product (GNP) A measure of the *market* value of goods and services produced by the labor and property of a nation. GNP includes receipts from the nation's business operations in *foreign* countries, as well as the share of reinvested earnings in foreign affiliates of domestic corporations. See also *gross domestic product.*

Gross rental value See *nonmoney income.*

Gross value of production A measure of *farm* costs and returns that is estimated for primary and secondary outputs. Crops are valued at their *harvest*-month *market* prices; *livestock* are valued at average market *prices* at the time of sale. Direct government payments (see *direct payments*) are excluded, but the measure does reflect the effects of government programs on market prices.

Gross value of sales All income during a year from the sale of crops, *livestock*, dairy, *poultry*, or other related agricultural products, including the landlord's share. When *commodities* are placed under *Commodity Credit Corporation (CCC)* loan, they are considered sold.

Ground limestone See *lime, agricultural.*

Groundnut See *peanut.*

Groundwater Water beneath the earth's surface between saturated *soil* and rock that supplies wells and springs (see *aquifers*).

Group "B" mill price See *price, raw cotton.*

Group Risk Plan (GRP) A federal crop insurance option based on the county expected *yield* rather than on individual *farm* yields (see *Federal Crop Insurance Program*). A policyholder receives an indemnity payment (see *crop insurance indemnity payment*) when—and only when—the county average yield is less than the yield guaranteed by the policy.

Grower price The unit *price* of the *commodity* paid to growers by *handlers.*

Growing-finishing house A building to house *pigs* while growing from 40–70 pounds per head to a *slaughter* weight of 220–240 pounds each.

Growout house A building used for *brooding* and rearing *broilers.* A typical growout house is 40 feet wide by 300 feet long and houses between 65,000 and 80,000 birds.

Growth hormones *Proteins* produced naturally in the pituitary gland of many animals. Virtually exact copies of naturally produced *hormones* can be produced using *recombinant DNA* techniques. Daily injections of these *synthetic hormones* (see also *synthetics*) have been used on a trial basis in *cattle* and hog (see *swine*) *production* to stimulate animal growth.

Growth inhibitor A chemical substance occurring naturally in plants or applied externally to retard new plant tissue growth. See also *growth stimulant.*

Growth regulator See *growth inhibitor* and *growth stimulant.*

Growth stimulant A chemical substance occurring naturally in plants or applied externally to hasten new plant tissue growth. See also *growth inhibitor.*

GSM See *General Sales Manager (GSM).*

GSM-5 See *direct (export) credit.*

GSM-102 See *Export Credit Guarantee Program* and *CCC commercial credit.*

GSM-103 See *Intermediate Export Credit Guarantee Program* and *CCC commercial credit.*

Guaranteed Export Credit See *Export Credit Guaranteed Program* and *Intermediate Export Credit Guarantee Program.*

Guaranteed Minimum Price (GMP) A mechanism that was used in Australia until 1991 to provide a *price* floor for *producers* of *grain* during a specific marketing season. It was intended to provide some degree of stability in growers' incomes.

Guava A tropical fruit native to the Caribbean. Guava is grown commercially in Hawaii and Florida.

Guayule A bushy, *drought*-tolerant *perennial* shrub native to the desert areas of north-central Mexico and southwest Texas. It produces a natural rubber than can be used in the production of tires, fan belts, surgical rubber, hydraulic hoses, resins, and other industrial and household products. In the United States, southern New Mexico and Arizona and western California also provide suitable climates for guayule *production.*

Gully erosion See *erosion.*

Gumbo soil A sticky, fine-textured *soil,* difficult to work when wet; usually high in clay content.

H

Habitat The physical *environment* in which a plant or animal *species* lives.

Hammer grinder A machine that pulverizes *grain* or other *seed* crops using thin steel knives or flails rotating at high speed. The hammer grinder is used to prepare *livestock* feed (see *feed*). Also called hammer mill.

Hammer mill See *hammer grinder.*

Hand 1: A unit of measurement of the height of a *horse* at the shoulders (withers).
2: A subjective measurement of the reaction obtained from the sense of touch created when handling fabric, reflecting the many factors that lend individuality and character to a material.

Handler A firm that prepares a *commodity* for shipment to *wholesalers* and retailers. In the case of *produce*, for example, handlers may be farmers, marketing *cooperatives*, or independent marketing (see *agricultural marketing*) companies. Preparations include arranging the picking, hauling, cleaning, *grading*, and packaging. For dairy, the term generally refers to fluid milk *processors*, including manufacturing plants (see *plants or establishments*) that also supply fluid milk *markets* (see *fluid milk products*).

Handler price The per-unit *price* of the *commodity* paid to *handlers* by *wholesalers* and retailers. If handlers, wholesalers, and retailers are competitive, the handler price approximately equals the grower or *producer* price plus the marginal total cost of marketing a unit of the commodity over time.

Handling technologies Technologies and equipment that are used to receive, dry, clean, store, convey, and transport *grain*.

Hard currency See *convertible currency*.

Hard fibers Comparatively stiff, elongated, woody *fibers* from the leaves or leaf stems of certain *perennial* plants. These fibers are generally too coarse and stiff to be woven (see *weave*) and are used chiefly in twine, netting, and ropes. Examples are *abaca*, *sisal*, and *henequen*. See *soft fibers*.

Hard manufactured dairy products Storable manufactured dairy products, including *butter*, *non-fat dry milk*, and *cheese*.

Hardpan A hardened sandy or claylike *soil* that may be cemented by iron oxide, silica, calcium carbonate, or other substances.

Hard red spring wheat *(Triticum aestivum)* A spring-seeded (see *seed*), high *protein* wheat (see *wheat*) produced in the upper *Great Plains* and used primarily to produce bread *flour. Varieties* include *dark northern wheat, northern wheat,* or *red wheat.* See *hard wheat.*

Hard red winter wheat *(Triticum aestivum)* A fall-seeded (see *seed*) *wheat* produced in the lower *Great Plains* and used primarily to produce bread *flour.* The wheat is medium to high in *protein* and may be either dark hard, hard, or yellow hard. See *hard wheat.*

Hardware disease A condition resulting when an animal eats metal objects that work their way through the digestive system and cause punctures through the stomach wall into the heart. Magnets are sometimes effective in preventing the *disease*, and surgery may save the animal.

Hard wheat *Wheat* with a vitreous *endosperm* suitable for making bread *flour* or *semolina.* A hard wheat *yields* coarse, gritty *flour* that is free-flowing and easily sifted. The flour consists primarily of regularly shaped particles of whole endosperm. See *hard red spring wheat* and *hard red winter wheat.*

Hard wheat flour See *flour.*

Hardwood A botanical group of trees, generally broad-leafed and *deciduous,* used to produce *lumber,* flooring, veneer, plywood, and ties. Temperate *varieties* include oak, maple, beech, birch, and walnut, while mahogany and teak are tropical varieties. The United States, Malaysia, and Indonesia are the major producers of hardwood.

Harmonized Commodity Description and Coding System The international classification system for goods, implemented by most major trading countries on January 1, 1988. The system is used for *tariff* classification, trade statistics, and transport documentation. It replaces the *Customs Cooperation Council Nomenclature*, also known as the *Brussels Tariff Nomenclature.*

Harmonized system (HS) See *Harmonized Commodity Description and Coding System.*

Harrow A *farm* implement used to perform shallow tillage (See *till*) operations. A harrow may be any of several types, including *disk, spike*, tine, *peg,* chain, spring tooth, and rotary.

Harvest To remove usable portions from a plant, to gather them, and to remove them from a field.

Harvested acres *Acres* actually harvested (see *harvest*) for a particular crop. This figure is usually somewhat smaller at the national level than planted acres because of abandonment due to *weather* damage, other disasters, or *market* prices (see *price*) too low to cover harvesting costs.

Harvester Any number of machines designed to cut, dig, or pick and separate the desired part of the plant, usually the *seed*, from the stems, roots, and leaves. See *combine, forage harvester, corn harvester,* and *cotton picker.*

Hatch Experiment Station Act A law passed in 1887 that provided states with federal grants to be used for agricultural experimentation.

Hatching eggs Fertile *eggs* used to produce chicks for egg and *broiler* production.

Hatching flock Breeding stock that produces *hatching eggs* and some *table eggs.*

Hay Grasses and other *commodities* mowed and cured (see *curing*) specifically for fodder. *Varieties* include *alfalfa*, Sudan Grass, *Timothy*, Bermuda Grass, prairie, mixed, *wheat*, *oats*, *barley*, *rye*, *straw*.

Hay baler See *baler*.

Hay conditioner A *farm* implement that crimps or crushes *hay* stems to allow *alfalfa* and other hay to dry faster in the field. It may be attached to a mower (see *hay mower*) or separate.

Haying and grazing A provision of *commodity* law that allows farmers to harvest (see *harvest*) *hay* or graze (see *graze*) *cattle* on *land* idled under *acreage reduction programs* for a limited time period and under specific circumstances.

Haylage *Legume* or grass *hays* that are *ensiled* (see *silo*).

Hay mower A *farm* machine that cuts *hay* crops, such as *alfalfa*. Several different types of mowers are used by farmers, including *sickle mowers*, drum mowers, and rotary mowers.

Hay rake An implement that collects loose *hay* and arranges it in rows across the field or in *windrows*, lengthwise along the field. Hay is later baled (see *bale*), stacked, or made into *silage*.

Hay silage Any grass or *legume* crop, most commonly *alfalfa*, that is harvested (see *harvest*) at a moisture content high enough to allow *fermentation* and preservation in a *silo*.

Hazard Analysis/Critical Control Points (HACCP) A system used throughout the food system to enhance food safety by identifying critical points in food *processing* where hazards might be introduced.

Hazelnut See *filbert*.

Header The part of the harvesting (see *harvest*) machine that cuts the stalks and sends them to the *threshing* mechanism of a *combine*.

Headgate A facility consisting of a pair of adjustable vertical bars, mounted in a passageway, that can be closed around the neck of *cattle* to restrict forward and backward movement. See *chute*.

Head rice Whole *kernels* of milled (see *milling*) *rice*. The kernel must be at least three-fourths the length of a whole kernel.

Heat-damaged kernels *Kernels* that have been affected by elevated temperatures. The damage, usually caused by mild spontaneous heating, may decrease the baking quality of the *flour* milled (see *milling*) from it.

Hectare A metric (see Appendix 6) unit equal to 10,000 square meters or 2.471 *acres*. See also *conversion factors*.

Hedging Entering a futures (see *futures contract*) or *options* position opposite to a cash position to reduce a firm's overall exposure to *price* variation. Hedging transactions can be viewed as temporary substitutes for cash transactions that will occur later. An example of hedging is an *elevator*'s selling of grain futures contracts to protect against price declines on *grain* it has bought from farmers and not yet sold.

Heifer An immature female *bovine* that usually has not given birth.

Hemp An *annual* Asiatic *herbaceous* plant whose *fiber* is used commercially to make *cloth*, floorcoverings, cordage, and specialty papers. Hemp can be grown in both temperate and tropical *climates*. Very small amounts of hemp are imported (see *import*) into the United States for use in rope, twine, and other cordage products. Worldwide, hemp is used for cordage and *cigarette* papers. Also known as common hemp or marijuana. See *soft fibers*.

Hen A mature female chicken or turkey (see *poultry*).

Henequen A *perennial* crop, related to *sisal*, that is grown in Mexico and other Central American countries. Because sisal and henequen are both used for the same end products (see *end use*), the term sisal is generally used to refer to both *fibers*.

Herb A plant used for culinary, medicinal, cosmetic, decorative, and fragrance purposes. Herbs are processed into spices, *essential oils*, and oleoresins for use in food seasonings, creams and lotions, teas, dyes, and other products. Fresh cut herbs are sold in *produce* markets (see *market*).

Herbaceous Nonwoody plants.

Herbicide Any substance used to destroy or inhibit plant growth. Herbicides are used mainly for killing weeds but in small doses may be used as growth regulators (see *growth inhibitor* and *growth stimulant*).

Herbivore An animal that eats only plant foods.

Herringbone milking parlor A milking (see *milk*) stall that allows the group milking of several *cows* at the same time. Because the parlor is raised, the person doing the milking does not have to bend, stoop, or walk a long distance as in an in-line parlor.

Heterozygous Having a mixed genetic constitution so that its offspring do not *breed* true. See also *homozygous*.

Hevea tree A tropical tree used to produce all of the world's natural rubber. The tree is native to the Amazon region, but most of the world's *production* now comes from Malaysia, Singapore, and Thailand. See also *guayule*.

Hide See *hides and skins*.

Hides and skins The outer coverings of *cattle* and hogs (see *swine*), which is used to produce *leather* and leather products. The United States, the former Soviet Union, Argentina, and Brazil are the major producers.

High-clearance tractor A *tractor* with tall wheels designed to run above the growing crop to avoid injury to plants.

High density The compression of a flat, modified flat, or *gin* standard *bale* of *cotton* to a density of about 32 pounds per cubic foot. Previously used for most exported (see *export*) cotton but currently replaced by a universal density compression of about 28 pounds per cubic foot.

High-fat soy flour *Flour* produced by adding back about 15 percent *soybean* oil (see *fats and oils*) or *lecithin* to *defatted soy flour*.

High fructose corn syrup (HFCS) A product of corn (see *corn*) *wet-milling* that serves as a substitute for *sugar* in food *processing*. The liquid sweetener is produced by treating *glucose* from *starch* with an *enzyme* to produce a combination of *dextrose*, *fructose*, and small amounts of other *saccharides*.

The typical composition (percentage) of commercially available HFCS products is as follows:

	HFCS-42	HFCS-55	HFCS-90
Fructose	42	55	90
Dextrose	52	40	7
Higher saccharides	6	5	3

Highly erodible cropland *Cropland* that meets specific conditions primarily relating to its *land* or *soil* classification and current or potential rate of *erosion*. The classifications, developed by the *Soil Conservation Service*, are used to determine eligibility of land for the *Conservation Reserve Program* and the *conservation compliance provisions*.

High-moisture grain Any *grain* containing enough moisture (usually 22 to 30 percent) to require the addition of preservatives or *fermentation* in a *silo* for storage.

High-protein animal unit An aggregate index that converts the *protein* requirement (44 percent *soybean meal* equivalent) for all animals over a base period to a common unit of *feed* consumption.

High-quality beef (HQB) *Carcasses* or any cuts from *cattle* not over 30 months of age that have been fed for 100 days or more on at least 20 pounds per day of nutritionally balanced, high-energy *feed* containing no less than 70 percent *grain*. As defined, the United States is the major producer. Australia and Japan are secondary producers. *U.S. Department of Agriculture (USDA)* Prime or Choice *grades* of meat automatically meet the definition of HQB.

High-value products (HVP) Products that range from highly processed (see *processing*), *value-added* goods to unprocessed but relatively expensive foods on a per-unit or per-volume basis, such as *eggs*, fresh fruit, and fresh vegetables. The *production* and *export* of HVPs stimulate more economic activity outside of production agriculture than *bulk products*.

Hired worker Anyone, other than an agricultural service worker, who was paid for at least one hour of *agricultural employment* on a *farm* or *ranch*. Worker type is determined by what the employee was primarily hired to do. Types of hired workers include:

Field worker: An employee engaged in planting, tending, and harvesting (see *harvest*) crops, including operation of farm equipment on crop farms.
Livestock worker: An employee tending *livestock*, milking (see *milk*) *cows*, or caring for *poultry*, including operation of farm equipment on *livestock* or poultry operations.
Supervisor: A hired manager, range foreman, crew leader, and so on.
Other worker: An employee engaged in agricultural work not included in the other three categories, such as a bookkeeper or pilot.

Hive Any container furnished by a *beekeeper* in which *honeybees* are kept.

Hobbyist beekeeper A person who keeps *honeybees* for pleasure. The industry generally considers hobbyists as maintaining fewer than 25 bee *colonies*. See also *apiarist* and *commercial beekeeper*.

Hog See *swine*.

Hog cholera A highly contagious, often fatal, viral *disease* of *swine* that is very similar to *African swine fever*. Hog cholera is not transmissible to humans.

Hog-corn price ratio See *corn-hog ratio.*

Hog cycle The general industrywide movement in *production* of hogs (see *swine*) from one peak or low of total *supply* to the next, historically averaging about four years due largely to biological lags in production adjustments but increasingly affected by the time needed to plan, finance, construct, and activate production facilities.

Hogshead A unit of liquid measure, usually a barrel or cask, equal to 63 gallons or approximately 239 liters. Also a large round wooden cask used for storing and aging *tobacco*. About 1,000 pounds of leaf can be stored in one hogshead.

Hold The interior of a ship or airplane used for carrying cargo.

Home delivery route A type of retailer who brings a variety of food directly to *consumers*' homes. In the past, this method of sales was often used by dairies.

Homestead 1: *Land* acquired and worked under state or federal laws that provide public land to settlers. The Homestead Act of 1862, for example, provided 160-acre tracts of public land to the head of a family. The law required the homesteader to clear and improve the land and live on it for five years.

2: A farmhouse, adjoining buildings, and land.

Homestead Act of 1862 The first of several public regulations designed to spur expansion into the western United States. The law provided 160-acre tracts of public *land* to the head of a family and required the homesteaders to clear and improve the land and live on it for 5 years. Other examples of "homesteading" legislation include the Desert Land Act of 1877, which gave fee simple title to 640-acre tracts of area lands in the West on condition that the land be reclaimed "by conducting water on it." The Land Reclaimation Act of 1902 authorized funds from the sale of public lands to be used to construct storage and power dams and canal systems for irrigable lands in the West. Settlers got the land free but had to pay for structures built on it over 10 years.

Homestead Protection A provision that allows *Rural Housing and Community Development Service* (RHCDS) borrowers who must convey property to the RHCDS containing the borrower's residence to lease and/or purchase their residence and up to 10 *acres* of adjoining *land*. The provision applies to RHCDS borrowers who lack the financial resources to make payments on a delinquent loan, are ineligible for a restructured loan, and are unable to buy out the loan at the net recovery value of the collateral property, and who convey the property to FMHA in lieu of loan payments.

Homogenized milk *Milk* that has been processed (see *processing*) to reduce the size of the fat (see *fats and oils*) particles (globules) and disperse them throughout the milk. In homogenized milk, the cream no longer rises to the top of the container.

Homozygous Having a pure genetic constitution so that the organisim breeds true for a specific hereditary characteristic. A plant that breeds true for flower color, for example, is called homozygous for this characteristic. See also *heterozygous*.

Honey The sweet, viscous fluid produced by *honeybees* from *nectar* obtained primarily from floral plants. Honey consists of about 17 percent water; two *sugars*, *dextrose* and levalose; and very small amounts of *sucrose*, mineral matter, *vitamins*, *protein*, and *enzymes*. Types of honey include *clover*, *alfalfa*, orange blossom, heather, and *rape*. The major honey producers are the former Soviet Union, China, the United States, Mexico, and Canada.

Honeybee Any of several social bees of the genus *Apis* that produce *honey*.

Honeycomb A group of hexagonal cells with three-faced bases built by *honeybees* from *beeswax* to raise their young and store *honey* and *pollen.*

Honeydew A sticky substance produced by *aphids* on the leaves of plants, sometimes found in cotton *lint*, that makes the *cotton* more difficult to process (see *processing*).

Honeydew melon A type of *muskmelon* with green or orange flesh. The United States, Chile, and Mexico are the major honeydew melon producers. California is the largest U.S. producer, with a 70–75 percent share. Texas and Arizona are the other principal domestic sources.

Honey flow The period when bees (see *honeybees*) are collecting *nectar* from plants in plentiful amounts.

Honey sac See *honey stomach.*

Honey stomach The area located between the end of the *honeybee*'s esophagus and the front of part of the bee's abdomen that expands when full of *nectar* or water. Also known as honey sac.

Hoof-and-mouth disease See *foot-and-mouth disease*.

Hops Ripe dried flowers from a vine of the mulberry family, which are added to beer and other *malt* liquors in the brewing process to impart the characteristic bitter flavor.

Horizontal integration A single management that controls, either through voluntary agreement or ownership, two or more firms performing similar activities at the same level or phase in the *production* or marketing (see agricultural marketing) process.

Horizontal merger Combining two firms producing the same or similar products in the same *market*. See *horizontal integration*.

Horizontal silo Any type of *silo* constructed horizontally on or below the surface of the ground; sometimes called a bunker silo (above ground) or trench silo (excavated).

Horizontal storage *Grain* storage in buildings constructed of metal, wood, or concrete that have flat floors and are filled by means of a portable incline belt or conveyors in the roof.

Hormone A discrete chemical substance secreted into the body fluids by the endocrine gland. Hormones have specific effects on the activities of other organs.

Hormone-type materials Materials that simulate the effects of a *hormone* and thereby cause the action of a hormone. *Growth stimulants* and *growth inhibitors* are an example.

Horse A large, hoofed animal domesticated and used for work (draft horses), racing, and pleasure (light horses). The different types of horses are classified according to size, build, and use. The major producers include the United States, the United Kingdom, France, Australia, Germany, China, and the former Soviet Union.

Horsepower Energy equal to 746 watts. For a *tractor*, horsepower is usually measured at the *drawbar*, but sometimes the term refers to the *power-take-off* or belt horsepower, as opposed to the engine horsepower measured for cars and trucks. See *drawbar horsepower.*

Horsepower hour The energy consumed by working at the rate of one *horsepower* for one hour; equal to 1,980,000 foot-pounds.

Horse Protection Act A law that prescribes care and treatment for *horses* in shows. The Act is enforced through inspections (see *inspection, federal*) by the *U.S. Department of Agriculture*'s (USDA) *Animal and Plant Health Inspection Service (APHIS)* personnel and by "Designated Qualified Persons," who are licensed by industry organizations and certified and monitored by APHIS.

Host A plant or animal invaded or parasitized by a *disease*-causing or inducing *agent* and from which the *parasite* or *pathogen* obtains its sustenance.

Host range The various kinds of plants that may be affected by a *parasite*.

H-2A temporary foreign workers *Foreign* workers who are authorized to work temporarily in *agricultural employment* in the United States and who have no intention of abandoning their home country.

Hull The outer covering of *grain* or other *seeds*.

Human Nutrition Information Service (HNIS) An agency of the *U.S. Department of Agriculture (USDA)* that conducts research on the *nutrient* composition of foods, nutrition monitoring, and nutrition education. Among other things, HNIS conducts national surveys to provide data on the dietary status of households and individuals, and publishes the *Dietary Guidelines for Americans*. HNIS was abolished under USDA's 1994 reorganization. HNIS's functions were merged with the *Agricultural Research Service*.

Humus Well-decomposed organic matter in *soil*.

Hundredweight (cwt) A unit of weight in the United States totaling 100 pounds or approximately 45.36 kilograms. In the British Imperial system, a hundredweight equals 112 pounds or 50.8 kilograms.

Hunger Prevention Act of 1988 (P.L. 100-435) An act that extended the *Temporary Emergency Food Assistance Program (TEFAP)* for two years and authorized an appropriation of $120 million per year for the *U.S. Department of Agriculture (USDA)* to purchase foods with a high *nutrient* value for distribution to low-income families through TEFAP. The act also authorized USDA to purchase $40 million worth of federal food for distribution to soup kitchens and shelters for the homeless in 1990.

Husk The green outer layer or envelope of many *seeds* and fruits. The green covering of an ear of *corn* is the most familiar example.

HVI (high volume, instrument) testing A process for determining *cotton* quality that utilizes instruments rather than sight and touch methods to determine quality characteristics. Almost all U.S. cotton is now tested using HVI.

Hybrid The first generation produced by breeding plants or animals of different *varieties* or *species*. See also *cross*.

Hybrid vigor The tendency of crossbred (see *crossbreeding*) offspring to perform better than the average of their parents in certain traits. Hybrid vigor decreases in each succeeding generation of inbreeding.

Hydrate dextrose See *dextrose*.

Hydrated lime See *lime, agricultural.*

Hydraulic pressing An intermittent batch-pressing operation using elevated temperatures in a hydraulic press that forces oil (see *fats and oils*) from *oilseeds*. The residual oil left in the oilseed cake is about 5–10 percent. This method is no longer used in the United States to any appreciable extent.

Hydrogenation A process of passing hydrogen gas through liquid oils (see *fats and oils*) in the presence of a catalyst (nickel) for the purpose of saturating (see *fatty acids, classification of*) the *fatty acids* (hardening them) for use in the manufacture of *shortening* and *margarine*. Hydrogenation makes a white hardened product that has a longer shelf life without refrigeration.

Hydrogen phosphide See *fumigant*.

Hydrolysis The splitting of complex molecules into simpler units by chemical reaction, usually by using a catalyst.

Hydroponics Growing plants in a liquid medium containing a *nutrient* solution. Also applies more generally to the process of growing plants in an inert medium, such as sand, *peat*, or vermiculite and adding a solution containing all essential elements needed by a plant for normal growth and development.

Hygroscopic water Moisture unavailable to plants because it is so closely held by the *soil* particles.

I

Idled acres *Land* planted in cover (see *cover crop*) and *soil* improvement crops, as well as completely idle *cropland.*

Illegal alien A person who has entered a country without legal documentation or who cannot legally accept employment in that country.

Imagery Photographic and digital satellite image data products used in the analysis of agricultural *production.*

Immunity The ability of a plant or animal to remain free from a *disease* by virtue of its inherent properties.

Imperfect information An economic characterization of firms that lack knowledge and an understanding of past, current, and future events that affect *market* outcomes.

Implement carrier A flatbed trailer used to haul *farm* machinery.

Import 1: To legally buy and receive goods from a *foreign* nation into a country.
2: The quantity or value of goods legally entering a nation.

Import barriers *Quotas*, *tariffs*, and *embargoes* used by a country to restrict the quantity or value of a good that may enter that country.

Importer A person, firm, or other entity that legally buys goods from *foreign* nations.

Import license A document required and issued by some governments permitting specified goods to be imported (see *import*) into their country.

Import Licensing Code See *Agreement on Import Licensing Procedures.*

Import prohibitions and restrictions Regulations that prohibit the importation (see *import*) of certain articles. If shipped to the United States, such items may be seized or forfeited.

Import quota The maximum quantity or value of a *commodity* allowed to enter a country during a specified time period. In 1994, for example, the United States imposed an annual *import* quota on raw *cotton* totaling 14.5 million pounds (about 30,000 bales) of short staple cotton having a length of less than 1-1/8 inches and a *quota* of 45.7 million pounds (about 95,000 bales) of long staple cotton having a length of 1-1/8 inches or more. A quota also may apply to amounts of a commodity from specific countries.

Import substitution A strategy that emphasizes replacing *imports* with domestically produced goods.

Import value The declared dollar value or invoice value of all shipments cleared by the *U.S. Customs Service* and reported in the official U.S. trade statistics.

Improved perennial pasture *Pasture*, covered with predominantly *perennial* grasses and/or *legumes*, managed relatively intensively by recurring reseeding, fertilization (see *fertilizer*), and/or mechanical or chemical weed control. Also applies to plant materials grazed (see *graze*) from intensively managed perennial pastureland (sometimes called tame pasture).

Incentive payments *Direct payments* made to *producers* of *wool* and *mohair*. Similar to *deficiency payments*, incentive payments are provided to producers when the *marketing year* is over if the average market (see *market*) *price* received is less than the support level. The support level is determined by a *cost-of-production* formula specified in legislation.

Incidence (of foodborne disease) The occurrence of one or more cases of *acute* or *chronic* illness. The rate of occurrence of specific illnesses is usually reported as the yearly incidence of foodborne disease.

Incline belt An endless belt used to convey *grain*, which is supported by rollers and driven by a shaft-mounted speed reducer motor.

Income insurance A variation of *revenue insurance* that would prevent income from falling below a specified level. An income insurance program has never been implemented.

Income support payment See *deficiency payment*.

Income support programs Government programs designed to enhance *farm incomes* and ensure adequate *resource* returns. Farmers receive *direct payments*, known as *deficiency payments*, if market (see *market*) *prices* fall below the established *target price*. See *price-support programs*.

Incomplete risk markets *Markets* characterized by an absence of a means for *producer*s, marketers, and *consumers* to exchange (conditional) promises today about something that they will do at a future date. Examples include the absence of *forward pricing* or *crop insurance* markets.

Inconvertible currency See *soft currency*.

Increasing economies of scale The range of output where the average total cost of producing a unit of output is falling.

Incubation The process of controlling the heat and moisture *eggs* are exposed to in order to get fertile (see *fertilization*) eggs to hatch. Chicks require 21 days and turkeys, 28 days, to hatch.

Indemnity The amount received from an insurer as settlement on a loss claim. In the case of crop insurance, it is calculated by multiplying the *price* election by the number of *bushels* of loss.

Independent grocer A food retailer or *foodservice* operator owning 10 or fewer stores or outlets. See also *chain*.

Index of Prices Paid See *Prices-Paid Index*. Technically known as the Index of *Prices* Paid by Farmers for *Commodities* and Services, Interest, Taxes, and Farm Wage Rates.

Index of Prices Paid by Farmers for Commodities and Services, Interest, Taxes, and Farm Wage Rates See *Prices-Paid Index.*

Index of Prices Received See *Prices-Received Index.* Technically known as the Index of Prices Received by Farmers.

Index of Prices Received by Farmers See *Prices-Received Index.*

Indian fig See *prickly pear.*

Individual activity A pattern of behavior of firms in which each firm operates without implied or express agreements with any other firm.

Individual cage A pen (see *lot*) used to hold one laying (see *layer*) *hen.* The pens are usually 8 to 12 inches wide, 18 inches long, and arranged in rows.

Individual profit maximization An attempt by a firm to maximize its profits regardless of the effect upon the profits of other firms.

Individual proprietorship A business organized and directed by a sole owner.

Industrial crops Crops, such as *industrial rapeseed, kenaf, crambe, meadowfoam, jojoba, lesquerella, guayule,* and *canola,* that have industrial applications. Meadowfoam, jojoba, and lesquerella *yield* oils (see *fats and oils*) that can be used by industry. See table on page 138.

Industrial fabrics Fabric (see *cloth)* used for nonapparel and nondecorative uses. These uses fall into several classes: (1) fabrics employed in industrial processes such as filtering, polishing, and absorption; (2) fabrics combined with other materials to produce a different type of product, such as tires, hoses, and electrical machinery parts; and (3) fabrics incorporated directly into a finished product, such as tarpaulins, tents, and awnings.

Industrial rapeseed An *oilseed* plant with *seeds* used to produce oil (see *fats and oils*) and high protein (see *protein*) *meal.* Much of the oil is used in the manufacture of plastic films and in automotive and industrial lubricants. Other potential uses include cosmetics, paints and coatings, plasticizers, and pharmaceuticals. See also *canola.*

Industrial users Users of a product who receive it directly from primary distributors for use in manufacturing or *processing.*

Industry Sellers of close-substitute products who supply a common group of buyers.

Infectious agent A *microorganism* that can cause *disease* infections in humans and animals.

Inflation rate The percentage change in a measure of the average *price* level. Changes are reported on a monthly basis and are stated as annual rates for longer-term comparisons. The two major measures of the average price level are the *Consumer Price Index* and *Producer Price Index.*

Infrastructure The transportation network, communications systems, financial institutions, and other public and private services necessary for economic activity.

Initial margin See *margin.*

Agriculturally Based Industrial Products

Crop	*Products*	*Major Current and Potential Uses*
Industrial rapeseed	High erucic acid oil	Plastic film, plasticizers, nylons, polyesters, cosmetics, lubricants, paints and coatings, surfactants, flotation agents, pharmaceuticals, nonconducting fluids
	High *protein* meal	*Livestock* or *aquaculture* feed
Canola	Edible vegetable oil	Cooking oil, food *processing*
	High protein *meal*	Livestock *feed*
Kenaf	Nonwood *fiber*	Newsprint, cordage, chicken litter, fire logs, cardboard, sewage treatment
Crambe	High erucic acid oil	See *industrial rapeseed*
	High protein meal	Beef *cattle* feed
Meadowfoam	Oil	Soap, hair care items, moisturizers, lubricants, polymers, waxes
	Meal	Livestock feed
Jojoba	Oil	Cosmetics, lubricants, transmission fluids
	Meal	
Lesquerella	Oil	Lubricants, plastics, protective coatings, pharmaceuticals, dietary *supplement* for beef cattle
	Meal	
Guayule	Rubber	Tires, fan belts, surgical items, hydraulic hoses, carbon char, resins
	Plant tissue	Cork, fiber

Source: Glynn, Priscilla. "New Crops and Old Offer Alternative Opportunities." *Farmline* XII, no. 1 (Dec.–Jan. 1991), p. 14.

Initial payment (IP) A payment made by the recipient country at the time of delivery under the *Public Law 480* program. See also *currency use payment.*

Inputs See *farm inputs.*

Insect growth regulators (IGRS) A class of *pesticides* that occur naturally in low concentrations at various points in the life cycle of an insect. Related chemicals (analogs) can disrupt a wide range of insect body functions when applied at a critical time during the life cycle.

Insecticide residue Trace amounts of *insecticide* on a food product resulting from *grain* fumigation, field crop treatment, or accidental environmental *contamination.* Sensitivity of testing methods means residues may be detected below the level harmful to human health. See *drug residue* and *pesticide residue.*

Insecticides Chemicals used to destroy *insect* pests.

Insoluble fiber See *fiber.*

Inspection (federal) Examination of animals, plants, and meat by the *U.S. Department of Agriculture*'s (USDA) *Food Safety and Inspection Service* to ensure health, cleanliness, and wholesomeness.

Inspection (meat) Examination of animals by the *U.S. Department of Agriculture*'s (USDA) *Food Safety and Inspection Service* to ensure health, cleanliness, and wholesomeness. Inspection includes live animals, *carcasses*, and processed (see *processing*) meat. In some cases, inspection is performed by states for meat not moving across state lines. See also *inspection, federal.*

Institute of Shortening and Edible Oils (ISEO) An association of edible vegetable oil (see *fats and oils*) and animal fat (see *fats and oils*) refiners (see *refining*) and manufacturers of *shortening*, salad, and cooking oils. The ISEO was founded in 1932, and its headquarters are in Washington, D.C.

Institutional wholesaler A *wholesaler* whose customers are restaurants, hotels, lunch counters, cafeterias, hospitals, colleges, and other institutions engaged in public or group feeding. Institutional wholesalers carry a broad line of products tailored for the *foodservice* industry and provide many services similar to a *service wholesaler.*

In-store bakery A *foodstore* department that prepares fresh-baked goods on the premises.

In-store promotion A short-term sales expansion scheme used in retail stores. Examples include sample tastings in local supermarkets (see *foodstore*), coupons, and rebates designed to provide *consumers* with temporary *price* incentives. In-store promotion activities require smaller budgets than media advertising and lend themselves to seasonal promotions for quick movement of perishable products.

Integrated chain A *chain* operator that also owns related purchasing, warehousing, and/or distribution facilities.

Integrated Crop Management An *agriculture* management system that integrates all controllable agricultural *production* factors for long-term sustained *productivity*, profitability, and ecological soundness.

Integrated Farm Management Program A voluntary program authorized by the *Food, Agriculture, Conservation, and Trade Act of 1990 (P.L. 101-624)* (see Appendix 3) to encourage participants to plant *resource*-conserving crops on 20 percent of their *crop acreage base*. The goal of the program is to enroll 3 to 5 million *acres* by 1995. To be eligible, *producers* must contract to develop and carry out an approved *farm management* plan over three to five years (can be extended another five years). Participating producers would not lose eligibility for *deficiency payments* on the portion of their base planted in resource-conserving crops.

Integrated management system A comprehensive, multiyear, site-specific system for planning and implementing a program to contain or control undesirable plant *species*. The methods may include education; preventive measures; physical, cultural, or mechanical methods; biological agents; *herbicides*; and general *land* management practices including manipulating *livestock* or wildlife grazing (see *graze*) strategies or improving wildlife or livestock *habitats*.

Integrated Pest Management (IPM) The control of pests using an array of crop *production* strategies, combined with careful monitoring of insect pests or weed populations, and other methods. Some approaches include selection of resistant *varieties*, timing of cultivation (see *cultivate*), biological control methods, and judicious use of chemical *pesticides* so that the natural enemies of pests are not destroyed. See *biological control of pests*.

Integrated resource management See *integrated crop management*.

Integration The combination (under the management of one firm) of two or more of the processes in the *production* and marketing (see *agricultural marketing*) of a particular product. The processes are generally capable of being operated as separate businesses. See *vertical integration* and *horizontal integration*.

Intellectual Property Rights (IPR) Ownership of the right to possess or otherwise use or dispose of products created by human ingenuity. Examples of IPRS include trademarks, patents, and copyrights.

Inter-American Development Bank (IADB) A regional financial institution established in 1959 to further the economic and social development of its Latin American member countries.

Inter-American Institute for Cooperation on Agriculture (IICA) An organization created by the *Organization of American States* to promote cooperation among the nations of the Western Hemisphere in *agricultural research*, training, and institution building.

Intercropping Planting one crop into another. Intercropping can be done either between rows or into the stubble of a previous crop.

Interest Payment Certificates A means available to the secretary of agriculture to refund interest charges on *nonrecourse loans*. The secretary may provide a negotiable certificate, redeemable for *Commodity Credit Corporation (CCC)* commodities (see *commodity*), to any *producer* who repays, with interest, a *price support* loan for *wheat*, *feed grains*, *rice*, or *upland cotton*.

Intermediary Relending Program A *U.S. Department of Agriculture (USDA)* program that provides grants to public bodies or nonprofit organizations to relend to businesses in *rural* areas. The program is administered by the *Rural Business and Cooperative Development Service*.

Intermediate Export Credit Guarantee Program (GSM-103) A program established by the *Food Security Act of 1985 (P.L. 99-198)* (see Appendix 3) that complements the *Export Credit Guarantee*

Program (GSM-102) but guarantees repayment of private credit for 3 to 10 years.

Intermediate handler A firm that operates at a stage of the *market* located between the *processing* or manufacturing and retail distribution levels. Intermediate *handlers* generally perform such functions as packaging, storing, transporting, promoting, and selling.

Internal growth Growth that is a result of a firm adding new stores, or increasing sales of existing stores, or both.

International Association of Meat Processors (IAMP) An international organization involved in meat and food *processing*, distribution, and manufacturing.

International Bank for Reconstruction and Development (IBRD) An international organization that makes long-term loans for economic development projects and programs in *developing countries*. The International Development Association, which is part of the IBRD, provides financing to very low-income countries to meet their development requirements. The terms of this assistance are more flexible and bear less heavily on the countries' budgets than conventional loans. Also known as the World Bank.

International Cocoa Agreement (ICCA) An agreement, originally signed in 1973, designed to stabilize world cocoa (see *cacao*) *prices* through various *buffer stock* mechanisms. A new International Cocoa Agreement with limited economic provisions was adopted to replace the previous ICCA, which expired on September 30, 1993. The new agreement, adopted in July 1993, focuses on *production* management and consumption promotion, rather than utilizing buffer stock schemes to influence global cocoa *market* prices. The United States, the world's largest cocoa consuming country, has yet to become a member of any ICCA. Similarly, some major producers such as Indonesia have also never become members.

International Coffee Agreement (ICA) An agreement, originally signed in 1962, designed to stabilize world *coffee* prices (see *world price*) through various *export* quota (see *export allocation or quota*) mechanisms. In June 1993, the International Coffee Organization (ICO) agreed to extend the existing ICA through September 30, 1994. The United States notified the ICO on September 27, 1993, that it could no longer justify participating in the ICA. The action resulted from lack of Congressional support for continued funding and the U.S. coffee industry's preference for a "free coffee market." The United States had been an active member of the ICO since its inception.

International Commodity Agreement(s) (ICA) Agreements, by a group of countries, that contain substantive economic provisions aimed at stabilizing world trade, *supplies*, and *prices*. The agreements may include various mechanisms and/or schemes, such as quota (see *export allocation or quota* and *import quota)* agreements among *exporters*, *buffer stock* systems, and multilateral (see *multilateral agreements*) *contracts* between exporters and *importers*. Agreements have been negotiated for cocoa, coffee, dairy products, *olive* oil, *sugar*, and *wheat*. See also *International Cocoa Agreement, International Coffee Agreement, International Dairy Arrangement, International Olive Oil Agreement, International Sugar Agreement*, and *International Wheat Agreement*. Largely because of conflicting goals between and among exporting and importing nations, it is not unusual for international commodity agreements to continue in existence but without economic provisions to limit supplies and stabilize prices, as has occurred with the International Cocoa and Coffee Agreements.

International Cotton Advisory Committee (ICAC) A worldwide association of governments that assembles, analyzes, and publishes data on world cotton *production*, consumption, *stocks*, and *prices*. ICAC closely monitors developments in the world cotton *market* and promotes intergovernmental cooperation in developing and maintaining a sound world *cotton* economy. The ICAC is headquartered in Washington, D.C.

International Dairy Arrangement (IDA) An arrangement under the *General Agreement on Tariffs and Trade (GATT)* signed by all 18 major dairy producing and exporting countries, with the objective of expanding and liberalizing world trade (see *trade liberalization*) in dairy products by improving international cooperation. The United States withdrew from the agreement in 1985 because the *European Community* violated a major objective of the IDA by selling *butter* and other basic dairy products at *prices* below the established minimum *export* prices.

International Development Association See *International Bank for Reconstruction and Development.*

International Finance Corporation (IFC) A separately organized member of the *International Bank for Reconstruction and Development* (also known as the World Bank) that encourages the flow of capital into private investment in *developing countries.*

International Food Policy Research Institute (IFPRI) One of the international research centers supported by the *Consultative Group on International Agricultural Research* established to identify and analyze strategies and policies for meeting world food needs. The focus of IFPRI is the reduction of malnutrition and hunger in low-income countries.

International Fund for Agricultural Development (IFAD) An agency of the United Nations that finances agricultural development projects, primarily for food *production* in *developing countries.*

International Institute for Cotton (IIC) A nonprofit organization of *cotton*-producing countries founded in 1966. Its purpose is to increase world consumption of cotton and cotton products through utilization research, *market* research, sales promotion, education, and public relations. The IIC is headquartered in Brussels, Belgium

International Monetary Fund (IMF) An organization established in 1946 to assist in expansion of stable world trade and monitor *exchange rate* policies of member countries. The IMF also acts as a banker of last resort for countries experiencing *foreign* exchange deficiencies.

International Office of Epizootics (IOE) An international veterinary organization formed in 1924. The organization consists of members from 130 countries and maintains a global animal *disease* reporting network. See also *International Plant Protection Convention* and *Codex Alimentarius Convention.*

International Olive Oil Agreement (IOOA) An agreement signed by 17 major *exporters* and *importers* to ensure fair competition, delivery of the *commodity* in accordance with *contract* specifications, and stabilization and development of *olive* oil (see *fats and oils*) *markets*. The United States is not a member of the IOOA.

International Organization for Standardization (ISO) A worldwide federation of national standards bodies located in Geneva, Switzerland. Formed in 1947, the ISO produces standards in all fields except electrical and electronic (which are covered by the International Electrotechnical Commission). The American National Standards Institute is the United States member of the ISO. See also *ISO* 9000.

International Plant Protection Convention (IPPC) A subsidiary of the *Food and Agriculture Or-*

ganization of the United Nations that was formed in the 1950s. The convention has members from about 90 countries and is concerned with issues involving plant pests and plant health. See also *International Office of Epizootics* and *Codex Alimentarius Convention.*

International Sugar Agreement (ISA) An agreement enacted in 1969 to stabilize international trade in *sugar* at equitable *prices* and to aid those countries whose economies are largely dependent on the *production* and *export* of sugar. The United States is not a member of the ISA. The current ISA, re-enacted in 1992, provides for only statistical functions and not for any actions that affect prices or trade.

International trade barriers Regulations imposed by governments to restrict *imports* from, and *exports* to, other countries. *Tariffs, embargoes, import quotas,* and unnecessary sanitary restrictions are examples of such barriers.

International Trade Centre An organization established in 1964 to provide assistance to *developing countries* in formulating and implementing *export promotion programs.* Since 1968, the Centre has been operated jointly by *GATT* and the *United Nations' Conference on Trade and Development (UNCTAD).*

International Trade Commission (ITC) An independent agency of the U.S. government established in 1916 to monitor trade, provide economic analyses, and make recommendations to the president of the United States in cases of *unfair trade practices.* Interest groups, such as growers or trade associations, can petition the ITC to investigate the trade practices of other countries to determine whether "material harm" has been done to U.S. *producers.*

International Wheat Agreement (IWA) An agreement to stabilize trade in *wheat* and other *grains.* This agreement contains two conventions: the Wheat Trade Convention and the Food Aid Convention. The Wheat Trade Convention, signed by 60 countries including the United States, provides a forum for the periodic exchange of information among member countries on the world grain situation. The Food Aid Convention, signed by 11 countries, commits signatories to minimum annual food aid contributions of edible grains or cash.

International Wool Secretariat (IWS) An organization funded by *wool* growers in Australia, New Zealand, South Africa, and Uruguay to expand the uses for wool throughout the world by promotion and wool *textile* research and development. The IWS is a non-trading organization, headquartered in London, but with branches in 33 countries. See also *Wool Bureau.*

Intervention See *intervention price (IP).*

Intervention price (IP) The price (set in *European currency units [ECUs])* that acts as a *price* floor in determining the price at which *European Union (EU)* farmers can sell certain *commodities* into EU storage—called intervention. *Grains,* beef, *butter,* and *nonfat dry milk* are the major commodities that may be sold into intervention as long as they meet certain quality and eligibility criteria. The IP is similar to the United States' *loan rate.*

In-the-beef An evaluation of a live animal's *carcass* value.

In-the-money option An *option contract* that would yield a positive return to the holder if exercised. An option is in-the-money if the *strike price* exceeds the market *price* for a *put option* or is less than the *market* price for a *call option.* The magnitude of this difference is the *intrinsic value* of the option.

Intrinsic quality Characteristics critical to the end use of *grain*. These characteristics are nonvisual and can only be determined by analytical tests. For example, the intrinsic quality of *wheat* is determined by characteristics such as *protein*, ash, and *gluten* content; the intrinsic quality of *corn* by its *starch*, *protein*, and oil (see *fats and oils*) content; and the intrinsic quality of *soybeans* by their protein and oil content.

Intrinsic value The amount that would be realized by exercising an *option* immediately and trading out of the resulting *futures* position. The intrinsic value is positive for an *in-the-money option* and zero otherwise.

Inventories The quantity of *commodities* on hand; equivalent to *stocks*.

Inventory (CCC) The quantity of a *commodity* owned by the *Commodity Credit Corporation (CCC)* at any specified time.

Inventory Reduction Program A discretionary program introduced in the *Food Security Act of 1985 (P.L. 99-198)* (see Appendix 3) to provide *producers* with *payments-in-kind (PIK)* if they reduce acreage (see *acre*) by half the required reduction and agree to forgo loans and *deficiency payments*. The inventory reduction payments, also known as producer option payments, are computed in the same way as *loan deficiency payments*. Inventory reduction programs have not been implemented to date.

Invert sugar The liquid sweetener that results when *sucrose* is treated with water and acid.

Invisible stocks Stocks (see *buffer stock*, *carryover*, and *Commodity Credit Corporation [CCC]*) held by wholesalers, retailers, and users as distinct from stocks of primary distributors. They are termed "invisible" because these quantities are not reported.

Invitation for bids A public announcement of the intention to buy or sell a *commodity* or service in accordance with specified terms regarding commodity specifications, terms of delivery and payment, the deadline for receipt of *bids*, and other eligibility requirements.

Iodine number (Iodine value) The number of grams of iodine that will react with 100 grams of fat (see *fats and oils*). It is a measure of the unsaturated (see *fatty acids, classification of*) *fatty acids* in a particular oil. Oils with relatively large quantities of unsaturated fatty acids are good drying oils. Higher iodine numbers indicate faster drying oils.

Irradiation The use of ionizing radiation on food to kill insects and *microorganisms* that cause spoilage or human health problems. Irradiation provides an alternative, nonchemical substitute to some preservatives and postharvest (see *harvest*) *fumigants*. In the United States, irradiation is legal for decontaminating spices, controlling trichinae (see *trichinosis*) in fresh pork, controlling insects in foods, and extending the shelf life of fruits and vegetables.

Irrigable land *Land* currently not irrigated (see *acres irrigated* and *irrigated farms*) but either has project works constructed by the *Bureau of Reclamation* and available water, or the Bureau of Reclamation has existing plans to provide water.

Irrigated farms *Farms* with any agricultural *land* (see *acres irrigated* and *irrigated farms*) irrigated in the specific calendar year. The acres irrigated may vary from a very small portion of the total acreage (see *acre*) in the farm to irrigation of all agricultural land in the farm.

Irrigation See *acres irrigated* and *irrigated farms.*

Irrigation furrow opener A type of *plow* or *disk* used to make a depression in the ground for irrigation (see *acres irrigated* and *irrigated farms*) water to flow between crop rows.

Irrigation levee plow A large *plow* used to pile dirt in long rows to form irrigation (see *acres irrigated* and *irrigated farms*) ditches.

Irrigation methods Any of several means of applying water regularly to crops. These include: (1) check irrigation in which water is applied to relatively level areas that are surrounded by levees, also known as border-strip or basin irrigation; (2) corrugation irrigation in which water is supplied to small, closely placed *furrows*, frequently used for *grain* and *forage* crops; (3) flood irrigation in which water is released from field ditches to flood *land*; (4) furrow irrigation in which water is applied in small ditches, commonly used for tree and row crops; (5) sprinkler irrigation in which water is sprayed over the *soil* surface and crops; (6) drip irrigation in which water is applied at very low pressure through plastic line, commonly used for tree and row crops. See also *acres irrigated* and *irrigated farms.*

Irrigation pipe Metal tubing, usually aluminum, used to apply water to crops.

Irrigation pump A motor-driven (electric or combustion engine) device to draw water from wells and ditches to irrigate (see *acres irrigated* and *irrigated farms*) crops. Pumps use impellers or other means to apply water under pressure for sprinkler systems.

Irrigation sprinkler head A device used to apply water as a mist, fog, or stream to crops.

Irrigation system, center pivot See *center pivot irrigation.*

Irrigation system, drip Applying water to crops, such as *grapes* in *vineyards*, using small plastic hoses running along the surface of the ground or buried. The hoses slowly release water to the plants. The drip irrigation system is more efficient than many other irrigation methods since less water is lost to evaporation. See also *acres irrigated* and *irrigated farms.*

Irrigation system, stationary gun Applying water to crops using a single, large, rotating nozzle and high pressure to force water through the air over a large area.

Irrigation water management The practice of limiting irrigation based on the water-holding capacity of the *soil* and the need of the crop in order to minimize *resource* losses.

ISO 9000 A series of international quality management and quality assurance standards published in 1987 by the *International Organization for Standardization (ISO).* The five parts of the ISO 9000 series (9000-9004) provide minimum quality management requirements and guidance for establishing and auditing quality management systems. The standards are generic in nature and may be applied to any type or size of business. ISO 9000 is becoming widely used for evaluating suppliers' quality management practices, particularly through third-party assessments and registration schemes.

Isoglucose *High fructose corn syrup.* The term isoglucose is favored in some countries.

Isolated soy protein *Protein* that has been greatly concentrated from its origin by chemical or mechanical means. It is produced by extracting white flakes or *flour* with water or a mild alkali. Isolate generally has a protein content of 90 percent.

Italian red lettuce See *radicchio*.

J

Jack bean A native *bean* of Central America and the West Indies that is relatively high in *protein*. The plant produces flat, straight, or scimitar-shaped pods that are among the largest of any domesticated *legumes*—up to 30 centimeters long and 3.5 centimeters wide. The Jack bean is not common in the United States.

Jackson-Vanik Amendment A law that requires satisfaction or presidential waiver of Section 402 of the Freedom of Immigration provision of the Trade Act of 1974 (P.L. 93-618) as a condition for the extension of *most-favored nation tariff* treatment to most nonmarket economy (see *centrally planned economy*) countries and for their participation in any U.S. government program extending credit, credit or investment guarantees, including the programs of the *EximBank.*

Jerusalem artichoke A small, pale brown *tuber* that tastes crunchy when eaten raw and has a hint of regular *artichoke* flavor when cooked. Also called sunchoke.

Jicama A popular Mexican root vegetable that resembles a large, brown *turnip*. The flesh has a texture and flavor similar to water chestnuts. Pronounced hik-ka-ma.

Jobber A synonym for a *wholesaler* or distributor.

Joint activity or collusion Joint decisionmaking by two or more firms as the result of express or implied agreement.

Joint Council on Food and Agricultural Sciences A group established by the Food and Agriculture Act of 1977 (P.L. 95-113) (see Appendix 3) to "bring about more effective research, extension, and teaching by improving the planning and coordination of publicly and privately supported agricultural science activities." Membership is drawn from government, university, foundation, farm, and *agribusiness* organizations.

Joint products The primary and all secondary products obtained by *processing* a raw material.

Jojoba A slow-growing, *perennial* shrub that produces an oil (see *fats and oils*) similar in chemical characteristics to sperm whale oil. Jojoba has been produced commercially for 10–15 years and is native to the Sonora Desert region of the United States and Mexico. Major producers include the United States, Mexico, and Israel. See *jojoba oil.*

Jojoba oil A liquid ester that provides a natural base for cream ointments and many other products in the cosmetic industry, which accounts for 90 percent of *jojoba* oil use. It is also suitable for lubricants and their activities.

Junta Nacional de Granos (JNG) An Argentinian government agency that regulates the *grain* industry in that country. It establishes *grading* standards (mandatory for *export* grain), conducts educational programs, licenses inspectors, and enforces regulations.

Jute An *annual*, tropical plant grown mainly in India and Bangladesh that yields a strong, coarse *fiber* used to make cordage. See *soft fibers*.

K

Karnal bunt A fungal (see *fungus*) *disease* of *wheat* affecting the quality of the wheat *kernel. Quarantines* are imposed on wheat from affected areas.

Kenaf An *annual*, nonwood, *herbaceous* plant whose *fiber* can substitute for newsprint and imported *jute* to produce cordage used in such items as carpet backing and padding, roofing, felt, and fine and coated papers. Kenaf is also used to make twine and bond paper. The plant's core is sold for use in *soil*-less potting mixes, animal litter and bedding, and *feed* products. Kenaf can be grown in moderate climates in the United States, from California to Florida.

Kennedy Round The sixth round of *multilateral trade negotiations* held in 1963–1967 under the *General Agreement on Tariffs and Trade (GATT)*. See *GATT Rounds*.

Kernel A *grain* or *seed* encased in a hard shell or *husk*.

Kilotex (Ktex) See *tex*.

Kiwano A fruit from New Zealand that is a *cross* between golden *cucumber* and melon. The fruit is spikey with a seedy (see *seed*), green, cucumber-flavored flesh.

Kiwifruit A small, fuzzy fruit of a vine native to Asia. Major producers include New Zealand, the United States, Italy, Australia, and Greece.

Knitting A method of constructing fabric (see *cloth*) by interlocking a series of loops of one or more *yarns*. The two major classes of knitting are *warp* knitting and *weft* knitting. In warp knitting, yarns run lengthwise in the fabric; in weft knitting, which includes circular knitting, yarns run transversely in the fabric. Warp knits are flatter and denser than weft knits. Tricot and Milanese are typical warp knit fabrics, often found in apparel use as lingerie. Jersey is a typical (circular) weft knit fabric, commonly seen in T-shirts.

Kobe beef Beef from Kobe *cattle* that is well-marbled and tender, making it highly prized in Japan and many other countries. Kobe cattle selected for fattening are generally mature females over three years of age that have never borne calves (see *calf*) or have done so only once. These animals are fed a concentrated (see *concentrate*) *feed*, mainly *barley* but also containing *wheat* bran (see *bran*), *rice* bran, and *soybean* cake.

Kohlrabi A member of the *cabbage* family with a delicate, *turnip*-like flavor. Kohlrabi is native to northern Europe.

Kumquat A fruit that looks like a miniature *orange*. Kumquats have edible sweet skin and a tart interior and are eaten fresh and used as a garnish.

Kwashiorkor A *disease* caused by a deficiency of *protein* and calories.

L

Labor coverage The minimum labor required at a store for customer service and to meet union agreements for staffing.

Labor force participant A person, 16 years or older who, at any time during the year, worked for any period of time, looked for work, or was laid off from a job.

Lactase See *lactose intolerance.*

Lactation period The length of time dairy *cows* and milk *goats* give *milk* following the birth of offspring.

Lactose A disaccharide contained in *milk.*

Lactose intolerance An inability to digest *lactose* that occurs because the human body has stopped producing an intestinal *enzyme* called lactase, which breaks lactose down into the simple sugars *glucose* and galactose during digestion. In the absence of the enzyme, sensitive people may suffer from uncomfortable gastrointestinal symptoms such as bloating, cramps, diarrhea, gas, and nausea.

Ladino clover A *variety* of white *clover* primarily used as a *pasture* crop.

Lamb A young *ovine* animal that has not cut the second pair of permanent teeth (less than one year old). The term includes animals referred to in the *livestock* trade as lambs, *yearlings,* or yearling lambs. The term also refers to meat from a young *sheep* of one year or less.

Laminitis See *founder.*

Land The total natural and cultural *environment* within which *production* takes place. Land is a broader term than *soil.* In addition to soil, its attributes include other physical conditions such as mineral deposits and water supply; location in relation to centers of commerce, populations, and other land; the size of the individual tracts or holdings; and existing plant cover, works of improvement, and the like.

Land capability A measure of the suitability of *land* for use in *agriculture* without damage. In the United States, it usually expresses the effects of physical land conditions, including *climate,* on the total suitability for agricultural use without damage. Arable *soils* are grouped according to their limitations in sustained *production* of common cultivated (see *cultivate*) crops without soil deterioration. Nonarable soils are grouped according to their limitations in the production of permanent vegetation and their risks of soil damage if mismanaged.

Land-capability classification A grouping of kinds of *soil* into special units, subclasses, and classes according to their capability for intensive use and the treatments required for sustained use.

Land classification The classification of units of *land* for the purpose of grouping *soil* of similar characteristics, in some cases showing their relative suitability for some specific use.

Land-grant A gift of public *land*, usually by the U.S. government, to promote homesteading (see *Homestead Act*) or a similar purpose.

Land-grant universities State colleges and universities established on *land* granted by the federal government to encourage further practical education in *agriculture*, homemaking, and the mechanical arts. The land-grant universities were established by the Morrill Acts of 1862 and 1890.

Land plane Equipment used on irrigated (see *acres irrigated* and *irrigated farms*) *land* after land has been leveled by heavy machinery to "touch up" land leveling operations, usually every two or three years following intensive cropping.

Land Reclaimation Act of 1902 See Homestead Act of 1862.

Land tenure The relationship of the *farm operator* to the *land* operated. The major land tenure categories in the United States are: (1) full owners, who own all the land they operate; (2) part owners, who own and rent the land they operate; and (3) tenants, who rent all of the land they operate.

Land-use planning The decisionmaking process to determine the present and future uses of *land*. The resulting plan is the key element of a comprehensive plan describing the recommended location and intensity of development for public and private land uses, such as residential, commercial, industrial, recreational, and agricultural.

Lanolin The wax secreted by *sheep* to protect *wool* fiber (see *fiber*). Lanolin is used extensively in health and beauty aids, such as cosmetics.

Lard The *rendered* fat (see *fats and oils*) from pork *carcasses.*

Large-bale baler A *baler* that makes *bales* of *hay* weighing from about one-half to a ton or more. Farmers handle far less bales than when using conventional-size bales but require more specialized, heavy duty equipment to pick up and move the bales. Also referred to as a square baler. (While the term square baler is used to refer to large-bale balers, the bales are actually rectangular and may be 2 feet by 3 feet or more in thickness and width and from 4 feet to 8 feet or more in length.)

Large-bale mover *Tractor*-mounted spikes or trailed implements designed to lift, transport, and unload large round (cylindrical) and large rectangular *bales* of *hay.*

Larva A white legless grub that lies curled up on the bottom of the wax cell of the *honeycomb*. The stage in the life of a bee (see *honeybee*) between *egg* and *pupa*.

Latency period The time from the introduction of a *disease*-causing *agent* into a host's body and the first *symptoms* of the illness.

Latin American Free Trade Association (LAFTA) See *Latin American Integration Association.*

Latin American Integration Association The umbrella organization under which *trade liberalization* takes place in South America. Known by its Spanish acronym, ALADI, the organization replaced the Latin American Free Trade Association (LAFTA). Member countries include Venezuela, Columbia, Ecuador, Peru, Bolivia, Brazil, Argentina, Uruguay, and Paraguay.

Lauric acid See *fatty acids (classification of).*

Layer A female chicken (see *poultry*) that produces *eggs* regularly. About 19 to 22 dozen eggs a year is the average for a good layer.

L-dopa A compound that occurs naturally in fava *beans* and velvetbeans; used to treat Parkinson's *disease.*

Leaching The removal of materials in solution by water passing through the *soil.*

Leasing of quota Payments for the right to grow and sell a specified quantity of *tobacco.* In most cases, quota tobacco can be grown on *farms* other than the farm to which it is assigned if the farms are in the same county. (However, cross-county leasing is allowed for *burley tobacco* in Tennessee.) Leasing is permitted for burley tobacco and some other types but is generally no longer permitted for *flue-cured.*

Least-cost feed formulation The combination of available *feed* ingredients that satisfies all of the *nutrient* and physical requirements of *livestock* at the least possible cost.

Leather An animal *hide* or skin that has been chemically treated to prevent its natural decomposition.

Lecithin The mixed phosphatides product obtained from vegetable oils (see *fats and oils*) in the degumming (see *degummed oil*) process. Lecithin is added to many food products for its emulsifying (see *emulsifier*) and antioxidant properties.

Leek A type of green *onion.* Leeks are larger and milder in flavor than regular green onions.

Legumes A family of plants, including *peas, beans, soybeans, peanuts, clovers, alfalfas,* and sweetclovers that convert *nitrogen* from the air to nitrates in the *soil* through a process known as *nitrogen fixation.* Many of the nonwoody *species* are used as *cover crops* and are plowed under for soil improvement.

Lemon A *citrus fruit* used for fresh consumption, juice, *essential oil,* peel, and *pectin.* Major producers include the United States, southern Europe, and Argentina. California dominates fresh output from December through September, while Arizona is the leading producer from October to December. Florida grows less than 2 percent of the U.S. fresh market *supply* of lemons. About 35 percent of the U.S. fresh *market* production is exported.

Lentil The dry, threshed *seed* of a widely cultivated (see *cultivate*) Eurasian, *annual,* leguminous (see *legume*) plant. The United States, India, Syria, Turkey, Pakistan, Iran, Spain, Ethiopia, the former Soviet Union, Morocco, and Canada are major producers.

Lespedeza A *legume* grown for *fodder.*

Lesquerella A small, oil-bearing plant that is a member of the *mustard* family. The oil (see *fats and oils*) in the *seeds,* which accounts for up to 25 percent of the seed weight, contains *fatty acids* that have potential uses in lubricants, plastics, protective coatings, and cosmetics. The fatty acids have a structure similar to ricinoleic acid in *castor oil.* The meal can be used as a *protein* supplement in beef *cattle* rations (see *ration*).

Less developed country (LDC) See *developing country.*

Letter of commitment A commitment by the *Commodity Credit Corporation (CCC)* under the *Public Law 480* program to reimburse a U.S. commercial bank for payments made or drafts accepted under *letters of credit* for the account of an approved *foreign* applicant.

Letter of credit A document issued by a bank to a buyer in a transaction that gives the buyer the financial backing of the issuing bank.

Lettuce A vegetable plant cultivated (see *cultivate*) for its edible leaves. Lettuce is used in salads and as a garnish. *Varieties* include Iceberg, Boston Butterleaf, and Romaine. The major lettuce producers are the United States, Spain, and Israel.

Leukosis A virus-produced form of cancer affecting *cattle* that causes trade restrictions in many countries. Only a small percentage of animals with positive serological tests develop clinical signs. Trade in animals from areas with leukosis may be restricted.

Leveler A machine used to flatten and smooth fields in preparation for surface irrigation (see *irrigation methods*).

Level terrace A broad surface channel or embankment constructed across sloping (see *slope*) *soil* on the contour, as contrasted to a graded (see *grade*) terrace, which is built at a slight angle to the contour. A level terrace can be used only on soils that are permeable enough for all of the storm water to soak into the soil so that none breaks over the terrace to cause gullies.

Leveraged buyout The purchase of common stock of a company largely through debt-financing, pledging the assets (see *assets, current* and *assets, fixed*) of the new company as collateral. Investor capital is minimized as a result.

Levy See *tariff.*

Liabilities Debts or other claims against assets (see *assets, current* and *assets, fixed*).

Liberalization The removal of policies and practices that restrict trade.

Liberal (trade) Trade policies that are relatively free from controls or restraints.

Licensing An *agreement on import licensing procedures* (licensing code) that requires that documentation (other than that required for *customs* purposes) be submitted to the relevant administrative body for approval before importing (see *import*) is allowed.

Lichee See *litchi.*

Licorice A European *legume*, the root or root extract of which is used to flavor *tobacco*, beverages, confectionery products, and pharmaceuticals. About 90 percent of the licorice extract used in the United States is for flavoring tobacco products. Major producers include the former Soviet Union, China, Iraq, Iran, and Turkey.

Light air-cured tobacco A thin, medium-bodied *tobacco* that is light tan to reddish brown, with a mild flavor. The tobacco is used chiefly in making *cigarettes*. *Burley tobacco* and *Maryland tobacco* are the two types of light *air-cured tobacco* grown in the United States.

Light horse See *horse.*

Light soil A *soil* that is easy to *cultivate* because it contains large amounts of sand or silt particles.

Lignan A compound occurring naturally in wild mayapples that has shown evidence of activity against certain types of cancers (such as lung cancer) and viruses (such as herpes). The cancer drug etoposide ("vepeside") is a semisynthetic derivative of one such lignan.

Lignin The resinous chemical that provides rigidity to woody plants. Lignin is chemically separated to release *cellulose* fibers (see *fiber*) to form *pulp.* Lignin has industrial applications. See *black liquor.*

Lima bean (large, baby) A *variety* of *bean* that can be harvested (see *harvest*) in fresh form or as a *dry edible bean.* There are two types, a large seeded bean (see *seed*) and a small seeded (baby) bean, originating in Central and South America. In the United States, limas are grown primarily in California.

Lime A *citrus fruit* used for fresh consumption, juice, *essential oil*, and peel. Major producers include Mexico, Brazil, Egypt, the United States, and Cuba. Florida is the primary U.S. producer. The United States is a large *importer* of limes, with more than half of U.S. consumption coming from Mexico and other *Caribbean Basin Initiative* nations.

Lime, agricultural Material usually composed of the oxide, hydroxide, or carbonate of calcium, or of calcium and magnesium. The most common forms used in *agriculture* are ground limestone, hydrated lime, burned lime, *marl*, and oyster shells.

Lime requirement The amount of standard ground limestone (see *lime, agricultural*) required to bring a 6.6-inch layer of an *acre* (about 2 million pounds in mineral *soils*) of acid soil to some specific lesser degree of acidity, usually to slightly or very slightly acid. In common practice, lime requirements are given in tons per acre of nearly pure limestone, ground finely enough so that all of it passes through a 10-mesh screen and at least half of it passes through a 100-mesh screen.

Limited cross-compliance See *cross-compliance.*

Limited partnership A business organized as a partnership in which one or more partners have limited liability and do not participate in management.

Limonene A compound occurring in *citrus fruits* and in *herbs* and spices such as caraway, *celery*, fennel, and peppermint. The compound is touted as a cancer preventive and has proven *bactericide*, *insecticide*, and sedative properties.

Linear reduction of tariffs A given reduction of *tariffs* by participating countries on a whole range of goods.

Linen *Cloth* woven from thread made from the *fibers* of the *flax* plant.

Linoleic acid See *fatty acids (classification of).*

Linolenic acid See *fatty acids (classification of).*

Linseed The *seed* of the *flax* plant used to produce *linseed oil* and *linseed meal.* Also called flaxseed. Major producers are Argentina, the *European Union*, the United States, and Canada.

Linseed meal The finely ground product that remains after the *oil* is extracted from *flaxseed.*

Linseed oil An edible, industrial oil (see *fats and oils*) extracted (see *extraction, mechanical* and *extraction, solvent*) from *flaxseed* used in the manufacturing of plants, inks, varnishes, lacquers, and linoleum.

Lint Raw *cotton* that has been separated from the *cottonseed* by ginning (see *gin*). Lint is the primary product of the cotton plant, while cottonseed and *linters* are *byproducts.*

Linters The fuzz or short *fibers* that remain attached to the *cottonseed* after ginning (see *gin*). Linters are usually less than 1/8 inch in length and are removed from the *seed* by a *delinting* process. Too short for usual textile use, they are used for batting and mattress stuffing and as a source of cellulosic (see *cellulose*) *fibers.* See also *lint.*

Liquid bulk carrier See *bulk carrier.*

Liquid feedlot wastes *Manure* and other residues from *cattle* feeding that contain enough water to be handled as a liquid.

Lister A *plow* that throws *soil* in both directions, leaving a *furrow* in the middle. Also called a middle breaker or a middle buster.

Listeria A bacterium (see *bacteria*) that can cause human illnesses ranging from mild to severe, including spinal meningitis. The immuno-compromised are the most susceptible to listeriosis. Soft cheeses and chicken (see *poultry*) have caused outbreaks.

Litchi A small, warty-skinned fruit with a texture and flavor similar to peeled Muscat *grapes.* Also known as lichee.

Litter A group of young born to the same female at one time.

Little leaf A *disease* of plants caused by a deficiency of zinc. The disease is fairly common and is characterized by pale yellow-green leaves and dwarfed shoots and leaves.

Liver fluke A *parasite* of the liver.

Livestock Domestic *farm* animals, such as *swine*, beef and dairy *cattle*, *horses*, and *poultry*, raised for pleasure or profit.

Livestock scale A device for weighing *cattle*, *horses*, and other *livestock.*

Livestock truck A truck (see *farm truck*) with high slatted sides for hauling *cattle*, *horses*, and other *livestock.*

Livestock worker See *hired worker.*

Liveweight The weight of an animal before it is slaughtered (see *slaughter*).

Living mulch Vegetation that adds organic matter to the *soil* and helps reduce *erosion* but does not compete heavily with the crop for water and *nutrients.*

Loader A *tractor* with a hydraulically operated bucket or fork used to move *hay, grain, manure,* dirt, or other material into trucks (see *farm trucks*) or *wagons*, or from place to place. A loader may be a self-contained unit specifically designed for loading (see *skid steer loader*).

Load spout The pipe that conveys a bulk *commodity* from the *elevator* into the hold of a vessel, truck (see *farm truck*), or railroad car.

Loam *Soil* having a moderate amount of sand, silt, and clay. Loam soils contain 7 to 27 percent clay, 28 to 50 percent silt, and less than 52 percent sand.

Loan deficiency payments A provision giving the secretary of agriculture discretionary authority to provide payments to *producers* who, although eligible to obtain *nonrecourse loans*, agree not to obtain loans for *wheat, feed grains, upland cotton, rice, oilseeds,* and *honey*. The payment is determined by multiplying the loan *deficiency payment* rate by the amount of *commodity* eligible for loan. The loan deficiency payment rate per *bushel* is the announced *loan rate* minus the repayment level used in the *marketing loan*. The amount of the commodity eligible for the loan deficiency payment is determined by multiplying the individual *farm* program crop acreage by actual *yield* on the farm.

Loan forfeiture The ability of *producers*, who pledged their crop as collateral for a *nonrecourse loan*, to forfeit their crop to the federal government whenever they are unable to earn enough on the *market* to profitably repay the loan.

Loan rate The *price* per unit (pound, *bushel*, *bale*, or *hundredweight*) at which the government will provide loans to farmers to enable them to hold their crops for later sale.

Local currency sale A direct or *Public Law 480* sale, made by the *Commodity Credit Corporation (CCC)*, in which payment is made in the recipient country's currency.

Lomé Convention of 1975 A trade agreement between the *European Union (EU)* and the former African, Caribbean, and Pacific (ACP) colonies of the EU member states. The agreement covers some aid provisions as well as trade and *tariff* preferences for the ACP countries when shipping to the EU. The convention also produced a provision known as STABEX, designed to stabilize earnings by the associated countries on certain *commodities*. The Lomé Convention derives its name from the capital city of Togo, where the latest agreement was signed.

Long A person or firm who owns a *commodity*, or *resources* committed to producing a commodity, or holds a *contract* to buy a commodity at a specified *price*.

Longan A small fruit native to India.

Long cut (tobacco) See *smoking tobacco*.

Long-grain rice *Rice* that meets a government-established standard for *grain* sizes and shape types. Long-grain rice has a length/width ratio of 3.0 or more. Long-grain rice accounts for about 70 percent of U.S. rice *production*. Most long-grain rice in the United States is grown in Arkansas, Louisiana, Mississippi, Missouri, and Texas. See also *medium-grain rice* and *short-grain rice*.

Long position The ownership of a *commodity* or *resources* committed to producing a commodity or the holding of a *contract* to buy a commodity at a specified *price*. See *open position*.

Longshoreman A worker who loads or unloads vessels. Longshoremen working on shore are known as dockworkers, while stevedores work on ships.

Long staple cotton *Cotton* fibers (see *fiber*) whose length ranges from 1 1/8 inches to 1 3/8 inches. Fibers whose length is 1 3/8 inches or more are known as *extra long staple (ELS) cotton.*

Long staple fibers See *staple fibers.*

Long ton A measure of weight equal to 2,240 pounds or 1,016 kilograms. See *metric ton* and *short ton.* See also Appendix 6.

Loom A machine that *weaves* fabric (see *cloth*) by interlacing a series of lengthwise (vertical) parallel threads, called *warp* threads, with a series of crosswise (horizontal) parallel threads, called *filling* threads.

Loose leaf (tobacco) See *chewing tobacco.*

Lot A relatively small, fenced enclosure for *cattle*; also refers to a pen or corral.

Low-calorie sweetener A product with a sweetness so intense that only a fraction is needed to provide the same degree of sweetness as *sugar. Aspartame, saccharin,* and *acesulfame-k* are the major low-calorie sweeteners.

Low-fat soy flour *Flour* produced either by partial removal of the oil (see *fats and oils*) from *soybeans* before grinding (see *grind*) or by adding back 5 to 6 percent *soybean oil* or *lecithin* to *defatted soy flour.*

Low-input sustainable agriculture (LISA) Alternative methods of farming that reduce the application of purchased *farm inputs* such as *fertilizer, pesticides,* and *herbicides.* The goal of these alternative practices is to diminish environmental (see *environment*) hazards while maintaining or increasing *farm* profits and productivity. Methods include *crop rotations* and mechanical cultivations (see *cultivate*) to control weeds; *integrated pest management* strategies such as introducing harmless natural enemies; planting *legumes* that transform *nitrogen* from the air into a form plants can use; application of *livestock* manures, municipal sludge, and compost for fertilizer; and overseeding of legumes into maturing fields of *grain* crops, or as postseason *cover crops* to reduce *erosion.*

Low-input Sustainable Agriculture (LISA) Program A *U.S. Department of Agriculture* (USDA) program designed to reduce farmers' risks of adopting low-input (see *farm inputs*) methods and to make reliable information readily available to farmers.

Low intensity animal production A system for raising animals for food that uses fewer purchased inputs (see *farm inputs*) and less energy than conventional systems. A *pasture* production system is one example.

Lumber A *hardwood* or *softwood* product that has been sawed, planed, sanded, resawed, or crosscut but not further manufactured. Major producers are the United States, Canada, the former Soviet Union, Sweden, and Finland.

Lumpers Specialized free *agents* at retailer warehouses who unload trucks for a fee.

Lupulin A fine, yellow, resinous powder in *hops.* Lupulin is valuable in preserving and flavoring beer and other malt liquors.

Luxury consumption The intake by a plant of *nutrients* in amounts beyond those required for normal growth and function. Thus, if potassium is abundant in the *soil*, *alfalfa* may take in more than is required.

Lychee A tropical fruit with firm, translucent, milky-white flesh that is eaten fresh, canned, preserved, or dried (see *drying*). Lychee is produced in Australia, China, India, Mexico, South Africa, Taiwan, and the United States (Hawaii and Florida).

M

Macadamia nut The round, hard-shelled nut of a tree native to Australia. Australia, the United States (Hawaii), and Central America are the major producers.

Macroeconomics A branch of economics focusing on the general economic environment in which the total economy or a sector, such as *agriculture*, operates. Macroeconomic policies include monetary policies that directly affect interest rates, money, and credit flows in financial *markets* and fiscal policies that involve government spending and taxation. See also *microeconomics*.

Maintenance margin See *margin*.

Maize Indian *corn* (*Zea mays*).

Major Land Resource Area (MLRA) A geographically associated area delineated by the *U.S. Department of Agriculture*'s (USDA) *Natural Resources Conservation Service (NRCS)* according to the pattern of *soil*, water, *climate*, vegetation, *land* use, and type of farming.

Make allowance The margin between the government *support price* for *milk* and the *Commodity Credit Corporation's (CCC)* purchase *price* for *butter*, *nonfat dry milk*, and *cheese*. This margin is administratively set to cover the costs of *processing* milk into butter and nonfat dry milk or cheese to reach the desired level of prices for milk in manufacturing uses (see *manufacturing-grade milk*).

Malanga A starchy root crop that is a staple in tropical countries and has uses similar to *potatoes*.

Malt Sprouted and steamed whole *grain* from which the *radicle* has been removed.

Mammary gland A *milk*-producing organ of females. See also *udder*.

Mandatory reporting Required reporting of transactions (*prices* and quantities) by law or regulation.

Mandatory supply controls A program that would prohibit farmers from producing or selling without penalty more than specified amounts of certain *commodities*. All *producers* of any controlled commodity would be required to participate, with fines or other legal penalties used to enforce restrictions.

Mango A tropical fruit borne by an evergreen tree of the sumac family. The ripe fruit is usually partly red or yellow and is eaten fresh and used in desserts. The major producers are the United States, Mexico, the West Indies, Central America, India, and Australia. The United States is a growing net *importer*, with Mexico supplying approximately 80 percent of U.S. *imports*.

Man-made fiber fabric Fabric (see *cloth*) made from synthetic *fibers*, or from a blend or mixture with *natural fibers* where the *synthetic* portion comprises more than 50 percent by weight or value.

Man-made fibers Industrially produced *fibers*, in contrast to *natural fibers* such as *cotton*, *wool*, and *silk*. Examples of man-made fibers are nylon, rayon, acetate, acrylic, polyester, and olefin. See also *man-made fiber fabric*.

Manufacturer deal A promotional strategy by which a manufacturer offers a retailer volume discounts on one or more of the manufacturer's products in return for specified advertising and merchandising obligations. Manufacturer deals often specify payment conditions, *wholesaler* notification provisions, and precise dates for *price* concession and retailer merchandising performance.

Manufacturing-dependent county See *county type classification.*

Manufacturing-grade milk *Grade B milk*, or *milk* used to make "soft" or "hard" manufactured dairy products. All milk not eligible for fluid (beverage) use.

Manure The refuse from stables and barnyards, including both animal excreta and *straw* or other litter.

Manure lagoon A structure to treat *livestock* wastes (see *manure*), which can be aerobic, anaerobic, or facultative (capable of living in either of the other two *environments*), depending on loading and design.

Manure spreader A *wagon* with a chain, *auger*, or other device to move *manure* to rapidly rotating blades that fling it evenly over the ground. See *fertilizer spreader.*

Marbling The fat (see *fats and oils*) interspersed with lean beef or pork muscle. More marbling is usually associated with higher palatability.

Margarine A fatty food product prepared by blending *fats and oils* with other ingredients, usually *milk* solids, salt, flavoring materials, and vitamins A and D. It is manufactured under *Food and Drug Administration* (FDA) *Standards of Identity* and must contain at least 80 percent fat. The fats used in margarine may be from either animal or vegetable origin, although vegetable oils are the most widely used.

Margin 1: See *marketing margin (or marketing spread).*

2: Required money or collateral that a customer deposits with a *commodity* brokerage (see *broker*) firm, or a brokerage firm deposits with a clearinghouse, to ensure fulfillment of obligations under a *futures contract*. The initial margin is the amount required to enter a futures position; the maintenance margin is the amount required to continue a futures position without receiving a *margin call.*

Marginal land *Land* that cannot be farmed profitably because the *soil* is too unproductive.

Margin call A request from a *commodity* or brokerage (see *broker*) firm to a customer for additional funds to set aside to ensure performance of the *contract* (*margin*) when the customer's equity is diminished by *price* movements. The customer's futures (see *futures contract*) position is eliminated within a specified time period if the margin call is not met.

Margin of preference The difference between the *tariff* that would be paid under a system of *preferences* and the tariff payable on a *most-favored nation* basis.

Marine insurance A *contract* whereby an insurance firm provides financial guarantees to the owner, purchaser, or shipper of specific goods against their loss or damage while in transit from one destination to another. An open policy covers all shipments of a trader and is used when the trader regularly *exports* or *imports*. A special type of marine insurance is used when the seller has only an occasional shipment or must supply evidence of insurance to bankers or buyers. Two types of marine insurance coverage are available: *free of particular average (FPA)* and *with particular average (WPA)* coverage.

Marjoram See *oregano*.

Market 1: A closely interrelated group of buyers and sellers.
2: A physical location where buying and selling or other forms of trade occur.
3: The process of selling a product.

Market access 1: The extent to which a country permits *imports*. A variety of *tariff* and *nontariff barriers* can be used to limit the entry of *foreign* products.
2: The degree to which a farmer can sell in a given *market*.

Market basket Average quantities of goods, including food, purchased per household for a given period. Used to compute an index of retail *prices*. See also *market basket of farm foods*.

Market basket of farm foods A statistic used by the *U.S. Department of Agriculture* (USDA) to track *price* changes for *commodities* farmers sell and the foods *consumers* buy in retail *foodstores*. The market basket of *farm* foods contains the average quantities of domestically produced food for at-home consumption purchased in a base period. It does not include fish, foods purchased in away-from-home eating establishments (see *foodservice*), or foods primarily imported, such as *coffee*, *tea*, *cocoa*, and *bananas*. See also *market basket*.

Market certificate A certification of compliance used in past national *wheat* programs that could be redeemed for a specified amount of money. *Market* certificates were a provision of the Agricultural Act of 1964 (P.L. 88-297) (see Appendix 3).

Market-clearing prices *Prices* that, in the absence of *monopoly* or government controls, equate *production* of a product with commercial use over a period of time.

Market concentration A measure of the share of business held by the leading firms in a *market*. For example, the four largest, general-line grocery wholesalers accounted for 26 percent of sales in 1987 compared to 53.8 percent for the 20 largest firms.

Market conduct The pattern of behavior that enterprises follow in adapting or adjusting to the *markets* in which they buy or sell.

Market development programs Government-funded *export promotion programs* intended to expand *demand* for a country's *exports* in a *foreign* market(s). Compensation is usually provided for such activities as advertising, public relations, exhibits, and other activities undertaken to promote a country's output. For these programs to be effective, the *commodities* promoted must be differentiated from those of other countries so that *consumers* form a mental association between the country of origin and the product. The *Market Promotion Program* is an example of a *market* development program.

Market entry The entry of a firm or business into a *market* where it did not previously operate.

Market exit When a firm or enterprise no longer operates in a *market* in which it had been actively doing business.

Market-extension merger A merger of firms producing the same product line but in different geographic *markets*. Market-extension mergers extend the marketing area of an acquiring firm to include the marketing area of the firm(s) acquired.

Market hog See *slaughter hog*.

Marketing agreement A marketing institution which, once activated, is binding on all signatory parties.

Marketing bill An estimate of the total charge for marketing all U.S. *farm* foods, including those consumed in restaurants and other eating places and those bought in retail stores. The figure is calculated as the difference between total civilian expenditures for these foods and total *farm value*. Marketing bill statistics are affected by changes in *prices*, volume, types of products marketed, and the quantity of marketing services per unit of product.

Marketing board A major form of government involvement in *commodity* marketing by some countries, such as Canada and Australia. These boards generally handle all *export* sales for the commodity. They may administer provisions to guarantee farmers a minimum *price* each year based on the *cost of production* and forecasts of world *market* conditions for the upcoming *marketing year*, or they may provide an initial minimum price with supplemental payments later based on export sales. Boards may oversee a *two-price plan* in which domestic prices differ from the export price. Australia uses marketing boards for selected *grains* and *wool,* while Canada uses marketing boards for *wheat*, *barley*, chicken, turkey, *eggs*, and dairy.

Canada provides a specific example of the operation of a marketing board. The *Canadian Wheat Board's (CWB)* initial payments or guaranteed prices for wheat and barley are based on the CWB's forecast of world market conditions for the upcoming marketing year. The CWB initial payments are set in regard to Canada's supplies (see *supply*), other exporting countries' supplies, importing countries' demands (see *demand*), and the level of export subsidies (see *subsidy*). CWB final payments are made after the conclusion of the marketing year, after the initial payments and administrative expenses are accounted for. The CWB discontinued its two-price wheat policy for domestic wheat in 1991, when it began pricing wheat for domestic human consumption on a daily basis in relation to the Chicago and Minneapolis *futures market* price.

Marketing certificate A certificate that may be redeemed for a specified amount of *Commodity Credit Corporation* (CCC) commodities. Such certificates may be generic or for a specific *commodity*.

Marketing contract A sales agreement made at least 10 days before delivery specifying *price*, or rules for setting the price, and other terms of sale.

Marketing loan See *marketing loan program*.

Marketing loan program A program first authorized by the *Food Security Act of 1985 (P.L. 99-198*) (see Appendix 3) that allows *producers* to repay *nonrecourse price support loans* at less than the announced *loan rates* whenever the *world price* or loan repayment rate for the *commodity* is less than the loan rate. Under the *Food, Agriculture, Conservation, and Trade Act of 1990 (P.L. 101-624)* (see Appendix 3), marketing loan programs are mandatory for *soybeans* and other *oilseeds*, *upland cotton*, and *rice* and discretionary for *wheat* and *feed grains*. Provisions of the Omnibus Budget Reconciliation Act of 1991 (see Appendix 3) made marketing loans mandatory for the 1993–1995 crops of wheat and feed grains.

Marketing margin (or marketing spread) The difference between *prices* at various levels of the marketing channel. Most often used for the difference between the prices paid by *consumers* and those obtained by *producers*.

Marketing order See *federal marketing orders and agreements*.

Marketing quota(s) *Quotas* authorized by the Agricultural Adjustment Act of 1938 (P.L. 75-430) (see Appendix 3) to regulate the marketing of some *commodities* when supplies (see *supply*) are or could become excessive. A quota normally represents the quantity the *U.S. Department of Agriculture* estimates to be required for domestic use, *exports*, and adequate *carryover stocks* during the year. Marketing quotas are binding upon all *producers* if two-thirds or more of the producers holding allotments for the *production* of a crop vote for quotas in a *referendum*. Marketing quotas are usually put into effect through *acreage allotments* or specific limits on marketing. Growers who plant more of a commodity than their *farm* acreage allotments or *market* more than their quota are subject to marketing penalties on the "excess" production and are ineligible for government *price-support* loans. Quota provisions have been suspended for *wheat*, *feed grains*, and *cotton* since the 1960s; *rice* quotas were abolished in 1981. Poundage quotas are still used for domestically consumed *peanuts* but not for exported peanuts. Marketing quotas are also used for major types of *tobacco*. The *Food, Agriculture, Conservation, and Trade Act of 1990 (P.L. 101-624)* (see Appendix 3) mandates marketing quotas for *sugar* and crystalline *fructose* when sugar imports for U.S. consumption are less than 1.25 million short tons, raw value.

Marketing research Research to develop high-quality, low-cost products through new science and technology. Research is also conducted to stimulate development, innovation, and testing of new concepts in marketing, transportation, *processing*, storage, and services.

Marketing spread See *farm-to-retail spread*.

Marketing year Generally, a specified one-year period that starts with the beginning of a new *harvest*.

Market institution Any established *market* custom, practice, rule, or law that affects how a product is marketed, including sanitary regulations, marketing orders and agreements (see *federal marketing orders and agreements*), organized exchanges, *price support* and other federal programs.

Market news service See *federal-state market news service*.

Market performance The end results in terms of *price*, output, *production* cost, selling cost, production design, and so forth, of decisions made by firms in a *market*.

Market power The potential for a firm or group of firms to affect market *prices* by controlling certain aspects of marketing, such as sales to a *market*, quality of a product sold, and funding for promotion and research. *Monopoly* is the most extreme form of market power, since one firm controls all aspects of marketing.

Market Promotion Program (MPP) An *export promotion program* authorized by the *Food, Agriculture, Conservation, and Trade Act of 1990 (P.L. 101-624)* (see Appendix 3) that replaces the Targeted Export Assistance (TEA) Program, authorized by the *Food Security Act of 1985 (P.L.*

99-198) (see Appendix 3). The MPP is designed to encourage development, maintenance, and expansion of commercial *farm* export *markets.* Unlike TEA, the MPP does not restrict assistance to U.S. *producer* groups or regional organizations whose *exports* have been adversely affected by a *foreign* governments's policies, although these cases receive highest priority. The program promotes exports of specific U.S. *commodities* or products in specific markets. Under the program, eligible participants received *generic commodity certificates* or payments for promotional activities approved by the secretary of agriculture.

Market share A measure based on the proportion of total *market* sales (purchases) by the largest sellers (buyers). Market shares usually are measured for the largest four, eight, or 20 sellers (buyers) in the market and provide an indicator of *market concentration.*

Market structure See *structure.*

Market support tools Activities of a *federal marketing order and agreement* or *research and promotion order* that attempt to influence *demand* by improving both buyers' and sellers' knowledge of and awareness of a product's availability and uses.

Marl An earthy, chalky deposit, consisting mainly of calcium carbonate commonly mixed with clay or other impurities. It is formed chiefly at the margins of freshwater lakes. It is commonly used for liming acid *soils.* See *lime, agricultural.*

Maryland tobacco A light, *air-cured tobacco* that is usually considered to have ideal burning qualities for use in *cigarette* blends. Maryland tobacco is similar to *burley* but is somewhat milder and lighter in taste. See *tobacco.*

Master marketers Private company sales and service contractors that sell crop insurance (see *Federal Crop Insurance Program)* policies as *agents* for the *Consolidated Farm Service Agency (CFSA).* Master marketers bear no risk of loss on the policies they sell and do not adjust claims. They are reimbursed for their selling expenses by the FSA at a percent of the total *premiums* they sell.

Mastitis Inflammation of a dairy cow's (see *cow*) *udder,* usually caused by infection by microorganisms. The infection may be caused by improper milking procedures and poor hygiene.

Maturing future A *futures contract* during or immediately before the period when the seller can elect to make or take delivery.

Maximum Acceptable Rental Rates (MARR) Rental rate guidelines for designated areas eligible for the *Conservation Reserve Program (CRP)* as determined by the *U.S. Department of Agriculture. Producer*s' applications for the CRP can be accepted if the yearly rental payment they would require (rental *bid*) does not exceed the established MARR.

Maximum Payment Acreage (MPA) The maximum acres on which producers participating in annual *commodity* programs may earn *deficiency payments.* MPA equals the *crop acreage base* minus *normal flex acres* minus any acres required to be removed from *production* under an *Acreage Reduction Program* (ARP).

Mayonnaise An emulsified, semisolid, fatty food that by *Food and Drug Administration* (FDA) *Standards of Identity,* must contain at least 65 percent (by weight) edible vegetable oil (see *fats and oils*) and 30 percent salad dressing. Mayonnaise also contains salt, *sugar,* spices, seasoning, vinegar, *egg* yolk, *lemon* juice, and other seasoning or flavoring ingredients.

Meadowfoam A winter annual, (see *annual* and *perennial*) *oilseed* plant that yields both oil (see *fats and oils*) and *meal*. Similar to *industrial rapeseed* and *crambe oils*, meadowfoam oil has potential uses for personal care applications, such as soap, hair-care products, and moisturizers. Meadowfoam oil's stability makes it an excellent source of feedstock material for industrial applications such as lubricants, polymers, waxes, and lactones. Meadowfoam is well adapted to *climates* similar to the poorly drained *soils* of the Pacific Northwest and has been grown in parts of the eastern United States, Alaska, Western Europe, and British Columbia.

Meal The coarsely ground (see *grind*) and unsifted *grains* of a *cereal* grass. Also refers to the solid residue left after extracting (see *extraction, mechanical* and *extraction, solvent*) oil (see *fats and oils*) from *oilseeds*.

Measurement ton A measure of volume generally equal to 40 cubic feet (1 cubic meter). Also known as cargo or freight ton.

Meat analog Material usually prepared from vegetable *protein* to resemble specific meats in texture, color, and flavor.

Meat and bone meal See *meat meal and meat and bone meal.*

Meat and poultry inspection A mandatory requirement that animals and birds used for human food (*cattle, calves, swine, goats, sheep, lambs, horses* and other equines, *chickens, turkeys,* ducks, geese, and guineas) be inspected at *slaughter* and during various stages of *production* for *disease* and microbial or chemical contaminants. Inspections are conducted by the *U.S. Department of Agriculture*'s (USDA) *Food Safety and Inspection Service*. See also *inspection, federal.*

Meat distribution The movement of meat from *packer* to *consumer*. The distribution functions include fabricating (see *fabrication*), *wholesaling*, and retailing. Retailers include both grocery and *foodservice* firms.

Meat extenders *Soybean* or other vegetable *proteins* used as partial substitutes for meat in processed (see *processing*) items such as patties, chili, and casseroles.

Meat Import Law See *U.S. Meat Import Law*.

Meat meal and meat and bone meal The finely ground (see *grind*), dry-rendered residue from animal tissue used in animal *feeds*. It is called meat and bone meal if it contains more than 4.4 percent phosphorus.

Meat packer A firm that slaughters (see *slaughter*) *livestock*. Some *meat packing* firms may also fabricate (see *fabrication*) products, process (see *processing*), or perform other functions.

Meat packing Converting *livestock* into meat products. The process includes slaughtering (see *slaughter*) and may also include cutting, wrapping into *primal* and *subprimal* cuts, and *processing*.

Meat price list A list of cuts and *prices* a *packer*-processor (see *processor* and *processing*) prepares for use in pricing products to their customers. The list may have geographic and/or transportation differentials.

Meat Products Export Incentive Program payments Payments made to guarantee equitable treatment of meat products in the *U.S. Department of Agriculture*'s (USDA) efforts to enhance *foreign* sales of U.S. agricultural products.

Mechanical bulk feeder A feeder to hold *silages* or *concentrates*, usually extending inside the *feedlot* or shelter barn. The feeder is filled by electrically powered devices.

Mechanical nests Nests that have a moving belt to deliver laid *eggs* to a central pickup point.

Mechanical stack mover A machine that loads, transports, and unloads mechanically formed *hay* stacks.

Medium-grain rice *Rice* that meets a government-established standard for *grain* sizes and shape types. Medium-grain rice has a length/width ratio of 2.0–2.9. Medium grain accounts for about 30 percent of U.S. *production.* Over half of all medium-grain rice comes from California, with Arkansas and Louisiana providing most of the remainder. See also *long-grain rice* and *short-grain rice.*

Mellorine A frozen dessert made from edible vegetable oil (see *fats and oils*) or animal fat (see *fats and oils*), *sugar* or other sweetening agents, *milk* solids, and flavoring. The fat content of mellorine averages about 8 percent.

Merchandising A sales technique used to stimulate quick sales responses in stores. Examples include end-of-aisle displays, banners and signs, in-store product demonstrations, recipe booklets, and menus.

Mercosur A group of Latin American countries working toward establishing a *common market* and creating a free trade area between Brazil and Argentina by 1995, extending to Uruguay and Paraguay in 1996. Also known as the Southern Cone Common Market.

Merger The legal transformation of two or more formerly independent firms to one control. See *horizontal merger.*

Merit pricing Product sales *prices* based on *grade and yield* specifications, with appropriate differentials for varying grades of products.

Meslin The ancestor of modern *wheat* varieties.

Mesquite bean A *drought*-resistant *legume* with edible *seeds* that contains nearly 40 percent *protein.* Mesquite beans are not commonly grown in the United States.

Metabolism The changes that take place in the *nutrients* after they are absorbed from the digestive tract, including (1) the building-up process in which the absorbed nutrients are used in the formation or repair of body tissues, and (2) the breaking-down processes in which nutrients are oxidized for the production of heat and work.

Metabolite The product resulting from conversion of an animal *drug* into a chemical derivative.

Methanol A liquid alcohol distilled from wood or made synthetically and used as a fuel, a solvent, and an antifreeze. Also known as methyl alcohol and wood alcohol.

Methyl alcohol See *methanol.*

Methyl bromide See *fumigant.*

Metric ton A measure of weight equal to 2,204.6 pounds or 1,000 kilograms. See also *long ton, short ton,* and Appendix 6.

Metric weights and measures See Appendix 6, *metric ton, short ton,* and *long ton.*

Metro areas See *Metropolitan Area (MA).*

Metropolitan Area (MA) A county or group of contiguous counties that contain either a city of at least 50,000 population, or some other urbanized (see *urban*) area with 50,000 or more people and a total MA population of at least 100,000. Additional contiguous counties are included in an MA if they meet certain criteria of metropolitan character or worker commuting. See also *nonmetro areas.*

Metropolitan Statistical Area (MSA) See *Metropolitan Area (MA).*

Mexican husk tomato See *tomatillo.*

Microbial contaminant or pathogen *Microorganisms* with damaging biological effects that can occur in the *environment* where food is produced. These contaminants or *pathogens* have damaging biological effects and can cause human *disease.*

Microbiology The branch of biology dealing with microscopic forms of life.

Microeconomics A branch of economics focusing on determination of *prices*, incomes, and the use of *resources* by individual quantities or units, such a households or *markets*, as opposed to the economy as a whole. See also *macroeconomics.*

Microingredients *Vitamins*, minerals, *antibiotics*, *drugs,* and other materials normally required in small amounts and measured in milligrams, micrograms, or parts per million (ppm).

Micronaire A measure of the fineness of *cotton* fiber (see *fiber*).

Micronaire reading The results of an airflow instrument used to measure *cotton* fiber (see *fiber*) fineness and maturity. See *cotton quality.*

Microorganisms Any microscopic or ultramicroscopic organism, such as *bacteria*, mycoplasma, and viruses. These *parasites* gain their sustenance from the material on which they grow, such as *grain.*

Middlebreaker See *lister.*

Middlebuster See *lister.*

Middling 1: The designation of a specific *grade* of *cotton* (see *cotton quality*). Grades are determined by the amount of leaf, color, and the ginning (see *gin*) preparation of cotton, based on samples from each *bale* of cotton. Middling is a high-quality white cotton.

2: A fine particle of *endosperm, bran,* and *germ.* Middlings are *byproducts* of *flour* milling. Middlings are normally used for *livestock* feed (see *feed*).

Middlings rolls A pair, or pairs, of smooth rolls used in the *milling* process to reduce *middlings* to *flour*-particle size. Also called reduction rolls.

Midwestern Association of State Departments of Agriculture (MASDA) See *National Association of State Departments of Agriculture (NASDA).*

Migrant and Seasonal Agricultural Worker Protection Act (MSPA) The major federal law dealing exclusively with agricultural employment. The MSPA, passed in 1983, was designed to provide seasonal and *migrant farmworkers* with protection concerning pay, working conditions, and work-related conditions and to require *farm labor contractors* to register with the *U.S. Department of Labor.* The law was also designed to provide necessary protection for farmworkers, *agricultural associations*, and agricultural employers.

Migrant farmworker A person who leaves his/her place of residence overnight to do *agricultural work* of a seasonal or other temporary nature. Exceptions are immediate family members of an agricultural employer or a *farm labor contractor* and *H-2A temporary foreign workers.*

Milanese See *knitting.*

Mildew A plant *disease* in which the causal *fungus* forms a superficial coating over the surface of plant parts; or the fungus causing such a disease. The coating, which is a mycelial (see *mycelium*) growth, is usually thin and whitish. The two types of mildew are downy mildew and powdery mildew.

Milk The white liquid produced in the mammary glands of mature females after giving birth. Milk contains on average: fat (see *fats and oils*), 3.9 percent; solids, not fat, 9.0 percent (albumin, 0.7 percent; *casein*, 2.5 percent; *lactose*, 5.1 percent; mineral matter, 0.7 percent), and water, 87.1 percent.

U.S. *consumers* primarily consume the milk of *cows* as fresh, canned, *condensed*, or evaporated milk; cream; *yogurt*; *cheese*; *butter* and *butteroil*; dried milk casein; *whey*; *buttermilk*; sour milk and cream; and ice cream. The major producers are the former Soviet Union, the United States, Europe, and India.

Milk contract A marketing or manufacturing agreement between a *producer* and purchaser of *milk.*

Milk Diversion Program A program in which *producers* signed *contracts* to reduce their *milk* marketing by 5 to 30 percent from levels of a specified base period. In return, they received payments from the *U.S. Department of Agriculture* (USDA) of $10 per *hundredweight.* Participating producers could only sell their dairy *cattle* for *slaughter* or to another producer in the program. The program ran from January 1984 through March 1985.

Milk equivalent A measure of the quantity of *fluid milk* used in a processed (see *processing*) dairy product.

Milkfat The fat (see *fats and oils*) normally occurring in *milk.* Milkfat consists of 12 or more short-chain *fatty acids.*

Milkfat content A measure of the amount of *butterfat* in *milk.*

Milkfat content

Type of milk or milk product	*Percentage butterfat*
Nonfat or skim milk	None
Lowfat milk	2 percent
Whole milk	Usually 3.5 percent
Half and half	10–12 percent
Coffee cream	About 20 percent
Whipping cream	32–40 percent

Milk-feed price ratio A measure of the value of 16 percent *protein* feed (see *feed*) *ration* to one pound of whole *milk* (see *milkfat content*).

Milk fever A *disease* of dairy *cows* caused by calcium deficiency. The animal becomes partially paralyzed, may lose consciousness, and die. The disease most often occurs soon after *calving*, with heavy milkers apparently the most susceptible.

Milking machine A device for milking dairy *cows*. A milking machine may have tubing that runs *milk* from the cow's *udder* to a bulk milk tank.

Milk marketing order A regulation issued by the secretary of agriculture or a state official that specifies minimum *prices* and conditions under which *milk* can be bought and sold within a specified area. See *federal marketing orders*.

Milk Production Termination Program A program, often called the Dairy Termination Program or the Whole-Herd Buyout, authorized by the *Food Security Act of 1985 (P.L. 99-198)* (see Appendix 3). The program was designed to reduce *milk* production. *Producers* whose *bids* were accepted by the secretary of agriculture agreed to *slaughter* or *export* all female dairy *cattle*, have no interest in milk *production* or dairy cattle for five years, and not to use their facilities for those purposes during that time. The program was in effect from April 1, 1986, through September 30, 1987. During that period, 1.6 million dairy cows were removed from production.

Milk solids, not fat The total solids (albumin, *casein*, *lactose*, and mineral matter) present in *milk*, not including *butterfat*.

Milkweed A weed that is cultivated (see *cultivate*) to a limited extent for its floss. Milkweed floss has properties similar to goose down.

Mill byproduct A secondary *milling* product obtained in addition to the principal product. Generally refers to *millfeed* (particularly *bran* and *middlings*). See also *byproduct*.

Mill consumption The quantity of *fiber* processed (see *processing*) in manufacturing establishments.

Millet Any of various small-seeded (see *seed*), annual (see *annual* and *perennial*) *cereal* and *forage* grasses used to produce cereals, *feeds*, and confections. Major producers include the former Soviet Union, Nigeria, India, and China.

Millfeed The material remaining after all the usable *flour* is extracted from the *grain*. The material is used by the *feed* industry to make animal feed and feed supplements.

Milling 1: The process of shaping or finishing a product, such as *cloth*, in a mill.
2: A process by which grain *kernel* components are separated either physically or chemically, and *grain* is ground (see *grind*) into *flour* or *meal*.

Millitex (mtex) See *tex*.

Mill mix A mixture of several types of *wheat* to produce a desired *flour*.

Mill run The ungraded (see *grades* and *grading*) and usually uninspected material that comes from the mill (see *milling*).

Mill (textile) A business concern or factory that manufactures *textile* products by *spinning*, *weaving*, or *knitting*.

Milo *Grain sorghum*; used in the United States.

Minimum import price (MIP) A measure used by the *European Union*. *Imports* priced (see *price*) at less than the MIP are charged a tax equal to the difference or may be subject to quantitative restrictions.

Minimum optimum size The smallest scale of plant or firm that is viable over time. More specifically, applies to the smallest size plant or firm that can realize most available *economies of size*.

Minimum-price contract An agreement between a farmer and a first-*handler* that guarantees the farmer a specified minimum *price* during a prescribed time interval while providing for a higher price if *market* conditions warrant. For example, an *elevator* might promise to pay a specified minimum price or the elevator's current *bid* price, whichever is higher when the farmer chooses to sell.

Minimum tillage The minimum *soil* manipulation necessary for crop *production*.

Mining-dependent county See *county type classification.*

Ministerial Decided or carried out by cabinet officers, such as the secretary of state, the secretary of agriculture, or the *U.S. Trade Representative*. Their counterparts in other countries are usually titled minister of foreign affairs, minister of agriculture, and minister of trade.

Minnesota-Wisconsin (M-W) price The average *price* per *hundredweight* paid to farmers for *Grade B milk* in Minnesota and Wisconsin as estimated by the *U.S. Department of Agriculture* (USDA). The M-W price provides the basis for minimum class prices under the *federal milk marketing orders*. The *Class III milk* price, for example, is set equal to the M-W price, while the *Class II milk* price uses a product price formula to update the M-W price and is generally about 10 cents higher than the Class III price.

Mist blower A machine that applies *pesticides* under pressure through nozzles, forming a fog or mist. See *sprayer.*

MIU The combined content of moisture (M), insoluble impurities (I), and unsaponifiable matter (U) in *tallow* and *grease*. The maximum level of these nonfatty materials is specified for different *grades* of tallow and grease.

Mixed-flow dryers The most prevalent type of large, continuous-flow, off-*farm* grain dryer (see *crop dryer*) used in countries outside the United States. In these dryers, *grain* is dried by a mixture of crossflow, concurrent flow, and counterflow drying processes, which dry grain more uniformly and produce a higher quality grain. These dryers are expensive to manufacture and require extensive air-pollution equipment.

Mixing Using agitation to combine two or more materials to a specific degree of dispersion.

Mixing concentrate (pre-mix) A mix of the ingredients necessary to provide essential *nutrients* that is added to *corn* and *soybean meal* to make a complete *feed* ration (see *ration).*

Mixing regulation A requirement in some countries that products sold domestically contain a designated portion of domestically produced materials.

Mobile grinder-mixer A *tractor*-powered machine that *grinds* and mixes *feed* and can be used to transport and distribute feed.

Modular A hydraulic press used to compress *seed cotton* into a compact block that is kept in the field until being sent to the *gin*.

Moduled seed cotton *Cotton* that has been mechanically compressed after *harvest* into large modules in the field so that it can be held temporarily on the *farm* or at the *gin* while awaiting ginning. About 60 percent of U.S. cotton is moduled. This practice is especially important in the Southwest and West.

Mohair The hair of the Angora *goat*. Major producers include South Africa, Turkey, and the United States. Mohair is generally used to produce specialty *yarns* and *textiles* usually blended with other materials such as *wools* or *synthetics*.

Molasses The edible *byproduct* of the manufacture of *sugar* when some, but usually not all, of the crystallizable sugar in the *sugarcane* juice is removed by the crystallization process. Also applies to the concentrated, partially dehydrated (see *dehydration*) juices from fruits.

Mold A superficial growth produced by fungi (see *fungus*) on damp or decaying matter that can be detrimental to crops. Mold growth on *grain*, for example, creates damaged *kernels*, deposits toxic substances in the grain, and results in a loss of dry matter.

Moldboard plow A *tillage tool* with a curved steel plate or blade to turn *soil* over.

Molting A short period of time during which chickens (see *poultry*) shed their feathers and reduce or stop *egg* production. Induced to improve egg shell quality and to prepare *hens* for an additional laying season.

Monetary compensatory account (MCA) A tax or *subsidy* applied to goods traded between members of the *European Union* to avoid possible trade disruptions caused by *support price* differentials brought about by differing *green rates of exchange*. For a member state whose green rate was below the *market* rate, the MCA applied as a *levy* on *imports* and a *subsidy* on *exports*; for a member state whose green rate was above the market rate, the MCA had the opposite effect. The MCA's were eliminated as a result of reforms of the green rate system introduced on January 1, 1993.

Monetary policy See *macroeconomics*.

Monoculture Crop *production* using a single plant *variety*.

Monogastric An animal that has one digestive cavity, including *swine* and *poultry*.

Monoglycerides See *emulsifier*.

Monopoly A *market* characterized by one seller of a product that has no close substitutes.

Monopsony A *market* characterized by one buyer of a product that has no close substitutes.

Monosodium glutamate (MSG) The monosodium salt of glutamic acid, an *amino acid*. MSG is used as a flavor enhancer in foods.

Monounsaturated fats See *fatty acids (classification of)*.

Morrill Acts of 1862 and 1890 See *land-grant universities*.

Most-favored nation (MFN) An agreement between countries to extend the same trading privileges to each other that they extend to any other country. Under a most-favored nation agreement, for example, a country will extend to another country the lowest *tariff* rates it applies to any third country. A country is under no obligation to extend MFN treatment to another country, unless they are both *contracting parties* to the *General Agreement on Tariffs and Trade (GATT)*, or unless MFN is specified in an agreement between them.

Motes *Cotton* waste material from the cotton ginning (see *gin*) process, primarily resulting from the *lint* cleaning operation. Motes can be reclaimed and sold for use in padding and upholstery filling, nonwovens, and some open-end *yarns*.

Mouton A fur made by chemically treating and processing *sheep* pelts.

Mower See *hay mower.*

Mower-conditioner A machine that cuts *forage* crops and conveys the mowed plant materials through rollers which crush or crimp plant stems to speed *drying*.

MTN (multilateral trade negotiations) agreements and arrangements Generally applied to the seven *nontariff trade barrier* agreements negotiated during the *Tokyo Round* of the *General Agreement on Tariffs and Trade (GATT)* negotiations. These include agreements on aircraft, *antidumping*, customs valuation, government procurement, *import licensing*, standards, and *subsidies*. In the *Uruguay Round,* a negotiating group was established to consider modifying or expanding the provisions of these agreements. See *GATT Rounds.*

Mulch A natural or artificial layer of plant residue or other material put on the *soil* surface to reduce *erosion*, conserve moisture, and provide the soil with organic matter. Common mulching materials include compost, sawdust, wood chips, *straw*, and *crop residue*.

Mulcher 1: *Tillage equipment* of various designs that leave *crop residue* on the surface to prevent *erosion*.

2: A machine used to blow grass *seed* and a mix of *straw* and crop residues or artificial materials on bare *soil* areas, such as ditches or newly constructed roads, to establish root growth to prevent soil erosion.

3: A residential lawn mower with a specially designed blade to chop grass clippings into very fine particles that serve as a "green" *fertilizer* for a lawn.

Mulch till A type of *conservation tillage* where the total *soil* surface is tilled (see *till*), but sufficient residue is left on the soil surface to reduce *erosion*. See *no-till*, *ridge-till*, and *strip-till*.

Mulch tillage A process of working the *soil* that leaves plant and *crop residues* on the surface to prevent soil *erosion*.

Mulch tillage equipment See *conversation tillage equipment.*

Mulch tillage plow See *coulter chisel plow.*

Multicars More than one car carrying the same product on a given train.

Multifactor productivity index See *productivity*.

Multilateral agreement An international compact in which three or more parties participate. See *General Agreement on Tariffs and Trade*.

Multilateral trade negotiations (MTN) In general, discussions of trade issues involving three or more countries. An example is the *General Agreement on Tariffs and Trade (GATT),* which serves as a forum for intergovernmental *tariff* negotiations. Also applied to any of the eight rounds of GATT negotiations since 1947. See *GATT Rounds*. See also *bilateral trade negotiations*.

Multiline A composite (blended) population of several, genetically related lines of a self-pollinated (see *pollination*) crop.

Multinational firms Firms with operations or offices in more than one country.

Multiplant economies See *economies of size*.

Multiple hatcher Commercial enterprises involved in the mass-*production* of chicks from breeding stock.

Multispecies slaughter plant A plant (see *plant or establishment*) that *slaughters* more than one *species*. For example, a *cattle* and hog (see *swine*) slaughter plant.

Multitier system The existence of two or more effective *exchange rates*. Also referred to as a dual exchange rate system.

Murrain See *shorn wool*.

Mushroom The fruit of an edible *fungus*. There are approximately 2,000 *species* of mushrooms. The most common species include Agaricus, Button, Chanterelle, Enoki, Morel, Oyster, and Shitake. Major producers include the United States, China, France, the Netherlands, Spain, and Indonesia.

Mushroom spawn The seeding material for propagation of *mushrooms*. Mushroom spawn is a mixture of sterilized *grains* innoculated with *mycelium* added.

Muskmelon A broad category of melon, including such *varieties* as casaba, crenshaw, honeydew, Juane Canari, Persion, Santa Claus, and Sharlyn.

Mustard green A dark, leafy vegetable. *Varieties* include regular and oriental or Chinese mustard greens.

Mustard seed The *seed* of plants in the mustard family used to produce mustard powder, edible mustard oil (see *fats and oils*), and mustard *meal* (toxic). India and Pakistan are the major mustard seed producers.

Mutton The meat of mature *sheep*.

M-W price See *Minnesota-Wisconsin price*.

Mycelium The white, thread-like plant often seen growing on rotting wood or moldy bread.

N

Napa See *cabbage*.

Naps Large tangled masses of *fibers* that often result from ginning (see *gin*) wet *cotton*. Naps are not as detrimental to quality as *neps*.

National Advisory Council on International Monetary and Financial Policies (NAC) A federal interdepartmental committee that coordinates the policies of all U.S. government agencies to the extent that they make *foreign* loans or engage in foreign monetary transactions, including *Public Law 480* agreements.

National Agricultural Advisory Board An organization created by the *Food and Agriculture Act of 1977 (P.L. 95-113)* (see Appendix 3) to review and assess the allocation of funds for *agricultural research* and extension made by the *U.S. Department of Agriculture (USDA)*, as well as evaluate the effectiveness of coordination of federal and private research initiatives and the private and public research and extension system. Membership is drawn from research, extension, and education users in *agribusiness*.

National Agricultural Library (NAL) A national public library that provides information on *agriculture* and related subjects and is a coordinator and primary resource for state land-grant (see *land-grant universities*) and field libraries. The NAL is operated by the *U.S. Department of Agriculture (USDA)*. In October 1994, the NAL became part of the *Cooperative State Research, Education, and Extension Service*.

National Agricultural Statistics Service (NASS) A *U.S. Department of Agriculture (USDA)* agency that conducts surveys and publishes reports detailing *production* and prospects for crops, *livestock*, dairy, and *poultry*. NASS also provides data on *stocks*, *prices*, labor, weather, and other information of interest to farmers, ranchers, and others associated with *agriculture*.

National American Wholesale Grocers Association (NAWGA) A national trade association of firms engaged in wholesale distribution of food and grocery products to retail and *foodservice* operations.

National Association of Conservation Districts (NACD) An organization established in 1946 to serve as the national voice for the nearly 3,000 *soil* and water *conservation districts* in the United States. NACD aids local conservation districts in the conservation, orderly development, and wise use of natural *resources*.

National Association of Margarine Manufacturers An organization of firms that manufacture *margarine* and spread products and those who supply the ingredients, packaging, and equipment materials or services to manufacturing companies. The Association serves the interests of industry members, its customers, and *consumers* of its products through a wide variety of activities and programs.

National Association of Meat Purveyors (NAMP) An association representing the interests of *red meat, poultry*, seafood, and game processors (see *processing*) who sell mainly to the *foodservice* industry. NAMP is headquartered in Reston, Virginia.

National Association of State Departments of Agriculture (NASDA) A nonprofit, nonpartisan organization comprised of the 50 state departments of *agriculture* and those from the trust territories of Puerto Rico, Guam, American Samoa, and the Virgin Islands. The members of the organization are commissioners, secretaries, and directors of the departments of agriculture in the 50 states and four trust territories. NASDA's purpose is to provide a voluntary, nonpolitical organization that promotes unity and efficiency in administration of agricultural statutes and regulations, establishes federal-state cooperative programs to promote agricultural interests, and ensures cooperation between the departments of comparable agencies, with the U.S. Department of Agriculture and its officials and with persons interested in or engaged in agriculture. The 54 members of NASDA are organized into four regional organizations: the Northeastern Association of State Departments of Agriculture (NEASDA), the Southern Association of State Departments of Agriculture (SASDA), the Midwestern Association of State Departments of Agriculture (MASDA), and the Western Association of State Departments of Agriculture (WASDA).

National Association of Wheat Growers (NAWG) An organization formed in 1950 primarily to keep U.S. *wheat* growers aware of the legislative and regulatory issues affecting their *farm* operations. The organization is engaged in political action and lobbying efforts affecting wheat *exports*. NAWG is comprised of 20 state associations representing Arizona, Arkansas, California, Colorado, Idaho, Kansas, Kentucky, Minnesota, Montana, Nebraska, New Mexico, North Carolina, North Dakota, Oklahoma, Oregon, South Dakota, Texas, Virginia, Washington, and Wyoming. NAWG is headquartered in Washington, D.C.

National Audobon Society The largest grassroots environmental (see *environment*) organization in the United States, coordinating the efforts of scientists, activists, lobbyists, teachers, and naturalists.

National Bank for Cooperatives (CoBank) See *Banks for Cooperatives*.

National Broiler Council (NBC) An organization of *producers* and *processors* of *broilers* and their suppliers formed in 1954 and incorporated in 1955. NBC represents the industry in communication with Congress and federal departments. The council also supports *consumer* education and public relations programs to increase consumer acceptance and sales of broilers.

National Cattlemen's Association (NCA) A nonprofit trade association formed in 1977 to represent the interests of all segments of the U.S. beef industry, including *cattle* breeders, *producers*, and feeders. NCA's headquarters is in Denver, Colorado.

National Center for Food and Agricultural Policy (NCFAP) A research center that studies interrelated national public policy issues involving *agriculture*, food, nutrition, international trade, natural *resources*, and the *environment*. The NCFAP was established by *Resources for the Future*, an independent, nonprofit research and educational organization.

National Corn Growers Association (NCGA) An organization of *corn* growers founded in 1957 to encourage corn consumption, marketing (see *agricultural marketing*), and *production*.

National Cotton Council of America (NCC) The central organization representing all seven sectors, or interests, of the raw *cotton* industry of the United States: *producers*, ginners (see *gin*), warehouses, merchants, *seed* crushers (see *crushing*), *cooperatives*, and manufacturers (spinners [see *spinning*]). NCC is a voluntary, private industry association established in 1939. NCC programs include technical services, *foreign* operations, communication services, economic services, and government liaison. The NCC is headquartered in Memphis, Tennessee.

National Cottonseed Products Association (NCPA) An organization, founded in 1897, of oil (see *fats and oils*) mills, *refiners* (see *refining*), dealers, brokers, chemists, and others involved in the manufacture of *margarine*, cooking fats (see *fats and oils*), soaps, lubricants, *feed* for *cattle*, and *fertilizer*. The NCPA maintains uniform trading rules regarding buying, selling, weighing, sampling, and analyzing *cottonseed* and its products and conducts extensive research to increase *processing* efficiency and improve the usefulness of cottonseed products.

National Council of Farmer Cooperatives A nationwide association of *cooperative* businesses owned and controlled by farmers. Its membership includes major *agricultural marketing*, *supply*, and credit *cooperatives*, plus 32 state councils of cooperatives.

National Dry Bean Council (NDBC) An organization of state and regional *bean* dealers, growers, and shippers associations founded in the early 1940s to promote the sale and consumption of dry beans (see *dry edible beans*) in domestic and international *markets*. Formerly known as the National Dried Bean Council.

National Environmental Policy Act (NEPA) A law requiring federal agencies to prepare environmental (see *environment*) impact statements for "every recommendation or report on proposals for legislation and other major federal actions significantly affecting the quality of the human environment."

National Farmers Organization (NFO) A nationwide collective bargaining and marketing (see *agricultural marketing*) agency for farmers that evolved in 1955 from a *farm* protest movement.

National Farmers Union (NFU) A bipartisan, general *farm* organization of family farmers (see *family farm*). Officially known as the Farmers Educational and Cooperative Union of America, the Farmers Union monitors legislative developments.

National farm program acreage The number of *harvested acres* of *feed grains*, *wheat*, *upland cotton*, and *rice* needed nationally to meet domestic and *export* use and to accomplish any desired increase or decrease in *carryover* levels. Program acreage for an individual *farm* is based on that farm's share of the national farm program acreage.

National Fisheries Institute (NFI) An organization of *producers*, distributors, processors (see *processing*), *wholesalers*, *importers*, *exporters*, and canners of fish and shellfish that was founded in 1945.

National Food Brokers Association (NFBA) A full-service international trade association serving *food brokers* and the manufacturers they represent. Founded in 1904, the NFBA membership is composed of approximately 1,500 *broker* firms employing 40,000 personnel representing a broad range of product categories.

National Food Processors Association (NFPA) The principal scientific and technical trade association for the food industry. The science-based association represents manufacturers of packaged and processed (see *processing*) fruits and vegetables, juices and drinks, meat and *poultry*, seafood, and specialty products on legislative, regulatory, and *consumer* issues. The NFPA was founded in 1907.

National forest A federal reservation dedicated to protection and management of natural *resources*, under the concept of multiple use, for a variety of benefits, including water, *forage*, wildlife *habitat*, wood, recreation, and minerals. National forests are administered by the *U.S. Department of Agriculture*'s *Forest Service*.

National Frozen Food Association (NFFA) An organization representing the frozen food industry whose purpose is to advance the frozen food industry and expand the purchase and use of frozen foods. Members include distributors, *packer*/processors (see *processing*), retail supermarket (see *foodstore*) *chains*, *foodservice* operators, *brokers*, suppliers, and warehousers of frozen foods. The NFFA was founded in 1945.

National Grain and Feed Association (NGFA) An organization of *grain* and *feed* companies working together to achieve mutual goals and resolve common problems. NGFA members include *country elevators*, *subterminal elevators*, *terminal elevators*, feed mills (see *milling*), *exporters*, *processors*, *corn* and *flour* mills, *cash grain* and feed merchants, *commodity* futures (see *futures trading*) commission merchants, as well as allied industries such as banks, railroads, grain exchanges, and equipment manufacturers.

National Grain Trade Council (NGTC) A national trade association founded in 1930 that advocates and defends, consistent with public interest, the principles and merits of open and competitive *markets* for the distribution of agricultural *commodities*. Regular, policymaking members include *grain* exchanges, boards of trade, and national grain marketing organizations. Associate members encompass a cross-section of the grain industry and related businesses, including individual grain companies, *milling* and *processing* firms, transportation companies, futures (see *futures trading*) commission merchants, banks, and ports.

National Grange A fraternal and social organization for farmers founded in 1867 as the Patrons of Husbandry. The Grange has strong interests in *farm* issues but is also involved in many community and family projects through local, county, and state Grange units. The National Grange is headquartered in Washington, D.C.

National grassland *Land* with mainly grass and shrub cover administered by the *Forest Service* as part of the National Forest System for promotion of grassland *agriculture*, *watersheds*, grazing (see *graze*), wildlife, and recreation.

National Grocers Association (NGA) A trade association of independently operated retailers, retailer-owned cooperatives, and voluntary wholesale distributors engaged primarily in the sale and distribution of food and related products. The NGA was founded in 1982.

National Institute of Oilseed Products (NIOP) An organization of *importers* and dealers of copra, *coconut*, and *palm oil* and their products that establishes and maintains trading rules. The NIOP was founded in 1934.

National Milk Producers Federation (NMPF) An association of dairy *cooperatives* and federations of dairy cooperatives accounting for most of the *milk* marketed (see *agricultural marketing*) by farmers. The NMPF was founded in 1916.

National Onion Association (NOA) An organization of growers, shippers, and suppliers in the *onion* industry that was founded in 1913.

National Organic Standards Board A group created by the *Food, Agriculture, Conservation, and Trade Act of 1990 (P.L. 101-624)* (see Appendix 3) to propose allowable and prohibited substances in food marketed as organically grown. The board is composed of organic farmers (see *organic farming*), organic food *processors*, retailers, environmentalists (see *environment*), *consumer* advocates, scientists, and certifying *agents*. The board will make recommendations concerning residue (see *insecticide residue*, *drug residue*, and *pesticide residue*) testing and *tolerance levels*.

National Pecan Marketing Council (NPMC) An association of U.S. commercial *pecan* growers and others interested in promoting pecans. The NPMC was founded in 1979 to help expand the use of pecans.

National Peanut Council (NPC) An association of *peanut* growers, shellers, *brokers, processors*, manufacturers, and businesses providing goods and services to the peanut industry. Founded in 1940, the NPC conducts research and collects information on peanut *production* and marketing.

National Pork Council Women (NPCW) An association of pork *producers*' wives and other women interested in pork *production*. The NPCW was founded in 1962 to support the pork industry, increase consumption of pork products, and address issues affecting the pork industry.

National Pork Producers Council (NPPC) A federation of state and local pork *producer* associations that promotes the interests of pork producers. NPPC was founded in 1954. Formerly known as the National Swine Growers Council. See also *National Pork Council Women*.

National Potato Council (NPC) An organization of commercial *potato* growers founded in 1948 to work on national potato legislative and regulatory issues.

National Program Acreage (NPA) The number of *harvested acres* of *feed grains*, *wheat*, *cotton*, and *rice* needed nationally to meet domestic and *export* use and to accomplish a desired increase or decrease in *carryover* levels. Program acreage for an individual *farm* is based on a *producer*'s share of *national farm program acreage*, except when an *acreage reduction program* has been announced.

National Resources Inventory A survey of *soil*, water, and related information conducted by the *U.S. Department of Agriculture*'s (USDA) *Natural Resources Conservation Service*.

National Restaurant Association (NRA) A leading national trade association for the *foodservice* industry, founded in 1919. The association provides members with a wide range of education, research, communications, convention, and government affairs services.

National Rural Electric Cooperative Association (NRECA) An organization founded in 1942 to represent the national interests of *rural* electric systems. NRECA also actively assists the development of rural electrification overseas.

National Rural Housing Coalition (NRHC) An organization of *rural* community activists, public officials, and nonprofit developers formed to work for better housing and communities for low-income rural people.

National Rural Telecommunications Cooperative (NRTC) An organization formed in 1986 to foster the development and growth of satellite technology in the *rural* United States.

National School Lunch Program (NSLP) The oldest and largest U.S. child-feeding program, which provides financial and *commodity* assistance for meal service in public and nonprofit private high schools, grade schools, and under, as well as public and private licensed nonprofit, residential, child-care institutions. All children may participate in the NSLP. Based on household income poverty guidelines, a child may receive a free, reduced-price, or full-price meal. Free and reduced-price meals are subsidized by the U.S. government.

National Soft Drink Association (NSDA) An organization of persons or firms engaged in the manufacture and marketing of soft drinks and equipment, products, supplies, or services to the soft drink industry. NSDA monitors legislative developments affecting the industry and provides technical, management, and leadership training.

National Soybean Crop Improvement Council (NSCIC) An association organized by the soybean *processing* industry to promote *production* and improvement of *soybeans.*

National Soybean Processors Association (NSPA) An association, founded in 1930, of *producers* of *soybean oil* and *soybean meal.* Member firms account for about 95 percent of the *soybeans* processed (see *processing*) in the United States.

National Sugarbeet Growers Association See *American Sugarbeet Growers Association (ASGA).*

National Swine Growers Council (NSGC) See *National Pork Producers Council (NPPC).*

National Turkey Federation (NTF) An association of turkey (see *poultry*) growers, hatcherymen, *egg* producers, *processors,* and marketers founded in 1939. The NTF works to increase turkey consumption, promote favorable legislation, and distribute information relevant to the turkey industry.

National Wildlife Federation The largest private, nonprofit, conservation education organization in the United States. The federation was founded in 1936 to "bring together all interested organizations, agencies, and individuals on behalf of restoration of *land,* water, forests, and wildlife *resources.*"

National Wool Act Legislation that provided *price support* for *shorn wool* and *mohair* at an incentive level to encourage *production.* The law also provided for a payment on sales of unshorn *lambs.* The *National Wool Act of 1954, Amendment (P.L. 103-130)* (see Appendix 3), passed in 1993, repealed the National Wool Act effective December 31, 1995, providing a two-year phase out period for the wool and mohair program.

National Wool Growers Association (NWGA) A federation of state and regional associations of *lamb* and *wool* ranchers and farmers. The NWGA sponsors relevant industry research and handles major legislative activities involving *producers* interests. See *National Wool Act.*

Native pasture Unimproved or nonintensively managed open (nonforested) *pasture.* Also applies to plant materials from predominantly native or escaped (introduced but unintentionally spread) *species,* grazed (see *graze*) from nonintensively managed pastureland.

Natural fibers *Fibers* of animal (such as *wool,* hair, or silk), vegetable (such as *cotton, flax,* or *jute*), or mineral origin (such as asbestos or glass).

Natural Resource Conservation Service (NRCS) A *U.S. Department of Agriculture* (USDA) agency created in 1994 by merging the *Soil Conservation Service* and the *Agricultural Stabilization and Conservation Service's* conservation *cost-sharing* programs. The NRCS is responsible for developing and carrying out national soil and water conservation programs in cooperation with landowners, *farm operators,* and others.

Navy bean A *variety* of *dry edible bean* used primarily in making canned baked *beans* and navy bean soup. Roughly 90 percent of domestic navy beans are canned.

Near futures First maturing *contract* month in *futures contracts*. See *distant futures*.

Near-infrared reflectance spectroscopy (NIRS) An analytical technique that can determine the structure of compounds and the composition of substances by examining them with a spectroscope that is designed to operate in the infrared region of the spectrum. One application of this technique is the measurement of moisture and *protein* percentages in *wheat*. See also *spectrophotometer*.

Necrosis Generally, the localized death of plant or animal tissues or organs. However, the condition can also be extensive.

Nectar The sweet secretion, primarily a solution of dissolved *sugars* in varying proportions, produced in the *nectaries* of many flowering plants and the basic raw product of *honey*. The function of nectar is to attract *honeybees* so that the flowers may be cross-pollinated (see *cross-pollination* and *pollination*).

Nectaries Special glands found primarily in flowers that secrete *nectar*.

Nectarine A smooth-skinned fruit similar to the *peach*. The United States, Canada, Mexico, the *European Union*, Argentina, Australia, and Chile are the major nectarine-producing nations.

Negotiated pricing When the buyer and seller of a specified product bargain to determine the *price*, quantity, and terms of trade.

Nematocide Any substance used to kill *nematodes*.

Nematode Generally, a plant *pathogen*, though some nematodes are animal or human *pathogens*. The nematodes that cause plant *disease* are worms that pierce the *cells* of plant roots and suck up the juices or transmit other pathogens, such as viruses. Plant nematodes are abundant in many *soils* and can destroy plant roots. See *nematocide*.

Neps Very small, snarled masses or clusters of *fibers* that look like dots or specks in the *lint* of *cotton* and are difficult to remove. If not removed, they will appear as defects in the *yarn* and fabrics (see *cloth*). See also *naps*.

Net cash farm income The cash available to a *farm* operation after all cash expenses have been paid (including interest). It is the income available for living expenses, principal repayment, income taxes, and other expenses or reinvestment in the farm business. It indicates the availability of funds to cover cash operating costs, finance capital investment and savings, service debts, maintain living standards, and pay taxes. Net cash farm income does not reflect the cash position of farm families because savings, wages paid to family members, and *off-farm income* are not included.

Net cash flow A financial indicator that measures the cash available to the *farm* sector in a given year. It indicates the ability to meet current obligations and provide for family living expenses. Net cash-flow is calculated as the sum of *gross cash income*, the change in loans outstanding, net rent to nonoperator landlords, and the net change in farmers' currency and demand deposits; minus gross cash expenses and gross capital expenditures.

Net cash household income A measure of the funds available to the *farm* household, after cash business and family living expenses are met, for business expansion, further consumption, savings, or other obligations. Net cash household income is calculated as family *off-farm income* plus *net* *cash farm income* minus an estimate of principal repayments and a family living allowance. Principal payments estimated for each operation are based on the amount of real estate and nonreal estate debt owed to each lender and are consistent with standard debt repayment schedules.

Net energy The amount of digestible energy actually used by an animal for maintenance and *production.*

Net exporter A country in which the value of *exports* exceeds the value of *imports.* The country has a positive *trade balance.* See also *net importer.*

Net farm income A measurement of the *farm* sector's profit or loss associated with a farm's *production* during a given year. It is an approximation of the net value of agricultural production, regardless of whether the *commodities* were sold, fed, or placed in inventory (see *inventories*) during the year. Net farm income equals the difference between *gross farm income* and total production expenses. It includes nonmoney items such as depreciation, the consumption of farm-grown food, and the net imputed rental value of operator dwellings. Additions to inventory are treated as income. Over time, net farm income shows a farm's ability to survive as a viable business.

Net importer A nation in which the value of *imports* exceeds the value of *exports.* The country has a negative *trade balance.* See also *net exporter.*

Net market price The sales *price* or other per-unit value received by a *producer* for a *commodity* after adjustments are made for a premium or a discount based on *grading* or quality factors.

New York Coffee, Sugar, and Cocoa Exchange, Inc. An organization in which world and domestic raw cane *sugar* contracts (see *contract*) are traded daily. The *world price* is the *No. 11 contract price* for raw cane sugar (*f.o.b.* Caribbean), and the domestic *price* is the *No. 14 contract price* for raw cane sugar (*c.i.f.*, duty/fee-paid, New York).

New York Cotton Exchange See *cotton exchange.*

New York dressed weight The *carcass* weight of a bird (see *poultry)* after only the feathers and blood have been removed. This weight is approximately 90 percent of *liveweight.* Sales of New York dressed poultry were common during the 1940s and 1950s.

Newcastle disease A highly contagious viral *disease* of *poultry*, characterized by respiratory and nervous disorders. Trade in live birds and poultry products from affected areas may be restricted.

Newly industrialized country (NIC) A country, previously considered less developed, that has achieved a significant level of economic development, largely through heavy industrialization. Examples include South Korea, Taiwan, and Brazil.

Ninety-Six-Degree Basis (96°) A computed weight of *sugar* determined by dividing the weight of its *sucrose* content by 96 percent. This is the standard basis for publication of most *U.S. Department of Agriculture* statistics.

Nitrogen A chemical element essential to life and one of the primary plant *nutrients.* Animals get nitrogen from *protein* feeds (see *feed*), plants get it from *soil*, and some *bacteria* get it directly from air.

Nitrogen cycle The sequence of biochemical changes undergone by *nitrogen* wherein it is used by a living organism, liberated upon the death and decomposition of the organism, and converted to its original state of oxidation. The nitrogen cycle is a critical part of the growth and metabolism of plants and *microorganisms*.

Nitrogen fixation A biological process primarily in plants by which certain *bacteria* convert *nitrogen* in the air to *ammonia* to form an essential *nutrient* for growth.

No. 11 Contract price An *f.o.b.*, Caribbean *price* for raw cane *sugar* as traded on the *(New York) Coffee, Sugar, and Cocoa Exchange*. It is usually referred to as the *world price* for sugar and is traded in both *spot markets* and *futures markets*.

No. 12 Contract price The *c.i.f.* duty/fee-paid (see *tariff*) New York *price* for imported or domestic raw cane *sugar* as traded on the *(New York) Coffee, Sugar, and Cocoa Exchange*. It stopped being traded on the *spot* market on May 31, 1985, and was no longer traded on the *futures market* after October 8, 1986. It had been used in conjunction with the market stabilization price to calculate *import* fees.

No. 14 Contract price The *c.i.f.* duty/fee-paid New York *price* for imported raw cane *sugar* as traded on the *(New York) Coffee, Sugar, and Cocoa Exchange*. It is traded only on the futures market (see *commodity futures exchange*) and commenced on July 8, 1985. It is now usually referred to as the domestic price for raw cane sugar.

N-OIL See *carbohydrate-based fat substitute*.

Nominal dollars A measure of *prices*, income, and other monetary values in the prevailing year. Because the measure does not consider the effects of inflation, current dollars cannot be meaningfully compared over time. See *real dollars*.

Nonairtight upright silo An unsealed (conventional), upright, cylindrical tower (*silo*) in which air circulation is retarded only by the density of the *ensiled* feedstuffs (see *feed*).

Nonarable soils See *land capability*.

Nonbin dryer An on-*farm*, high-capacity, high-temperature *grain* drying system. See also *crop dryer* and *on-farm dryer*.

Noncellulosic fibers The group of *fibers* that includes *synthetic* and inorganic fibers. Such fibers are made from petroleum-derived chemicals. The major types are polyester, nylon, acrylic, and polypropylene.

Noncentrifugal sugar Crude *sugar* made from the *sugarcane* juice by evaporating and draining off the *molasses*.

Noncompetitive imports Agricultural products purchased from *foreign* countries because they cannot be grown profitably on a large scale in the United States. These products include *coffee*, *cocoa*, rubber, and *bananas*. Also referred to as complementary imports.

No net cost A provision requiring that the *sugar* and tobacco *price support programs* be operated at no cost to the federal government. Under the *Food, Conservation, and Trade Act of 1990 (P.L. 101-624)* (see Appendix 3), the sugar program must be operated so that there are no forfeitures of sugar to the *Commodity Credit Corporation*. The *No-Net-Cost Act of 1982* requires that, to be eligible for price support, *producers* of all kinds of *tobacco*, beginning with the 1982 crop of tobacco, must pay assessments to an account established by the *cooperative* association that makes federal support loans (see *nonrecourse loans*) available to producers. The funds are collected to cover potential losses in operating the price support program.

No-Net-Cost Act of 1982 See *no net cost.*

Nonfarm foods Foods not originating on U.S. *farms*, such as imported (see *import*) products and seafood.

Nonfarm income See *off-farm income.*

Nonfat dry milk (NDM) Dried (see *drying*) skim milk (see *milkfat content*) containing no more than 1.5 percent fat (see *fats and oils*) and 5 percent moisture. The term includes *buttermilk* powder but not *whey* powder. Also abbreviated NFDM or SMP (skim *milk* powder).

Nonfed cattle *Livestock* that do not enter *feedlots* prior to *slaughter* but are kept on *forage* for *feed.*

Nongrade-determining factors Factors that influence the quality of *grain* but that are not taken into account in the *grading* of grain. They must be reported as information whenever an official inspection is made.

Nonindustrial private forest lands Privately owned *rural* lands (see *land*) with existing tree cover or suitable for growing trees.

Nonindustrial rapeseed See *canola.*

Noninsured Assistance Program (NAP) A standing disaster assistance program for crops not yet covered by a federal crop insurance (see *Federal Crop Insurance Program*). The NAP provides coverage similar to catastrophic crop insurance coverage, but is triggered by a 35 percent area loss. Once the area-loss threshold is met, farmers are paid for their crop losses in excess of 50 percent at a rate of 60 percent of the average *market* price. NAP is new with the Federal Crop Insurance Reform Act of 1994 and will be in place for 1995 crops. It is permanent rather than ad hoc (see *Ad Hoc Disaster Assistance*).

Nonmarket economy See *centrally-planned economy.*

Nonmetro areas Counties not within metro areas. See *Metropolitan Area (MA).*

Nonmoney farm income A statistical allowance used in *farm income* compilations to credit farmers with income for the value of *farm* products produced and consumed on the farm (instead of being sold for cash), including the rental value of services from farm dwellings. It assumes farmers otherwise live rent-free on their farm business enterprises.

Nonpoint-source pollution Pollutants that cannot be traced to a specific source, such as storm water *runoff* from *urban* and agricultural areas.

Nonprice competition Competition among sellers (buyers) involving factors other than *price*, such as *product differentiation.*

Nonprogram crops Crops, such as *potatoes*, vegetables, fruits, and *hay*, that are not included in federal *price-support programs*. Also includes crops other than *program crops* (*wheat*, *feed grains*, *upland cotton*, and *rice*) that are eligible for price support benefits, such as *soybeans* and minor *oilseeds.*

Nonreal estate debt *Farm* debt not secured by real estate.

Nonrecourse loans The major *price-support* instrument used by the *Commodity Credit Corporation (CCC)* to support the *price* of *wheat, feed grains, cotton, honey, peanuts, tobacco, rice, oilseeds,* and *sugar.* Farmers or *processors* who agree to comply with each *commodity* program provision may pledge a quantity of a commodity as collateral and obtain a loan from the CCC. The borrower may elect either to repay the loan with interest within a specified period and regain control of the collateral commodity, or default on the loan. In case of a default, the borrower forfeits without penalty the collateral commodity to the CCC. The loans are nonrecourse because the government has no option (or recourse) but to accept forfeiture as full satisfaction of the loan obligation, including the accumulated interest, regardless of the price of the commodity in the *market* at the time of default.

Nonruminant A simple-stomached animal that does not ruminate (see *ruminant*). Examples are *swine, poultry, horses,* dogs, and cats.

Nontariff trade barriers Regulations, other than traditional customs duties (see *tariff*), used by governments to restrict *imports* from and *exports* to other countries. *Embargoes, import quotas, licensing, variable levies, state trading,* and unnecessary sanitary labeling and health and sanitary standards are examples of the types of nontariff trade barriers that have increased since the end of World War II, while tariff rates have declined significantly.

Nontillable pasture *Forages* for grazing (see *graze*) produced on *land* that cannot be tilled (see *till*) for crop *production.*

Nonwoven fabrics Material made primarily of randomly arranged *fibers* held together by an applied bonding agent, by fusion, or by entanglement (interlacing).

Normal climate The average of the measured site-specific daily *weather* elements, such as temperature and rainfall, for a 30-year period; used by the World Meteorological Organization to describe climatic conditions. See *climate.*

Normal crop acreage (NCA) The acreage (see *acre*) on a *farm* normally devoted to a group of designated crops. A farm's total planted acreage of these designated crops, plus any *set-aside,* cannot exceed the NCA if a farmer wants to participate in the federal farm programs. NCA has not been used as the basis for *commodity* programs since 1980.

Normal flex acres A provision of the Omnibus Budget Reconciliation Act of 1990 *(P.L. 101-508)* and the *Food, Agriculture, Conservation, and Trade Act of 1990 (P.L. 101-624)* (see Appendix 3) that requires a mandatory 15-percent planting flexibility for program participants. Under this provision, producers are ineligible to receive *deficiency payments* on 15 percent of their *crop acreage base* (not including any acreage removed from *production* under any production adjustment program). Crops on normal flex acreage are still eligible for *nonrecourse loans* and *marketing loans.* Producers are allowed to plant any crop, except fruits, vegetables, and any crop prohibited for planting by the secretary of agriculture, on these acres. Also referred to as *triple base.* Producers may plant alternative crops on up to 10 percent additional base acreage if they agree to forego *deficiency payments.* See *optional flex acres.*

Normal soil A *soil* having a profile in near equilibrium with its *environment.*

Normal yield The average historical *yield* established for a particular *farm* or area. Normal *production* is calculated as the normal acreage harvested (see *harvest*) of a *commodity* multiplied by the normal yield.

Norm price See *indicative price.*

No-roll Ungraded (see *grades* and *grading*) steer or *heifer* beef. Some *carcasses* are not graded, and others that are graded are not *rolled* or marked with the grade received. (When carcasses are graded, the quality and *yield* grade mark is rolled the length of the carcass.) Most no-rolls would be Select (formerly Good) grade. See *Prime, Choice,* and *Select.*

North American Export Grain Association (NAEGA) An association, founded in 1920, of U.S. and Canadian *grain* exporters. The NAEGA headquarters are in Washington, D.C.

North American Free Trade Agreement (NAFTA) A trade agreement between Canada, the United States, and Mexico implemented on January 1, 1994. See Appendix 4.

Northeastern Association of State Departments of Agriculture (NEASDA) See *National Association of State Departments of Agriculture.*

Northern wheat See *hard red spring wheat.*

No-till A farming practice that involves seeding (see *seed*) the ground in one operation, leaving residue (see *crop residue*) from the previous crop on the surface to prevent soil *erosion*. No other *soil* tillage (see *till*) operations are conducted from the time of harvesting (see *harvest*) the last crop until the current seeding operation. Also called slot tillage, zero cultivation, or no-tillage. See *conservation tillage* and *minimum tillage.*

No-tillage See *no-till.*

No-till drill A *drill* of heavy construction, designed to sow *grain* in *soil* not tilled since the last crop was harvested (see *harvest*). The drill is generally either an *air drill* or a specially constructed drill that uses some variation of *double-disk openers*. Heavy construction is needed to penetrate the ground, which may become packed and hard due to lack of tillage from harvesting to seeding operations.

Nurse bees Three- to ten-day-old worker *honeybees* that work inside the *hive*. In particular, the nurse bees feed the *larvae.*

Nursery plants Plants at all stages of growth that consist of trees, shrubs, ground covers, vines, and fruit and nut plants. Some nursery plants are field grown and sold as "bare-root" or "balled and bagged" nursery stock, while some are grown and sold in containers. Sales from plants bought and resold within 30 days by landscape businesses or garden center outlets and *cut Christmas trees* are excluded. Examples of nursing plants are *deciduous* shade trees, flowering trees and shrubs, evergreen trees and shrubs, *citrus fruit* and other fruit and nut trees, plants for commercial and home *orchards*, rose bushes, ornamental grasses, waterlilies, and ground covers such as ivy, phlox, and portulaca.

Nutrient *Carbohydrates*, fats (see *fats and oils*), *proteins*, minerals, *vitamins*, water, and other chemical elements or compounds essential for normal body metabolism, growth, and *production.*

Nutrient, plant Any element taken in by a plant, essential to its growth, and used by it in elaboration of its food and tissue.

Nylon See *synthetic fibers.*

O

Oats A grass whose *seeds* are used for fodder and *foods,* such as *cereal.* The major producers of oats include the former Soviet Union, the United States, Canada, and the *European Union.*

Oca A hardy tropical plant grown in Africa and Asia. It can be served boiled, baked, or fried.

Ocean freight differential The difference in freight costs between two carriers or destinations. For the *Public Law 480* (P.L. 480) program, the amount by which the ocean freight for the portion of *commodities* required to be carried on U.S. flag vessels exceeds the cost of carrying the same amount on *foreign* flag vessels. Under P.L. 480, that difference is paid by the *Commodity Credit Corporation* (CCC).

Offal The edible or inedible meat from organs (liver, kidney, etc.) or from parts of the thoracic and abdominal cavities of animals. More generally, offal is the *byproduct* of preparation of some specific product, minus the valuable products and byproducts of *processing.*

Off-beat trades Trades of *commodities* or products not normally associated with general trading practices.

Offer See *bid.*

Off-farm dryers High-capacity, high-temperature, commercial *grain* dryers (see *crop dryer*) that are used away from the *farm.* These fall into three categories: *crossflow*, *concurrent-flow*, and *mixed-flow dryers.*

Off-farm income All income from nonfarm sources received by *farm operator* households.

Office of Agricultural Biotechnology An office of the *U.S. Department of Agriculture* (USDA) that promotes and coordinates *biotechnology*-related regulatory and research activities within the Department.

Office of Energy (OE) An office of the *U.S. Department of Agriculture (USDA)* that coordinates the Department's energy policy, including all issues dealing with agricultural energy use and the impacts on *agriculture* of energy development and use. In 1994, OE was incorporated into USDA's *Economic Research Service.*

Office of International Cooperation and Development (OICD) A *U.S. Department of Agriculture* (USDA) agency that coordinated agricultural training and research programs carried out in cooperation with other nations. OICD sponsored international research projects and scientific and technological exchanges with other nations on topics of interest to U.S. farmers and *agribusiness.* In

1994, OICD was incorporated into the USDA's *Foreign Agricultural Service.*

Office of Management and Budget (OMB) A federal agency responsible for evaluating, formulating, and coordinating management procedures and program objectives within and among federal departments and agencies. OMB also controls the administration of the federal budget and provides the U.S. president with recommendations regarding budget proposals and relevant legislative enactments.

Office of Public Affairs (OPA) A *U.S. Department of Agriculture (USDA)* agency that provides leadership, expertise, and counsel for the development of public affairs strategies necessary for the overall formulation, awareness, and acceptance of USDA programs and policies.

Office of Small-Scale Agriculture An office of the *U.S. Department of Agriculture* (USDA) created as part of the *Cooperative State Research Service (CSRS)* to increase and improve the flow of information about small-scale farming (see *small-scale farms*) to agricultural *producers* and consumers. The office functions as a liaison to other USDA agencies, such as the *Extension Service* and the *Agricultural Research Service (ARS)*, in identifying and directing research and education programs to improve the usefulness of information received by all small-scale farmers.

Office of Technology Assessment (OTA) A nonpartisan support agency serving Congress established in 1972 to provide analysis of the many impacts of technological developments, including the social, biological, and economic effects.

Office of the Consumer Advisor An office of the *U.S. Department of Agriculture (USDA)* that coordinates USDA actions on problems and issues of importance to *consumers*. The Office of the Consumer Advisor represents the Department in policy discussions related to consumer issues before Congress, in meetings with other departments and agencies, and in various public forums. In October 1994, the Office of the Consumer Advisor was merged into the newly created *Food and Consumer Service Agency.*

Office of the U.S. Trade Representative See *U.S. Trade Representative.*

Office of Transportation (OT) A *U.S. Department of Agriculture (USDA)* agency created in 1978 to handle transportation concerns and issues affecting farmers and *rural* communities. OT represents agricultural interests on transportation matters before regulatory agencies and performs related research and other activities. In 1991, OT was incorporated into the *Agricultural Marketing Service (AMS)* to improve the coordination of USDA's transportation and marketing policies.

Offset requirement A *countertrade* whereby an importing (see *import*) country requires an *exporter* to purchase a specific amount of locally produced goods or services from the importing country. Alternatively, the exporter may be required to establish manufacturing facilities in the importing country.

Offsetting compliance A program that requires that a *producer* participating in a *paid land diversion* or *acreage reduction program* must not offset that reduction by planting more than the *acreage base* for that crop on another *farm* under the same management control. Offsetting compliance provisions were used as recently as the late 1970s. The *Food, Agriculture, Conservation, and Trade Act of 1990 (P.L. 101-624)* (see Appendix 3) does not require offsetting compliance as a condition for program eligibility.

Oil cake See *oilseed meal.*

Oil palm An African palm with fruits whose flesh and *seeds* yield *palm oil.* Major producers of oil palm include Malaysia, Indonesia, West Africa, and China.

Oils See *fats and oils.*

Oilseed A *seed* or crop grown largely for its *oil.* See *oilseed crops.*

Oilseed crops Primarily *soybeans*, *peanuts*, *cottonseed*, *sunflower seeds*, and *flaxseed* used for the *production* of edible and/or inedible *oils*, as well as high *protein meals.* Other oil crops include *rapeseed*, *safflower*, *castorseed*, *sesame*, and *mustard seeds.*

Oilseed hulls The outer covering of *oilseeds.*

Oilseed meal The product obtained by grinding (see *grind*) the cakes, chips, or flakes that remain after most of the *oil* is removed from *oilseed crops.* Oilseed meals are mainly used as a *feed* for *livestock* and *poultry.* They are also used as a raw material in *processing* edible vegetable-*protein* products.

Oilseed processing Removing oil (see *fats and oils*) from *oilseeds.* There are three basic types of *processing*—solvent extraction (see *extraction, solvent*), mechanical processing, and *hydraulic pressing.*

Okra A long, thin, green vegetable popular in the southern United States. Florida dominates U.S. *production* of okra, while Texas, Georgia, and other southern states *market* small quantities in the spring and fall. *Imports* account for 80 percent of U.S. supplies. Mexico is the primary supplier, accounting for 90 percent of U.S. imports.

Old turkeys (breeders) Fully matured birds held for *egg* production, usually over 15 months of age.

Olein The liquid portion of *palm oil*; *stearin* is the solid portion.

Oligopoly A *market* characterized by a few large sellers with a high degree of mutual interdependence. See also *oligopsony.*

Oligopsony A *market* characterized by a few large buyers with a high degree of mutual interdependence. See also *oligopoly.*

Olive The small fruit of the semitropical evergreen tree, *Olea europaea.* Olives are used as a food and as a source of edible oil (see *fats and oils*). The Spanish-style (green) olive and the California-style (black) olive are the two main types. The major producers are Italy, Greece, Spain, and Tunisia.

Omnivore An animal that eats foods of animal and plant origin.

Once-over harvester A harvesting (see *harvest*) method that allows harvesting the entire crop with only one trip through the field. In the case of *tobacco*, for example, all the leaves are removed at one time and then *cured* together. This method has been used for *flue-cured tobacco*, but little is currently harvested in this manner.

Once-refined oil Oil (see *fats and oils*) that has been treated by alkali or caustic wash.

One-person baling The use of field pickup hay (see *hay*) *balers*, with self-tying attachments and *bale* ejectors, that allow one person to *harvest* hay crops.

One-way disk A *farm* implement with concave rotating steel plates, all designed to enter the *soil* at the same angle and turn or throw the soil in the same direction (one way). A one-way *disk* does not usually completely turn the soil over, leaving some stubble or *mulch* on the surface. Also referred to as one-way.

On-farm dryers *Grain* dryers (see *crop dryer* and *drying*) used by farmers to dry grain. On-farm dryers fall into three categories: *bin dryers*, *nonbin dryers*, and *combination dryers*.

On-ground pile storage Storage of *grain* placed in piles directly on the ground or on pads, either covered by a tarp or left uncovered. Piles can be contained by fixed or movable sloping walls or circular rings. Grain stored by this method is difficult to load, unload, aerate, and fumigate (see *fumigant*).

Onion A *bulb* or bulblike member of the lily family. *Varieties* include green onions (sometimes called scallions), *leeks*, shallots, and the red and white type (often referred to as *"dry" onions*). Major producers include the United States, Canada, Japan, Korea, and Italy.

Open contract A *contract* that has been bought or sold without the transaction having been completed by subsequent sale, repurchase, or actual delivery of the *commodity*.

Open-end spinning *Processing* fibers directly from a fiber *supply* (typically sliver) to the finished *yarn*, in contrast to *ring spinning*. "Open-end" refers to the fact that *fibers* are separated from the feedstock and individually reunited on the tail of the forming yarn. Advantages over ring-spun yarns include increased *production* speed, reduced labor, and floorspace requirements. Drawbacks are mainly related to lower yarn strength and a crisper feel to fabrics (see *cloth*). Of the three basic methods (mechanical, electrostatic, and fluid), only the mechanical principle of rotor spinning has achieved commercial success. Friction spinning, also based on the mechanical system, has had limited success.

Open-front shed A building that has walls on the ends and one side with *cattle* free to go through an open side to an adjacent *lot*.

Opening line A sequence of machines, particularly in *cotton* mills (see *milling*), used to blend and progressively reduce the size of the tufts of cotton (opening), allowing trash and other impurities to be released (cleaning). Ideally, the fibers should be sufficiently disentangled to permit the next process (*carding*) to occur without *fiber* damage.

Open-lot system A *feedlot* that does not provide buildings for the *livestock*.

Open policy See *marine insurance*.

Open-pollinated See *cross-pollinated crop*.

Open position A *futures contract* that has been entered into by an individual and not yet fulfilled by delivery or canceled by an opposite trade. The total number of long open positions (see *long position*) among all traders equals the total number of short open positions (see *short position*) for each *contract* month. This total is the total open interest for the contract month.

Open upright silo See *nonairtight upright silo*.

Operator (farm) See *farm operator.*

Option A right, without obligation, to buy or sell a *commodity* or asset (see *assets, current* and *assets, fixed*) at a specified *price* over a specified time period. See *commodity option.*

Optional flex acres A planting flexibility provision of the *Food, Agriculture, Conservation, and Trade Act of 1990 (P.L. 101-624)* (see Appendix 3) that allows *producers* to plant up to 25 percent of their *crop acreage base* to another crop (except fruits and vegetables) without losing base but to receive no *deficiency payments* on these *acres*. The Omnibus Budget Reconciliation Act of 1990 (P.L. 101-508) (see Appendix 3) made a 15 percent planting flexibility mandatory (see *normal flex acres*). The remaining 10 percent are the optional flex acres.

Option contract See *option.*

Option grantor (writer) A person who sells an *option contract*, receives the *premium*, bears the obligation to buy (sell) the asset (see *assets, current* and *assets, fixed*) at the *strike price*, and guarantees performance on the obligation by making and maintaining a *margin* deposit with a *broker*.

Option premium The amount an *option* buyer pays the *option grantor* for an *option contract.*

Orange A *citrus* fruit, which may be eaten fresh or processed (see *processing*) into juice. Oranges are also used for such diverse products as a beverage base, peel, *pectin*, *essential oil*, and animal *feed*. Major producers include the United States, Brazil, Spain, Italy, Egypt, Mexico, Israel, and Morocco.

Orchard *Land* used for growing fruit or nut trees.

Order buyer An *agent* who purchases *commodities* for a fee according to the specifications of the buyer.

Orderly marketing agreement (OMA) An agreement between governments to limit *exports* to each other. Such agreements are used to prevent injury to domestic industries and are generally undertaken to avoid imposition of unilateral *import* restrictions.

Oregano A *perennial* plant used as a seasoning. Also known as wild marjoram and organy. Major producers include Mexico, Greece, and Turkey.

Organic farming A *production* system that completely or mostly excludes the use of synthetically compounded *fertilizers, pesticides,* or *growth regulators*. Instead, *crop residues*, animal *manures*, *legumes*, *green manure*, off-*farm* organic wastes, mechanical cultivation (see *cultivate*), mineral-bearing rocks, and aspects of biological pest control are used to maintain *soil* productivity, supply plant *nutrients*, and control insects, weeds, and other pests.

Organic soil A *soil* or a soil horizon that consists primarily of organic matter, such as *peat* soils, muck soils, and peaty soil layers.

Organization for Economic Cooperation and Development (OECD) An organization founded in 1961 to promote economic growth, employment, a rising standard of living, and financial stability; to assist the economic expansion of *developing countries*; and to further expand world trade. The member countries are Australia, Austria, Belgium, Canada, Denmark, Finland, France, Germany, Greece, Iceland, Ireland, Italy, Japan, Luxembourg, the Netherlands, New Zealand, Norway,

Portugal, Spain, Sweden, Switzerland, Turkey, the United Kingdom, and the United States.

Organization of African Unity (OAU) A group of all independent countries in Africa except the Republic of South Africa. The OAU was founded in 1963 to promote self-government, respect for territorial boundaries, and social progress.

Organization of American States (OAS) A regional organization established in 1948 to provide for the peaceful settlement of disputes, common action against aggression, and economic and social collaboration among members. Members include the United States, Mexico, and most Central American, South American, and Caribbean nations.

Organy See *oregano.*

Oriental tobacco A small-leaved *tobacco* with a distinctive aroma used primarily to produce Oriental or aromatic types of *cigarettes.* Oriental tobacco was the first type of tobacco used in cigarettes. About half the world *production* of Oriental tobacco enters international trade. The former Soviet Union (mainly Russia), Turkey, Greece, Bulgaria, and Italy are the major producers of Oriental tobacco. The United States *imports* most of its Oriental tobacco from Turkey.

Out-of-the-money option An *options contract* that cannot be profitably exercised at the current market *price.* This occurs when the market price exceeds the *strike price* for a *put option* or is less than the strike price for a *call option.*

Output The marketable *cash crops, livestock* products, or breeding stock of a farming (see *farm*) operation.

Overfinished *Cattle* with excessive fat-to-lean ratios. These cattle are often *price*-discounted.

Over-order payment A payment charged by a producer's (see *producer*) *cooperative* in excess of the minimum *price* specified by a *marketing order.* The payment is negotiated between buyers and sellers to cover the cost of providing *market* services or attracting *milk* away from manufacturing plants (see *plants or establishments).* Over-order payments could also result from market power. Over-order payments usually apply to *Class I milk.*

Overproduction See *surplus.*

Overseas Private Investment Corporation (OPIC) A U.S. government agency that assists U.S. investors in making profitable investments in *developing countries* while encouraging projects that enhance the social and economic development in those countries.

Ovine An animal of the *sheep* family.

Ovum A female *germ cell* (see also *cell*).

Owala See *pod seed.*

Oysterplant See *salsify.*

Oyster shells See *lime, agricultural.*

P

Package bees *Honeybee*s produced for sale, supplied by the pound, and transported in a box with a wire screen on two opposite sides. The most popular size packages contain 2 to 5 pounds of adult bees, without *brood* or *comb*. Also called combless package.

Packer 1: A firm that packages *commodities*, such as field-harvested (see *harvest*) fresh produce, and ships it to receivers, including *wholesalers*, retailers, institutions, and *foodservice* organizations.

2: A firm that slaughters *livestock*. These firms may or may not fabricate (see *fabrication*) products, process (see *processing*), or perform other functions.

3: A series of wheels, usually attached to another implement such as a *drill*, that presses *soil* on top of *seed*. The wheels may be steel, rubber, or rubber mounted on steel.

Packer brand A product that is packaged under the brand name of a *packer* and usually advertised.

Packers and Stockyards Act A law, administered by the *U.S. Department of Agriculture*'s (USDA) *Grain Inspection, Packers and Stockyards Administration*, that assures free and open competition and prohibits unlawful and deceitful employment practices as well as unfair business practices in the *livestock*, meat, and *poultry* trade. Specifically included are livestock markets (*terminal markets* and *auction markets*), livestock *market* agencies, livestock dealers, *meatpackers*, and live poultry dealers. Livestock markets, buying stations, dealers, *packers*, and poultry *processors* subject to the act must maintain accurate scales and weigh livestock, poultry, and meats accurately.

Packers and Stockyards Administration (PSA) The name of the former *U.S. Department of Agriculture* (USDA) agency responsible for enforcing provisions of the *Packers and Stockyards Act*, the Truth in Lending and Fair Credit Billing Acts, and the Equal Credit Opportunity Act with respect to firms subject to the Packers and Stockyards Act. The PSA was abolished in October 1994 and its functions transferred to the newly created *Grain Inspection, Packers and Stockyards Administration.*

Packer sales office 1: A local *wholesaling* organization for *packers*. A packer sales office does not physically handle meat but acts as a salesforce and as troubleshooters.

2: The sales office at a packing plant.

Packer-to-packer sales Sales from one *packer* to another, which may be necessary to fill an order. Some packers may also process (see *processing*) more than they *slaughter* or need *carcasses* of specific quality and size not met from their own slaughter.

Packer-shipper A firm that receives field-harvested (see *harvest) produce*, sorts, cleans, packs, and sometimes cools and ships to receivers, including retailers, *wholesalers*, institutions, and *foodser-*

vice establishments.

Paid land diversion A program that offers payments to *producers* for reduction of planted *acres* of a *program crop*, if the secretary of agriculture determines acres planted should be reduced. Farmers are given a specific payment per acre to idle a percentage of their *crop acreage base*. The idled acreage is in addition to an *acreage reduction program*.

Palatability The relative attractiveness of food or *feed*, including such factors as appearance, odor, taste, texture, and temperature.

Palmitic acid See *fatty acids (classification of)*.

Palm kernel The *seed* of the *oil palm*, used to produce palm kernel *meal* and palm kernel oil (see *fats and oils*) used in baking and industry. Major producers include Malaysia, Nigeria, Brazil, and Indonesia. See *palm oil*.

Palm oil The edible oil (see *fats and oils*) extracted (see *extraction, mechanical* and *extraction, solvent*) from the fruit of the *oil palm*. Malaysia, Indonesia, and West Africa are the major palm oil producers.

Pampas An extensive, generally grass-covered plain that is part of temperate South America east of the Andes.

Papain A proteinase in the juice of unripe *papaya* that is used as a meat tenderizer, stabilizer in beer, and in pharmaceutical products. Major papain producers include Uganda, Zaire, Tanzania, Sri Lanka, and India.

Papaya A fruit-bearing giant *herb* grown in tropics of the United States (Hawaii), Mexico, Brazil, the Philippines, Indonesia, India, and Zaire.

Parasite An organism that obtains its *nutrients* wholly or in part from another living organism. Examples include *bacteria*, viruses, and *yeasts*.

Parboiled rice *Rough rice* soaked in warm water under pressure, steamed, and dried before *milling*. In the United States, all parboiled rice is southern *long-grain rice*. Parboiled rice has superior *milling* qualities—fewer *kernels* are broken in the process—compared to regular, milled white rice. Parboiled rice is also easier to cook, is fluffier, and sticks together less than regular, milled white rice, although it takes longer to cook. Parboiled rice is desired by *consumers* who like a chewy, wholesome taste. Restaurants favor parboiled rice because it retains its shape, texture, and taste longer after cooking than regular, milled white rice.

Parity A measurement of the purchasing power of a unit (*bushel, pound, or hundredweight*) of *farm* product. Parity was originally defined as the *price* that gives a unit of a *commodity* the same purchasing power today as it had in the 1910–1914 base period. In 1948, the parity price formula was revised to allow parity prices for individual commodities to reflect a more recent relationship of farm and nonfarm prices by making the base price dependent on the most recent 10-year average price for commodities. Except for *wool, mohair*, and certain minor *tobaccos*, parity is not currently used to set price-support levels (see *price-support programs*) for any *program crops*. However, parity remains part of *permanent legislation*.

Parity index See *prices-paid index*.

Parity ratio A measure of the relative purchasing power of *farm* products. It is the ratio between the *prices-received index* by farmers for all farm products and the *prices-paid index* by farmers for *commodities* and services used in farm *production* and family living. The parity ratio measures relationships between prices received and prices paid.

Partial house brooding The use of only a portion of a *growout house* for *brooding*. The remainder of the house is closed to conserve energy until the birds are large enough to need the entire house, usually at around 2 to 3 weeks old.

Participation See *compliance*.

Partnership A business organized under the direction of two or more joint owners.

Passion fruit A tropical, *egg*-shaped fruit with a thick, hard shell and gelatinous, yellow orange *pulp*. The flavorful juice is used as a topping for desserts and as an addition to beverages.

Pasta Products made primarily from *durum wheat*, such as macaroni, spaghetti, and noodles.

Pasteurization A process of heating *milk* or other foods to kill all harmful *bacteria*. Milk, for example, is pasteurized by holding it at 140° F for 30 minutes or at 161° F for 15 seconds.

Pasture *Land* used for grazing (see *graze*); also deliberately established or naturally occurring plant material grazed. See also *native pasture, rangeland, rotation pasture,* and *temporary pasture*.

Pasture production system Hogs (see *swine*) produced on *pasture* with portable facilities.

Pasture seeder A *drill* designed to sow grass *seed;* also *broadcast seeders*.

Patent flour The "cut" of *flour* from the front of the mill (see *milling*). It is low in ash and *protein* and highest in value of all *grades* of flour.

Pathogen A *disease*-causing organism, such as a *bacterium, fungus*, parasite, or virus.

Pathology The study of the nature of *disease* and diseased conditions of the *host*.

Patrons of Husbandry See *National Grange*.

Pause A nonproductive period in *egg* layers (see *layer*) that usually lasts at least 15 days.

Payment-in-kind (PIK) 1: A payment made to eligible *producers* in the form of an equivalent amount of *commodities* owned by the *Commodity Credit Corporation* (CCC). Payments-in-kind were first used in the 1930s to reduce government-held surpluses of *cotton*. A PIK program in 1983 offered *surplus* agricultural commodities owned by the government in exchange for agreements to reduce *production* by cutting crop acreage.

2: Benefits, such as food, lodging, and clothing, furnished to agricultural workers by their employers for services performed.

Payment limitation The maximum amount of *commodity* program benefits a person can receive by law. The payment limitation was first imposed by the Agricultural Act of 1970 (P.L. 91-524) (see Appendix 3). Separate payment limitations are set for the *honey, wool*, and *mohair* programs (see *National Wool Act*). Persons are defined under payment limitation regulations, established by the *U.S. Department of Agriculture (USDA)*, to be individuals, members of joint operations, or entities such as limited partnerships, corporations, associations, trusts, and estates that are actively engaged in farming.

Pea The rounded green *seed* of an *annual* vine used as a vegetable. *Varieties* include smooth green, smooth yellow, Austrian winter, wrinkled, and mixed. The major producers are the United States, Canada, the former Soviet Union, China, India, Pakistan, Australia, France, the United Kingdom, Ethiopia, Hungary, and New Zealand.

Peach The soft, juicy fruit of a peach tree. The fruit has yellow to red, fuzzy skin and yellow flesh. The United States, Canada, Mexico, the *European Union*, Argentina, Australia, and Chile are the major producers.

Peanut The edible *seed* of an *annual* tropical vine that is cultivated (see *cultivate*) in semitropical regions. Also called ground nut. Peanuts are used for food and as a source of peanut oil (see *fats and oils*) and *peanut meal*. The major producing countries are India, China, the United States, Brazil, Senegal, and Indonesia.

Peanut harvester See *harvester*.

Peanut meal A coproduct of *peanut* oil (see *fats and oils*) extraction (see *extraction, mechanical* and *extraction, solvent*) used for animal *feed*.

Pear The edible fruit of the pear tree. Major producers include the United States, Canada, Mexico, Europe, Argentina, Australia, and Chile.

Pearl onion An *onion* less than 16 millimeters (0.63 inches) in diameter.

Peat Unconsolidated *soil* material consisting largely of undecomposed or only slightly decomposed organic matter accumulated under conditions of excessive moisture.

Pecan The smooth, thin-shelled nut of the pecan tree. *Varieties* include Caddo, Cherokee, Cheyenne, Choctaw, and Mohawk. The major pecan producers are the United States, Mexico, Australia, Brazil, and South Africa.

Pectin A gelatinous, water-soluble substance found in ripe fruits, which causes jam or jelly to set.

Pedigree A recorded line of ancestry.

Peg harrow A *farm* implement pulled over the ground to break up *soil* clods and prepare the seedbed. A peg harrow consists of a steel frame with short (4- to 6-inch) vertical *spikes*, which are sometimes adjustable to change the angle at which they enter the ground. Also called a spike-tooth harrow.

Peewee A lightweight beef *carcass*, usually not graded (see *grading* and *grades*).

Pellet binders Materials, such as sodium bentonite, that enhance the firmness of *pellets*.

Pellet machine A machine used to compress *alfalfa* hay (see *hay*) into compact, high-density *pellets*.

Pellet mill A machine used in a *feed* mill (see *milling*) to compress the finished *feed* mixture into small units the size of a *kernel* of *corn*. This process insures a uniform *ration* for feeding since loose feed mixtures separate or otherwise allow ingredients to settle out.

Pellets *Feed* formed by compacting and forcing by a mechanical process through die openings.

Pen See *lot.*

Pepino A melon native to South America and grown in New Zealand and California on a small-scale, commercial basis. The skin is marked with purple and greenish yellow stripes, and the flesh is yellow green to yellow orange.

Pepper, bell The podlike, seedy fruit of several different types of vine and woody plants. Peppers can vary in size, shape, and pungency, depending on the *variety.* Varieties include bell pepper, *pimento*, and cherry and chili peppers, with the latter types being the most pungent. The United States and Mexico are the major bell pepper producers.

Pepper, black The dried, blackish fruit of a woody vine native to the East Indies and used as a spice. Pepper is sold both whole and ground. The major producers are India, Brazil, Malaysia, and Indonesia.

Peppermint A plant with leaves that yield an edible oil (see *fats and oils*) used as a flavoring. Producers include the United States, China, Paraguay, and Brazil.

Per capita income The mean income computed for each individual in a particular group. It is calculated by dividing the total income of a particular group by the total population in that group.

Percolation (soil water) The movement of water through *soil* due to gravity.

Perennial A plant that has a life span of more than two years. See also *annual* and *biennial.*

Performance Industry- or firm-level attributes, such as profitability, capacity use, efficiency, and equity.

Pericarp The ovary wall in plants.

Perishable Agricultural Commodities Act A law establishing a code of trading ethics and encouraging fair trading in the marketing (see *agricultural marketing)* of fresh and frozen fruits and vegetables. It prohibits unfair and fraudulent business practices and provides a forum to resolve *contract* disputes. The law is administered and enforced by the *U.S. Department of Agriculture*'s (USDA) *Agricultural Marketing Service.*

Perishable commodities *Farm* goods that are not processed (see *processing*) and cannot be stored for a substantial period of time without deterioration or spoilage. Examples include meat, *poultry*, and fresh fruits and vegetables.

Perishable manufactured dairy products Manufactured dairy products with limited storage life, including ice cream, cottage cheese, *yogurt*, and sour cream.

Permanent legislation Legislation that would be in force in the absence of all temporary amendments and temporarily suspended provisions. The Agricultural Adjustment Act of 1938 (P.L. 75-430) and the Agricultural Act of 1949 (P.L. 89-439) (see Appendix 3) serve as the principal laws authorizing the major *commodity* programs. These laws are frequently amended—provisions are added, suspended, and repealed. For the past several decades, periodic omnibus *agriculture* acts have provided for specific fixed-period commodity programs by adding temporary amendments to these laws and suspending conflicting provisions of those laws for the same period. The temporarily suspended provisions of the 1938 and 1949 Acts go back into effect if current amendments lapse and new legislation is not enacted.

Permanent pasture A *pasture* of *perennial* or self-seeding (see *seed*) plants kept indefinitely for grazing (see *graze*). See also *rotation pasture.*

Permanent press See *durable press.*

Permanent wilting point The point at which a plant is too dry to recover even if watered and put into a humid atmosphere.

Permeability, soil The quality of a *soil* horizon that enables water or air to move through it. It can be measured quantitatively in terms of the rate of flow of water through a unit cross section in unit time under specified temperature and hydraulic conditions.

Permitted acreage The maximum acreage of a crop that may be planted for *harvest* by a program participant. The permitted acreage is computed by multiplying the *crop acreage base* by 1 minus the *acreage reduction program (ARP)* requirement (announced by the *Commodity Credit Corporation* each year). For example, if a *farm* has a crop acreage base of 100 acres and a 10-percent acreage reduction is required, the permitted acreage is 90 acres (100 x (1 –.10) = 90).

Persistent poverty county See *county type classification.*

Pesticide A substance used to kill a pest. Pesticides include *insecticides*, *fungicides*, *herbicides*, *nematocides*, and growth regulators (see *growth inhibitor* and *growth stimulator*).

Pesticide residue Trace amounts of *pesticides* on food products resulting from crop treatment or accidental environmental *contamination.* Sensitivity of testing methods means residues may be detected below the level harmful to human health. See *drug residue* and *insecticide residue.*

pH A measure of the acidity or alkalinity of *soil* or water. A pH of 1 to 6 is the acid range; 8 to 13 is alkali; and a pH of 7 is neutral. Each plant has its own ideal pH range, with 6 to 7.5 considered best for most agricultural crops. See *acid soil* and *alkali soil.*

Pheromone A chemical secreted by certain insects and other animal *species* to elicit responses from the same species. Pheromones may be produced synthetically for use in insect control.

Phosphate A *fertilizer* that supplies phosphorous, a major element in fertilizers.

Photosynthesis The process, in living plants, of converting energy from the sun into chemical energy by transforming *carbon dioxide* from the air into simple *carbohydrates.* The process is conducted by the *chlorophyll*-containing *cells* in plants and results in the release of oxygen into the atmosphere, making plants critical for human survival.

Phylloxera An insect that attacks the roots of grapevines.

Physical quality A *grain* characteristic associated with the outward appearance of the grain *kernel*, including kernel size, shape, color, moisture, damage, and density.

Physiology The science of the function of organs, systems, and the whole living body.

Phytosanitary Plant health. See *sanitary and phytosanitary regulations.*

Phytosanitary certificate A document issued by a government to an *exporter* that certifies that the *commodity* is free from pests or *disease*, in accordance with the importing country's standards.

Phytosanitary regulations Regulations to minimize the threat of insect or *disease* infestation of domestic *production* areas. While phytosanitary regulations protect the food *supply* from new pests, they can also be used as unjustified *trade barriers* if not founded on legitimate scientific evidence or if they place greater controls on *imports* than those applied to domestic *commodities*. Phytosanitary regulations may also complicate trade patterns and represent an additional source of risk to *exporters* because each country determines its own standards.

Pick See *filling*.

Picker A self-propelled or pull-type machine for harvesting (see *harvest*) crops, such as *corn* and *cotton*.

Piece work A method of payment, where pay is determined by multiplying the units produced (boxes, crates, *bales*, etc.) by the rate per unit.

Pig A young *swine* weighing less than 120 pounds.

Piggyback Hauling truck trailers cross-country on railroad flatcars.

PIK and Roll A procedure by which *producers* attempt to profit from situations where certificate exchange values (posted county prices) are below *nonrecourse loan rates*. With this procedure, a producer places the eligible *commodity* under *nonrecourse loan* at the *loan rate* and uses *generic certificates* to exchange for *Commodity Credit Corporation (CCC)* commodities. If the posted county price is below the nonrecourse loan rate, the producer is able to acquire the quantity placed under loan for less than the proceeds of the nonrecourse loan, in addition to saving interest and storage charges.

Pile The cut or uncut loops that make the surface of a pile fabric (see *cloth*). Some common pile fabrics include velvet, corduroy, terry toweling, furniture covering, and rugs and carpets.

Pimento A red bell *pepper*, which usually is canned in brine and used as a condiment. Spain and the United States are major producers.

Pineapple The large, edible fruit of a tropical plant. The fruit is characterized by a tough outer shell with spiky leaves and sweet, juicy, yellow flesh. Hawaii is the sole producer of fresh pineapples in the United States and controls approximately 60 percent of U.S. fresh *market* supplies. The United States is a significant *net importer*, with Mexico accounting for 60 percent of U.S. supplies.

Pineapple guava See *feijoas*.

Pinto bean A *variety* of *dry edible bean* used to make canned refried beans, as well as three-bean salad and soups such as minestrone, stews, and casseroles. About three-fourths of pintos are sold dry (see *drying*) in bags, with the remainder canned.

Pistachio The edible *seed* of a tree in the sumac family. Major producers include Turkey, Iran, Italy, and the United States.

P.L. 480 See *Public Law 480*.

Plant breeding The development of plants with certain desirable characteristics. *Grain* breeding programs generally aim to improve *yield* and harvestability (see *harvest*), increase *disease* resistance, and satisfy apparently desirable, intrinsic quality goals.

Planter An implement used to sow *seed* in rows. Used for crops such as *corn*, where the rows need to be wide enough to allow a *cultivator* between rows, as opposed to a *drill*, which is used to sow close-grown crops such as *wheat*. See *air planter* and *corn planter*.

Planter, no-till A machine designed to cut through the heavy *mulch* of a previous crop to seed *corn* and other crops. Most no-till planters are of heavy construction and use *coulters* to cut through *crop residue*. See *planter*.

Plant or establishment A single production (see *production*), *processing*, or distribution facility.

Plant-specific economies See *economies of scale*.

Plant Variety Protection Act A law extending patent-type protection to developers of plants that reproduce through *seeds*. Developers of new *varieties* of such plants as *soybeans, wheat, corn,* and marigolds apply to the *U.S. Department of Agriculture (USDA)* for certificates of protection. USDA examiners determine whether the variety is actually novel and entitled to protection. The holders of certificates can turn to the courts to protect their "inventions" from exploitation by others.

Plow A *farm* implement used to lift and turn *soil* over. Includes a number of different types of devices, including a smooth, curved blade with a cutting edge; a large, concave steel *disk* that rotates at an angle to the direction of travel; and a *chisel point* mounted on a *shank*. See *coulter chisel plow, lister*, and *moldboard plow*.

Plowing Breaking up *soil* with a *plow*.

Plow pan A compacted layer of *soil* reaching 6 to 8 inches below the surface. Plow pan is generally caused by compression at the bottom of a *plow* or other *tillage equipment*.

Plow point The hardened, cutting edge of a *plow* that slices the *soil* being plowed.

Plow share The smooth, curved blade of a *plow* designed to turn *soil* over.

Plug (tobacco) See *chewing tobacco*.

Plum The smooth-skinned, edible fruit of any of several *varieties* of small shrubs or trees. The fruit is deep reddish purple to dark purple. *Prune*-type plums are used primarily for *drying*, but some varieties are canned or shipped to fresh *markets*. Major producers include the United States, Canada, the *European Union*, Australia, Chile, and South Africa.

Plurilateral Negotiations among a select few countries, as opposed to *multilateral trade negotiations* that involve a larger number of participating parties.

Ply The number of single *yarns* twisted together to make a composite yarn. When applied to *cloth*, it means the number of layers of fabric combined to make the composite fabric.

Pod seed A large forest tree found in tropical Africa that bears long bean-type pods containing large *seeds*. The seeds are locally processed (see *processing*) for the oil (see *fats and oils*), which is used to produce soap and candles. Also known as the owala oil tree.

Point Used when quoting the *price* of raw *cotton*. One point is equal to 1/100 of a cent.

Point source pollution A pollutant that can be traced to a specific source, such as a factory smokestack or chemical spill.

Point-to-point rate A transportation charge from a single point of origin to a single point of destination.

Polariscope See *polarization.*

Polarization A measure of *sucrose* concentration based on its ability to rotate the plane of polarized light. The degree of polarization is determined by a saccharimeter (commonly referred to as a polariscope) and is indicative of the percentage of sucrose in high-purity products such as raw cane *sugar* and white *refined sugar.*

Pollen Dustlike material produced in the male parts of flowering plants and necessary on the female parts of the flower for *seed* production. Pollen is the *protein* food essential to *honeybees* for raising *brood.*

Pollen basket An area on the hindleg of a *honeybee* adapted for carrying *pollen* which has been gathered from flowers, to the *hive.*

Pollination The transfer of *pollen* from the male parts (anther) of a flower to the female parts (stigma) of the same flower or another flower of the same *species.* See also *cross-pollination.*

Polyculture Growing many crops at once in the same field.

Polyester See *synthetic fibers.*

Polyglycerol esters See *fatty-acid-based fat substitute.*

Polypropylene See *synthetic fibers.*

Polysaccharide Compounds formed by the chemical union of two or more simple *sugars.*

Polyunsaturated fats See *fatty acid (classification of).*

Pomace See *pulp.*

Pomegranate A fruit with a leathery, deep red to purple rind. Only the *seeds* of the fruit are edible.

Pomology The cultivation (see *cultivate*) of fruit crops.

Porcine Pertaining to *swine.*

Porcine Somatotropin See *bovine Somatotropin.*

Poromerics Plastic materials containing microscopic holes that allow for the passage of air and moisture and used as a substitute for *leather.*

Portable auger An *elevator* or conveyer on wheels for *grain* and other crops that can be pulled to different locations. See *auger.*

Portable bunk feeder A container designed to hold and provide *livestock* access to *feed* that is situated inside a *lot* or *pasture* and that can be moved with a *tractor* or *farm truck.*

Portable housing Shelters that can be moved from one location to another with a *tractor* or *farm truck.*

Portion-controlled product A product that has been individually packaged to meet certain weight and other specifications.

Posted county price (PCP) A county-wide *market* price used in the calculation of loan redemptions with generic *payment-in-kind* certificates and *marketing loan payments* for *wheat, oilseeds,* and *feed grains.* The PCP reflects changes in *prices* in the 18 major terminal grain markets, corrected for the cost of transporting *grain* from the county to the *terminal elevator.* The *Farm Service Agency (FSA)* calculates the PCP daily.

Post-hole digger A tool or machine used to dig holes in the ground for fence posts. It can be manual or power-driven. Manual diggers usually consist of a double-bladed device attached to a long handle that can be rotated to remove *soil* and form a hole in the ground. Others may be *augers* or simple bars (crowbars). Power-drive units usually consist of an auger mounted on the back of a *tractor* and driven by *power-take-off (PTO).*

Postmortem condemnations *Carcasses* or parts of slaughtered (see *slaughter*) animals condemned by an inspector because of *disease* or mishandling and removed from the slaughter line and destroyed. See *inspection, federal* and *meat and poultry inspection.*

Potash A *fertilizer* that supplies potassium, an essential *nutrient* for plant growth; expresses the percentage of potassium oxide in potash salts and mixtures.

Potato The edible *tuber* of the potato plant. Common *varieties* include the Russet, Round white, Red, and Yellow. The United States and Canada are the major potato producers.

Potato digger A machine used to dig, shake dirt loose, and lay *potatoes* on top of the ground.

Potato eye The bud on a *potato* that produces stems or roots.

Potato harvester A machine used to dig, shake dirt loose, and collect *potatoes* in a bin or sacks.

Potato planter A machine used to place seed (see *seed*) *potatoes* in the ground.

Potted flowering plants Plants produced for their decorative shape, size, color, and stem characteristics and usually sold by the pot or hanging basket. House plants for indoor or patio use and large specimens used for the interiors of hotels, restaurants, and offices are included. Examples are poinsettias, African violets, Easter lilies, and kalanchoes. See also *foliage plants.*

Poult A young *turkey.*

Poultry Chickens, turkeys, ducks, geese, and guinea fowl, which *yield* meat, *eggs*, and feathers as products. The United States, China, the former Soviet Union, Brazil, Japan, Hungary, and France are the major poultry producers.

Poultry Products Inspection Act A law requiring inspection (see *inspection, federal*) of *poultry* in privately owned poultry *processing* plants in the United States to ensure that domestically produced poultry sold in interstate and *foreign* commerce is safe, wholesome, and accurately labeled. The *U.S. Department of Agriculture*'s (USDA) *Food Safety and Inspection Service (FSIS)* is responsible for carrying out federal inspection laws.

Pound A unit of weight equal to 16 ounces, 7,000 *grains*, or 453.592 grams.

Poundage quotas A quantitative limit on the amount of *tobacco* and *peanuts* that can be produced (see *production*) and/or marketed (see *agricultural marketing*) under federal *price support programs.*

Poverty status The status of families and unrelated individuals whose annual income falls below an income cutoff or "poverty threshold." These cutoffs vary by family size, number of children, and age of the household head. Poverty thresholds are updated each year to reflect changes in the *Consumer Price Index.*

Poverty threshold See *poverty status.*

Power-take-off (PTO) A device, primarily found on *tractors* but also on some *farm trucks* and other vehicles, used to drive other equipment, either pull-type equipment or mounted on a power unit.

Powdery mildew See *mildew.* See also *downy mildew.*

Power tooth harrow A *harrow* with vertical *spikes* or tines that till *soil* in circular patterns. A power tooth harrow is driven by tractor (see *tractor*) *power-take-off (PTO).*

Precleaning Removing *foreign material* such as weeds, *seeds*, dirt, stems, and cobs from the *grain* before it is dried (see *drying* and *crop dryer*), resulting in more uniform moisture content. Precleaning is generally not practiced by U.S. dryer operators. See *cleaning.*

Precooked rice *Rice* that has been cooked and dehydrated (see *dehydrate*) after *milling.* This reduces the time required for cooking. Precooked rices include quick-cooking rices, instant rices, and boil-in-the-bag rices.

Precooling The rapid cooling of fruits or vegetables immediately following *harvest.*

Predator Any animal, including insects and *microorganisms*, that preys upon other animals, resulting in injury and/or death. The predator may devour part or all of the prey.

Preferences Special trade advantages given by governments to trading partners in order to promote *export* growth and development. These advantages include admitting goods at *tariff* rates below those imposed on goods of competing *exporters* or exempting the favored countries from certain *nontariff trade barriers.* Preferences are often granted to *developing countries* by developed nations. See *tariff preference.*

Preferential trade agreement See *bilateral trade agreement.*

Premium 1: An amount paid for insurance.

2: The *price* paid by an *option* buyer and received by the option seller for an option *contract.* See also *commodity option.*

Premix A uniform mixture of one or more *microingredients,* such as *vitamins*, trace minerals, or *drugs*, mixed with a carrier. Premixes are used to distribute *microingredients* evenly throughout formula *feed*. Premix is usually added at a rate of less than 100 pounds per ton of finished feed.

Prepared flour mixes *Flour* blends that are tailored for the *production* of specific items; examples are biscuit, cake, doughnut, and pancake mixes.

Prevented planting acreage *Land* on which a farmer intended to plant *program crop(s)* but was unable to because of a natural disaster, such as a *drought*.

Price The amount of goods or money required to be exchanged for goods and services. See *price competition, price coordination, price discrimination, price fixing, price index, price-later contracts, price leader, price leadership, price pooling, prices-paid index, prices-received index, price support level, price support programs, price umbrella,* and *world price*.

Price competition Competition between sellers or buyers based on open *price* rivalry.

Price coordination Joint activity among firms in determining selling *prices*.

Price discrimination Charging a higher *price* in one or more segments of a *market* than in others for similar but not necessarily identical goods. Charging different prices can allow a firm to realize higher profits. A seller is able to price discriminate if it can divide or segment the market and if *consumers* differ in their sensitivity to price changes. For example, a seller may charge less for a product in *foreign* markets. *Dumping* is an example of price discrimination.

Price fixing An agreement among buyers or sellers to set the *price* of their product at a certain level, generally to reduce competition. The Sherman Antitrust Act made price fixing illegal in the United States.

Price index An indicator of the average *price* change for a group of *commodities* or products that compares prices for the same items in some other period, commonly called the *base period*. See *Consumer Price Index, Prices-paid index,* and *Prices-received index*.

Price-later contracts See *delayed pricing*.

Price leader A firm that is first among a group of firms to initiate a *price* change.

Price leadership A pattern in which a number of firms in a *market* follow the pricing policies of one firm, termed the *price leader*.

Price pooling A policy tool used by *state marketing boards* to guarantee *producers* a minimum *price* and provide price stability. All income received by the boards from *grain* sales is placed or "pooled" in one fund. If the average price exceeds the minimum price established, producers are paid a pool bonus. Otherwise, the government assumes the cost of the differential.

Prices-paid index An indicator of changes in the *prices* farmers pay for goods and services used for producing *farm* products and those needed for farm family living. In addition to prices paid for *farm inputs* like *fertilizer*, the index includes interest, taxes, and farm wage rates. It is referred to as the *Parity Index* when 1910–1914 is used as the base period.

Prices-received index A measure computed on the basis of *prices* farmers received for their products at the point of the first sale (usually the *farm* or local *market*).

Price support level The *price* for a unit *(pound, bushel)* of a *farm* commodity (see *commodity*) that the government will support through *nonrecourse loans*, direct purchases, and payments. Price support levels are determined by law and are set by the secretary of agriculture.

Price support programs Government programs that keep *farm* prices received by participating *producers* from falling below specific minimum levels. Price support programs for major *commodities* are carried out by providing *nonrecourse loans* to farmers so that they can store their crops during periods of low *prices*. The loans can later be redeemed if commodity prices rise sufficiently to make the sale of the commodity on the *market* profitable, or the farmer can forfeit the commodity to the *Commodity Credit Corporation (CCC)*. In the latter case, the commodity is stored and is not available to the market until prices rise above statutory levels that allow the CCC to sell the commodities. Other price support mechanisms include direct purchases and other payments. Commodities supported in the United States include *wheat, corn, grain sorghum, barley, oats, rye, rice, soybeans, peanuts, tobacco,* certain dairy products, *wool, mohair,* and *sugar.*

Price umbrella The concept that agricultural *price* supports (see *price support programs*) tend to hold *world prices* of the supported *commodities* at high enough levels to stimulate *production* in other countries.

Prickly pear The red fruit of a cactus plant native to Mexico. Also known as cactus pears, Indian figs, and tunas.

Primal cuts Major divisions of the *carcass* (chuck, rib, loin, and round) and rough-cuts (brisket, plate, and flank).

Primary feed manufacturing The *processing* and *mixing* of individual *feed* ingredients, sometimes with the addition of a *premix*. *Feed grains*, mill (see *milling*) *byproducts*, *oilseed* meals (see *meal*), and animal *proteins* are examples of feed ingredients.

Prime, Choice, Select *U.S. Department of Agriculture* (USDA) quality *grade* designations applied to qualifying young, grain-fed (see *grain-fed animal*) *steers* and *heifers*. See *quality grades for steer and heifer slaughter (Prime, Choice, Select, Standard, Commercial, Utility, Cutter, and Canner).*

Prime farmland *Land* where the *soil* quality, growing season, and moisture make it ideally suited to producing food, *feed*, *forage*, *fiber*, and *oilseed* crops.

Priming Removing ripened leaves from *tobacco* plants by hand (also referred to as cropping). *Flue-cured tobacco* and *cigar* wrapper *tobacco*s (see *cigar classes of tobacco*) are harvested (see *harvest*) by priming.

Priming aid A machine that permits workers to ride as they manually break off *tobacco* leaves. See *priming*.

Principal See *agent.*

Principal supplier The country that is the most important source of a particular product imported by another country.

Prior import deposit A requirement in some countries that *importers* seeking an *import license* deposit local funds with the central bank, often up to 100 percent of the value of the imported (see *import*) goods.

Private label brand A product sold under a retailer's brand name. The retailers buy the product processed (see *processing*) to their specifications and packaged in their packages. Also referred to as a store label brand or controlled brand.

Private trade A transaction in which the buyer and seller agree not to report the sale to *market* news reporting services. See *market news service.*

Private voluntary organization (PVO) A nongovernment, nonprofit institution that provides economic and social assistance to countries or people in need. PVOS may operate development assistance programs and play an important role in distributing U.S. food aid under *Public Law 480,* including emergency relief.

Prizing Packing *tobacco* into *hogsheads.*

Processed fresh vegetables Vegetables that have been shredded, diced or sliced but not heated, frozen, or dried (see *drying).*

Processed vegetables Vegetables that have been heated, frozen, or dried (see *drying*) and perhaps sliced or diced.

Processing The conversion of raw *farm* commodities (see *commodity*) into intermediate or final products.

Processor A firm that converts raw *farm* commodities (see *commodity*) into intermediate or final products. A dairy processor, for example, uses raw *Grade A milk* to produce *fluid (milk) products.*

Produce Fresh fruits and vegetables.

Produce grower A firm that produces either fruits or vegetables or both.

Producer 1: A person or firm that grows or manufactures goods to sell or offers services for a fee. 2: More specifically for *agriculture*, a person who, as owner, landlord, tenant, or *sharecropper*, is entitled to a share of the crops available for marketing from the *farm* or a share of the proceeds from the sale of those *commodities*.

Producer allotments A quantity provision of *federal marketing orders* that assigns a maximum quantity that a producer/handler can provide to the *market* in a single season.

Producer option payments See *Inventory Reduction Program.*

Producer Price Index (PPI) A measure of the average *price* received by *producers* during a specific period compared against a benchmark period. Formerly called the Wholesale Price Index.

Producer subsidy equivalent (PSE) An estimate of the effect of government policy by measuring the amount of the cash *subsidy* or tax that would be needed to hold farmers' incomes at current levels if all government agricultural programs were removed. PSE's and *Consumer Subsidy Equivalents (CSE's)* are used to compare different policy tools and their effects on farmer revenue and *consumer* costs across countries. As a result, most *General Agreement on Tariffs and Trade (GATT)* proposals for *trade liberalization* hinged on the use of measures such as PSE's and CSE's in negotiating lower protection levels.

Produce wholesaler A firm that receives *produce* from *packers* and distributes the produce to food retailers, institutions, *foodservice* organizations, and restaurants.

Product competition Competition among sellers on the basis of product design, promotion or branding, packaging, or other ancillary product characteristics.

Product differentiation A form of *nonprice competition* where firms focus on differences in such factors as quality, product characteristics, packaging, credit arrangements, or *supply* reliability to persuade *consumers* that their product is different from or better than that of their rivals.

Production 1: The process of growing or manufacturing goods or services to sell.

2: For statistical purposes, an estimate of the quantity of a product domestically grown and marketed, as reported by some official statistical reporting service. In the United States, for example, the *U.S. Department of Agriculture*'s (USDA) *National Agricultural Statistics Service* is one source of agricultural production information.

Production contract An agreement between a crop or *livestock* grower and a *processor* that specifies *production* activities to be carried out, the product or service to be delivered, and the method for determining the payment from the processor to the grower. In some cases, a production *contract* calls for the processor to provide certain inputs.

Production controls Any government program or policy intended to limit *production*. These programs or policies have included *acreage allotments*, *acreage reduction programs*, *set-aside*, *paid land diversion*, quantity and acreage *marketing quotas*, *payment-in-kind*, production termination, and *soil bank*.

Production Credit Associations (PCA) Organizations that provide short- and intermediate-term loans to farmers and ranchers for up to 10 years with funds obtained from investors in the money *markets*.

Production expenses The total cash outlays for *production* of *commodities* and services in the *farm* sector. Capital expenses are figured on annual depreciation rather than on yearly cash outlays for capital items.

Production flexibility The ability of a farmer to vary the crops mixture in response to prevailing *market* conditions. This capacity is hampered to some extent by the need to maintain base acreage (see *crop acreage base*) in the *commodities* for which program payments are profitable, which restricts shifts to *nonprogram crops* (see *program crops*), such as *soybeans*.

Productive capacity The amount of goods that could be produced if all the *resources* currently available were fully employed using the best available technology. Productive capacity increases whenever the available resources increase or the *productivity* of those resources increases.

Productive soil A *soil* that is favorable for the *production* of the crops suited to a particular area.

Productivity The relationship between the quantity of inputs (*land*, labor, *tractors*, or *feed*, for example) employed and the amount of output produced (see *production*). An increase in productivity means that more output is produced with equal or reduced inputs. Both single-factor and multifactor indexes are used to measure productivity. Single-factor measures examine the output per unit of one input at the same time other inputs may be changing. Examples of single-factor measures include crop *yield* per *acre*, output per workhour, and *livestock* production per breeding animal. Multifactor productivity indexes consider productive *resources* as a whole, netting out the effects of substitution among inputs. The "Total Farm Output per Unit of Input" index is a multifactor measure. See *productivity growth.*

Productivity growth A measure of the rate of growth of output, relative to the growth of the *inputs* (labor, capital, and materials) used to produce that output. See *productivity.*

Productivity (of soil) The present capability of a kind of *soil* for producing a specified plant or sequence of plants under a defined set of management practices. It is measured in terms of the outputs or *harvests* in relation to the *farm inputs* of *production* factors for a specific kind of soil under a physically defined system of management.

Product-specific economies See *economies of size.*

Product stream Any one of 125 to 150 mill streams in the *flour* manufacturing process.

Product weight The weight of a product as it is sold at the retail level. In the meat trade, product weight is differentiated from *carcass weight equivalent* and may or may not include the weight of bone, fat (see *fats and oils*), or additional water.

Program (agricultural) Government activities aimed at accomplishing a certain result. Federal *price support programs* include *nonrecourse loans* and purchases and payments. Other agricultural programs include *commodity* storage, transportation, *exports*, and *acreage reduction.*

Program costs These costs may be either (1) gross or net expenditures of the *Commodity Credit Corporation (CCC)* on a *commodity* or all commodities during a fiscal year or other period; (2) the realized loss on disposition of a commodity, plus other related net costs during a fiscal year or other period; or (3) the net costs attributed to a particular year's crop of a commodity during the *marketing year* for that commodity.

Program crop A crop for which federal *deficiency payments* are available to *producers*. Program crops include *wheat, corn, barley, grain sorghum, oats, upland cotton*, and *rice*. Some *nonprogram crops*, such as *rye*, ELS cotton, *soybeans,* minor *oilseeds*, *tobacco, peanuts, sugar, honey,* and *milk* are eligible for federal *price support programs*, but not deficiency payments, except for ELS cotton.

Program slippage See *acreage slippage.*

Program yield The farm *commodity* yield of record determined by a procedure outlined in legislation. Under the *Food, Agriculture, Conservation, and Trade Act of 1990 (P.L. 101-624)* (see Appendix 3), program *yields* were set at the same levels determined for *program crops* in 1986. The law also allows the *U.S. Department of Agriculture (USDA)* to update program yields at the average of the preceding 5 years' harvested (see *harvest*) yield (after dropping the high and low years). However, this provision, known as proven yields, has not been implemented. The *farm* program yield applied to eligible acreage determines the level of *production* eligible for *direct payments* to producers.

Promotion, research, and package regulations The set of regulations authorizing grower funding for *generic advertising* and promotion, *production* and *marketing research*, and establishment of package and container standards.

Propolis An orangey brown to red resinous substance obtained by *honeybees* from certain trees and used to close small openings or cover objectionable objects within the *hive*. Also called "bee glue."

Prorate A quantity provision in a *federal marketing order* designed to even out weekly (or occasionally some other specified time period) shipments.

Proso millet A *grain* used primarily for birdseed in the United States. More widely grown in Africa and Asia as a food grain.

Protectionism A *tariff, subsidy*, or *nontariff trade barrier*, for example, imposed (see *import*) by a country in response to *foreign* competition, in order to protect domestic *producers*. This distorts the world trading system by impairing the operation of *comparative advantage* and provides incentive for inefficient domestic *production*.

Protein A naturally occurring combination of *amino acids*, containing the chemical elements *carbon*, hydrogen, oxygen, *nitrogen*, and sometimes sulphur. Protein is one of the essential constituents of all living things and of the diets of humans and animals.

Protein-based fat substitute A *fat substitute* composed of a mixture of *protein* and water and shaped into particles that yield a taste and texture similar to fat (see *fats and oils*). The particles can be used as a partial fat substitute in manufacturing food products.

Protein premium The amount of money per *bushel* that a high-*protein* wheat normally commands over *wheat* of the same *grade* specification with lower protein content.

Protein quality The effectiveness of *protein* in promoting growth. The main factor affecting protein quality is the adequacy of the amounts of *essential amino acids* in the protein to meet human requirements.

Prune A plum dried (see *drying*) whole without *fermentation* at the pit and characterized by a high *sugar* content. Major producers include the United States, Argentina, Australia, Chile, France, and southeast Europe.

Pruning Removing or cutting wood from trees or vines.

Pseudorabies A contagious *livestock* disease that is mostly associated with *swine*. Although the *disease* does not harm humans, it can affect *cattle*, *sheep*, dogs, cats, and other animals.

pST See *bovine Somatotropin (bST)*.

Public Law 480 (P.L. 480) The common name for the Agricultural Trade Development and Assistance Act of 1954 (P.L. 83-480) (see Appendix 3), which seeks to expand *foreign* markets for U.S. agricultural products, combat hunger, and encourage economic development in *developing countries*. Also called the Food For Peace Program.

Title I of P.L. 480 makes U.S. agricultural *commodities* available through long-term dollar credit sales at low interest rates for up to 30 years. Donations for emergency food relief and nonemergency assistance are provided under title II. Title III authorizes "food for development" projects. The *Food, Agriculture, Conservation, and Trade Act of 1990* (P.L. 101-624) (see Appendix 3) made fundamental changes in the U.S. food aid program, including shortening the maximum repayment term of Title I loans from 40 to 30 years. The 1990 legislation also authorized a new Title III Food for Development program that provides government-to-government grant food assistance to least developed countries.

Pulled wool *Wool* removed from the skins and pelts of slaughtered (see *slaughter*) *sheep* or *lambs*.

Pullet A young, female chicken (see *poultry*).

Pulp 1: Material prepared by chemical or mechanized means from various materials, such as wood or *kenaf*, used in making paper or *cellulose* products.
2: The solid residue remaining after extracting juices from fruits. Also called bagasse or pomace.

Pulpwood Wood used in the manufacture of paper, fiberboard, and similar products. See *pulp*.

Pulses The edible *seeds* of various leguminous (see *legumes*) crops, such as *peas*, *beans*, and *lentils*.

Pummelo An exotic, subtropical *citrus fruit* believed to be native to Malaysia and Indonesia. The pummelo fruit, highly prized in the Orient, is light yellow, sweet, and solid. The sphere-shaped fruit typically weigh about 3 pounds each, measure 6 to 8 inches in diameter, and have a smooth, half- to three-quarters-inch thick, *lemon* yellow rind. As of 1994, the fruit was not grown commercially in the United States. Also known as shaddock.

Punta del Este Declaration A declaration written by the *contracting parties* to the *General Agreement on Tariffs and Trade (GATT)* in September 1986 to launch the *Uruguay Round* of *multilateral trade negotiations*. The declaration included a statement of objectives and an agenda for the negotiations to follow.

Pupa The third stage in the life of a developing *honeybee* when the larval (see *larva*) body is inactive in its cell.

Purchase authorization (PA) A document that authorizes an importing government to procure *commodities* and, in certain instances, ocean freight under *Public Law 480 (P.L. 480)*. The PA specifies the *grade* and type, approximate quantity, and maximum value of the commodities. It also states the timespan for their purchase and delivery, method of financing, and other provisions and limitations.

Pure conglomerate merger The merger or consolidation of firms that produce substantially different products or product lines.

Purveyor *Wholesalers* that do some *processing* of the products they buy and resell them to other firms. Purveyors may handle a variety of product lines and perform other services.

Put option The right, without obligation, to sell a *futures contract* at a specified *price* during a specified time period.

Pyrethrum An extract, powder, or dried (see *drying*) flower from any of several chrysanthemum *species* used in *insecticide* products. Kenya, Tanzania, Rwanda, Zaire, Ecuador, and Papua New Guinea are the major pyrethrum producers.

Q

Quality grades for steer and heifer slaughter (Prime, Choice, Select, Standard, Commercial, Utility, Cutter, and Canner) The *U.S. Department of Agriculture* (USDA) system of *grading* based on the maturity of the animal, the amount of *marbling*, and other palatability characteristics.

Quantitative restriction (QR) An explicit limit on the quantity or value of a product permitted to enter or leave a country. Examples include quotas (see *export allocation or quota, marketing quota(s),* and *import quota*), *embargoes*, restrictive *licensing*, and other means of limiting *imports*.

Quarantine Any restriction, including a total ban, on the entry or departure of cargo or persons that may harbor exotic insects, *diseases*, or weeds to prevent their transmission.

Quarantine, sanitary and health laws and regulations Government measures to protect human, plant, and animal health, such as restricting the use of potentially injurious preservatives and other *additives* in food products or denying entry of plants or animals from countries where specific *diseases* are present. While largely imposed for health reasons, in some cases these laws and regulations may be used to restrict *foreign* competition. See also *African swine fever* and *foot-and-mouth disease*, for example.

Queen bee A fully developed female bee (see *honeybee*). Under normal conditions, the queen bee is the mother of all the other individuals in the *colony*. See *honeybee*.

Queen cell The cell in which the *queen* bee develops. The queen cell is the largest cell built and hangs vertically in the *hive*, while the others are horizontal. See *honeybee*.

Quince A hard, yellow, acidic fruit of a tree in the rose family. The fruit is shaped like a *pear* and has a fuzzy, *peach*-like skin. The raw fruit is inedible.

Quinoa A small, tropical *grain*.

Quintal A metric measure of weight equal to 100 kilograms, or 220.46 pounds. See Appendix 6.

Quota See *export allocation or quota*, *marketing quota(s)*, and *import quota*.

Quota-exempt sugar *Sugar* imported (see *import*) into the United States that is exempt from a *quota* (see *export allocation or quota*, *marketing quota(s)*, and *import quota*) charge. This sugar is entered under bond to be reexported (see *reexport*), used as *feed* for *livestock*, or for the *production* of polyhydric alcohol.

Quota tobacco See *leasing of quota.*

R

Rack jobber A type of *wholesaler* that usually handles nonfood lines, such as health and beauty aids or housewares and displays and services items on a rack, typically in retail *foodstores*. Rack-jobber salespersons usually service racks, rearrange displays, replenish stock from their trucks, sell new items, and so on. Rack jobbers are a variation of *wagon jobbers*.

Radicchio A salad green with a flavor similar to *escarole* and *Belgian endive*.

Radicle A rootlet; the first stem of a plant or both the first stem and the root. The radicle forms the primary root of a young seedling.

Radish A root vegetable that comes in various sizes, shapes, and colors from red button radishes to long, thin, mild-flavored varieties.

Railed out When meat *carcasses* are evaluated, sorted, or put aside for later *inspection* (see *meat and poultry inspection* and *inspection, federal*) by being moved onto another rail in the slaughtering (see *slaughter*) plant.

Raisin A dried (see *drying*) *grape*. *Varieties* include Thompson seedless and Muscat. The United States, Australia, Greece, South Africa, Turkey, and Afghanistan are the major producers.

Rake Any number of tools and machines with tines used to pick up or move *hay*, leaves, and other materials. *Farm* rakes are power-driven implements used to collect hay into rows or *windrows*. See *dump rake* and *side-delivery rake*.

Ram A breeding male *sheep*.

Rambutan A red to yellow, golf-ball-sized, tropical fruit with long spikes. The flesh is pearly white, sweet, crunchy, and juicy, and it can be eaten fresh, canned, stewed, or in jams and jellies. The fruit is native to Malaysia, while Thailand is the world's largest *exporter*. In the United States, rambutan is grown in Hawaii. Rambutan is a relative of *lychee*.

Ramie The *fiber* or bast of an Asian plant. Ramie is used to make *cloth* and cordage. See *soft fibers*.

Ranch Used mostly in the western United States to describe a tract of *land*, including facilities, used for *production* of *livestock*. Accepted usage generally refers to the headquarters facilities, *pastures*, and other land as the ranch, as distinguished from *rangeland*. Loosely defined, a ranch may also be a small western *farm*, such as a fruit ranch or chicken (see *poultry*) ranch.

Range See *rangeland*.

Range drill A specially constructed, heavy duty *drill* used to seed grass on *rangeland.*

Rangeland *Land* that is predominantly grasses, grasslike plants, or shrubs suitable for grazing (see *graze*) and browsing. Rangeland includes natural grasslands, savannahs, many *wetlands*, some deserts, tundra, and certain shrub communities. It also includes areas seeded to native or introduced *species* that are managed like native vegetation.

Rape A European *herb* of the mustard family, grown as a *forage* crop and for its oil-producing (see *fats and oils*) *seeds*. Also called *canola*. Canada, the *European Union*, China, and India rank among the major producers.

Rapeseed See *industrial rapeseed.*

Raspberry The edible, dark purplish red fruit of the thorny raspberry shrub.

Ration The amount of food supplied to an animal for a definite period, usually for a day; also the ingredients that comprise the daily amount of *feed.*

Ratoon The second and subsequent crops grown from the root systems of previous plantings of some crops, such as *sugarcane* and rice. Usually one or more ratoon crops are harvested (see *harvest*) before the fields are plowed (see *plowing*) and replanted.

Raw fibers Natural *textile* fibers before any manufacturing activity has taken place. *Cotton* as it comes from the *bale* is an example of raw *fibers.*

Raw sugar Any *sugar* that is to be further refined (see *refining* and *refined sugar*) or improved in quality to produce liquid or crystalline sugar.

Ready-to-cook poultry Dressed *poultry*, without the feathers, head, feet, and most internal organs, ready for the *consumer* to cook. Includes neck and giblets.

Ready-to-eat Foods that require no cooking. Most cold breakfast *cereals*, for example, are ready-to-eat.

Real exchange rate The *trade-weighted exchange rate* adjusted by relative rates of inflation as measured by *consumer price indexes.*

Real dollars A measure of the value of *prices*, income, and other monetary amounts in a base period. The *nominal dollar* values are "deflated" by a *price index* to take into account the effects of *inflation* and allow comparisons of a series over time. Real Gross National Product *(GNP)*, for example, measures the output of the economy in any one period at the prices of a selected base year. The actual GNP in 1993 was $6.4 trillion, while the GNP measured in 1987 dollars was $5.1 trillion.

Receival standards Used in Australia for the standards that *grain* has to meet during inspection and *grading* at the point of first sale, when grain passes from the grower to the *Australian Wheat Board (AWB).*

Reciprocal baking An arrangement in which a company's plants (see *plants or establishments*) within a region specialize in the *production* of different bakery products. These plants ship items through regional centers for final assembly, distribution, and sale to *consumers.*

Reclamation The process of converting unproductive *lands* to productive uses.

Recombinant DNA Techniques involving the incorporation of DNA fragments into a suitable *host* organism. By growing the host in a culture to produce clones with multiple copies of the incorporated DNA fragment, the genetic makeup of plants can be altered to enhance various desirable characteristics. See *biotechnology* and *genetic engineering.*

Reconstituted milk Fluid *milk* recombined from ingredients (*nonfat dry milk*, *condensed* milk, and *butterfat*) or concentrated milk.

Red chicory See *radicchio.*

Red clover See *clover.*

Red kidney bean A *variety* of *dry edible bean* sold both dry (see *drying*) and canned and used in making chili and other Mexican dishes. Most kidney beans are canned.

Red meat Meat from beef, veal, pork, and *lamb* and *mutton.*

Red wheat See *hard red spring wheat* and *hard red winter wheat.*

Redrying Preparing *tobacco* for storage in *hogsheads.* Redrying involves removing tobacco moisture below a critical level, followed by application of a uniform moisture content throughout all the leaf.

Reduction rolls See *middlings rolls.*

Re-export *Export* of imported goods or *commodities* without substantial *processing* or transformation. See also *entrepot*, *foreign trade zone*, and *sufferance.*

Re-export sugar The process whereby program participants import *sugar* exempt from *quota* and then process it for *export* either as *refined sugar* or in a sugar-containing product.

Reference price The minimum *import* price for certain *farm* products under the *European Union's (EU) Common Agricultural Policy.* The reference price is normally based on an average of EU *market* or *producer* prices over a given period.

Referendum The referral of a question to voters to be resolved by balloting. For example, *marketing quotas* and *acreage allotments* have been subject to *producer* referenda.

Refined and further processed Oils (see *fats and oils*) that have been subjected to alkali or soda ash or other *refining* processes, and also may have been bleached (see *bleaching*), deodorized (see *deodorizing*), hydrogenated (see *hydrogenation*), or winterized (see *winterizing*).

Refined corn syrup See *dextrose.*

Refined, not further processed Oils (see *fats and oils*) that have been refined (see *refining*) by any one of several processes. This excludes oils that have been bleached (see *bleaching*), deodorized (see *deodorizing*), hydrogenated (see *hydrogenation*), or winterized (see *winterizing*).

Refined sugar A *sugar*, with most of the undesirable impurities removed, that is used primarily for human consumption.

Refining The process of removing impurities from a *commodity*, such as *sugar* and *fats and oils.*

See *refined and further processed, refined corn syrup, refined, not further processed*, and *refined sugar.*

Refining loss The loss from the original quantity of unrefined oil (see *fats and oils*) resulting from various *refining* processes. This loss varies considerably depending on the product, other substances removed during refining, and the method of refining.

Regular auction market A *market* facility that receives *livestock* from the seller and sells to buyers for a fee at an *auction* open to the public.

Reinsured company Private companies that sell, service, and settle claims on crop insurance policies. Policy terms and *premiums* are set by the *Consolidated Farm Service Agency (CFSA),* and are the same as for the policies sold by *master marketers.* Unlike master marketers, reinsured companies bear risk on the policies they sell, although FSA reinsures them against extraordinary losses. They are reimbursed for their selling expenses, at a percent of the total premiums that they sell.

Release price See *farmer-owned reserve program.*

Remote sensing The acquisition of information on planting, harvesting (see *harvest*), water supplies, and other crop developments, usually by satellite, through the use of infrared and other sensing apparatus.

Render To extract (see *extraction, mechanical* and *extraction, solvent*) *fats and oils* from *livestock* or *poultry* by melting down or reprocessing (see *processing*) meat, bone, feathers, or *byproducts.*

Rendering Separating animal fat (see *fats and oils*) from tissue and cellular (see *cell*) structure by applying heat, pressure, solvents, or a combination of these.

Renewable natural resources *Resources,* such as forests, *rangeland*, *soil*, and water, that can be restored and improved.

Renting quota Payment for the right to grow and sell a specified quantity of *tobacco.* Generally, the tobacco is grown on the *farm* to which the quota is assigned.

Reopening signup payments *Payments-in-kind* made to *producers* participating in a *production control* or loan program to encourage additional planted acreage (see *acre*) to be diverted prior to *harvest.* The secretary of agriculture has the option to reopen signup and accept *bids* from producers willing to divert additional acreage if domestic or world *demand* or *supply* conditions change substantially after normal signup, resulting in burdensome and costly surpluses (see *surplus*).

Replacement heifers Immature female *cattle* selected at or after *weaning* to be bred and added to the brood cow (see *beef cow*) herd.

Replenishment agricultural worker (RAW) program A provision of the Immigration Reform and Control Act of 1986 allowing *farmworkers* to enter the United States between 1990-1993 to replace newly eligible *special agricultural workers (SAW)* who might have shifted to other occupations if there were a shortage of workers to produce and *harvest* "perishable" crops. Replenishment agricultural workers are required to work in crops included under the SAW program for at least 90 worker-days (defined as a person working at least four hours a day) a year in each of the first three years of residency in order to remain in the United States and become permanent residents. If they continue to work 90 worker-days in the covered crops for two additional years, they can qualify for U.S. citizenship.

Request/offer A negotiating approach whereby requests are submitted by a country to a trading partner identifying the *concessions* it seeks through negotiations. Compensating offers are similarly tabled and negotiated by delegates of the countries involved.

Reseal Federal loans extended past the original maturity date during which time farmers receive storage payments.

Reserve pool A quantity provision in a *federal marketing order* that requires that some marketable supplies be withheld from the primary (fresh) *market* for sale in a secondary (e.g., frozen or processed) food market, for sale in a nonfood use, or for *stocks* to be sold in a future *marketing year.*

Reserves See *buffer stock*, *carryover*, and *Commodity Credit Corporation.*

Residual fertilizer The amount of *fertilizer* that remains in the *soil* after one or more cropping seasons.

Residual market The *market* serviced after *demand* has been saturated in more profitable markets.

Residual supplier A country that furnishes supplies to another country only after the latter has obtained all it can from other preferred sources.

Resources The available means for *production*, including *land*, labor, and capital.

Restitution An *export subsidy* on agricultural products; used by the *European Union.* Specifically, a restitution is a *subsidy* calculated to offset the difference between the EU *prices* and *world prices.* In contrast, a subvention is a subsidy given without regard to *market* prices.

Retail beef or retail cuts The form in which beef cuts are sold by retailers, such as grocery stores, to *consumers.*

Retailer-owned wholesaler See *cooperative wholesaler.*

Retail price The per-unit or per-item *price* for the *commodity* paid to retailers by *consumers.*

Retaliation An action taken by one country against another for imposing a *tariff* or other *trade barrier.* Forms of retaliation include raising tariffs, imposing *import* restrictions, or withdrawing previously agreed upon trade *concessions.*

Retirement-destination county See *county-type classification.*

Returns to operators The return to management after all other factors of *production* have been covered.

Revenue insurance A program that would provide farmers with a guaranteed revenue. Farmers would select and purchase insurance for a specific percentage of their normal *yield* at a set *price.* A revenue insurance program has not been implemented. See also *income insurance.*

Revenue pool With a classified pricing system such as that used in federal and state *milk* marketing orders (see *federal marketing orders and agreements*), processors pay for milk at different *prices* for each use category. *Producers* are paid a weighted average, or *blend price,* for all uses of milk in a particular order or *market. Processors* pay into the pool on the basis of their uses of milk and the respective class prices in each use; these are the pool revenues. Producers participating in the pool receive identical uniform blend prices, with adjustments for *butterfat* content and location of the *farm.* See also *classes of milk.*

Reverse osmosis filtration A membrane separation technique used to remove water from fluid *milk,* yielding a *concentrate* for shipping and recombining at the final destination. The process can *yield* a 50 percent concentrate without altering the milk's taste and *nutrient* characteristics.

Reverse preferences *Tariff* advantages offered by *developing countries* to *imports* from certain developed nations that granted *preferences* in return.

Rhizobium A genus of cylindrical, aerobic, mesophilic *bacteria* that live on the roots of leguminous (see *legume*) plants causing the formation of nodules. Rhizobia remove *nitrogen* from the air and *soil* and convert it to forms that plants can utilize for growth.

Rhubarb A vegetable with edible stalks, which are used in pies, compotes, and other desserts.

Rice A *grain* used for milled (see *milling*) rice, bakery products, beer, and wine. *Varieties* include Indica, Japonica, Basmati, and glutinous. The United States, Thailand, China, India, Indonesia, and Burma are the major rice producers.

Rice bran The outer cuticle layers and *germ* directly beneath the *hull.* This is removed during the *milling* process. Rice bran is rich in *protein* and B *vitamins.* Rice oil (see *fats and oils*) is extracted (see *extraction, mechanical* and *extraction, solvent*) from the rice bran.

Rice Millers' Association (RMA) An organization of independent and farmer-*cooperative,* rice milling operators that provides data on rice *production* and *milling* and serves as legislative liaison on *rice* issues.

Rice oil See *rice bran.*

Ridge-till A *conservation tillage* method where the *soil* is left undisturbed prior to planting. *Crop residue* is placed in valleys between ridges to reduce soil *erosion*, provide shelter for wildlife, reduce evaporation, and increase moisture. *Seeds* are planted in the ridges, usually with one row of seeds per ridge. See also *no-till*, *mulch-till*, and *strip-till.*

Ridge-till cultivator A machine used for *conservation tillage.* The ridge-till *cultivator* piles *soil* in long rows, or ridges, to control *erosion* and weeds and to conserve moisture.

Rinderpest A usually fatal, viral *disease* of *ruminants* most often characterized by profuse diarrhea. The disease is transmittable by contact with an infected animal, though not to humans.

Ring spinning The traditional method of *yarn* formation in which *fibers,* having been roller-drafted, are twisted together as they emerge, by the rotation of the yarn about itself caused by the rotation of the vertical bobbin upon which it is to be wound. The yarn is directed onto the bobbin by passing around a small wire clip (traveller) that it pulls around a vertically moving ring concentric with the bobbin.

Riparian areas Transitional areas between water bodies and uplands.

Riparian rights See *water rights.*

Ripper An implement with long, heavy, steel *shanks* mounted on a sturdy steel frame. Designed to dig or loosen hard *soil*, using *chisel points* or other types of *shovels*.

Risk assessment The systematic estimation of the probability of harm caused by a hazard and the distribution of likely severities.

Risk aversion Preference for a certain payoff over a random payoff with equal expected value.

Risk premium A return paid or earned for bearing risk.

Risky Subject to randomness in outcomes that are not equally desirable to the decisionmaker.

RNA The acronym for ribonucleic acid, a molecule that decodes instructions for *protein* synthesis carried on by the genes. See also *DNA*.

Roaster A 12- to 16-week old *poultry* bird, usually weighing 4 to 6 pounds dressed (see *dressed animal*), used for pan roasting.

Roasting pig A *milk*-fed *pig* weighing from 60 to 100 pounds.

Robusta See *coffee tree*.

Rock picker A heavily constructed implement designed to lift rocks from the surface of the *soil* and store them in a hopper or bin, to be discarded later in a rock pile. Construction varies from forks on the front of a loader to a fork and chain *elevator* mechanism on a unit pulled behind a *tractor*. Rock removal is necessary in some parts of the country to reduce damage to *drills*, *combines*, and other equipment.

Rodweeder An implement with a rotating horizontal rod drawn below the *soil* surface to kill weeds and help prepare the seedbed.

Roll-away nest A nest constructed with a sloping floor so that *eggs* roll away from the hen after laying. Roll-away nests are used to ensure cleaner eggs and make collection easier.

Rollbar See *roll-over protection system (ROPS)*.

Rolled beef Beef stamped (using a roller) the length of the *carcass* with a *grade* designation visible to the buyer.

Roller mill A machine, usually electrically powered, designed to crush and prepare *grains*, such as *oats* for *feed*. Grains are crushed between rollers under pressure and may be softened with steam.

Rolling 1: Changing the shape and/or size of *grain* particles by compressing them between rollers. It may entail *tempering* or conditioning. See *roller mill*.

2: The act of putting a *grade* designation on beef *carcasses*. See *rolled beef*.

Rollout or switchout or pullout The procedure to remove one or several *carcasses* to a center rail from a long rail of carcasses in the cooler. This is done by a mechanical device or use of a switch without moving all the carcasses.

Roll-over frame See *roll-over protection system (ROPS)*.

Roll-over protection system (ROPS) A frame or bar mounted over the seat of a *tractor*, constructed of steel tubing or bars of sufficient strength to bear the weight of the tractor and protect the operator if the tractor accidently turns over.

Romaine A variety of *lettuce* with upright clusters of large, crunchy leaves.

Rooster A mature male chicken (see *poultry*).

ROPS See *roll-over protection frame.*

Rot The breakdown or decomposition of plant tissue by *bacteria* and fungi (see *fungus*).

Rotary hoe A tool pulled behind a *tractor* used to control weeds by removing them from the *soil.* See *spider wheel cultivator.*

Rotary tiller *Tillage equipment* used to break up *soil*, chop surface residue, and mix the materials into the soil.

Rotation, crop The growing of different crops in recurring succession on the same *land.*

Rotation pasture A *pasture* used for a few seasons and then plowed (see *plowing*) for other crops.

Roughage Vegetable plants that are relatively bulky, high in crude *fiber*, and low in total digestible *nutrients*, in contrast to *concentrate*. The major groups of roughage include *pasture* and *range* plants, *hays*, and *silages*.

Roughage consuming animal units (RCAU) An aggregate index that converts *roughage* consumption for different *species* to a common unit. *Livestock* and *poultry* numbers are weighted by *pasture* and roughage consumption factors for a base period.

Rough rice Harvested (see *harvest*), whole-*kernel* rice with the *hull* remaining. Rough rice is sold to mills (see *milling*) for dehulling and polishing.

Round A *multilateral trade negotiation* under the auspices of the *General Agreement on Tariffs and Trade (GATT)* that results in trade agreements among participating countries. Eight rounds of negotiations, including the most recent *Uruguay Round*, have been conducted since the GATT was initiated in 1947. See *GATT Rounds.*

Round baler A *baler* that makes large cylindrical *bales* weighing 500–2,000 pounds.

Round bean See *bambara groundnut.*

Routine hedging *Hedging* according to standard rules without attempting to anticipate changes in the futures price.

Roving The penultimate stage of the *ring spinning* process of *yarn* manufacturing in which *sliver* is attenuated and the resultant assembly is slightly twisted, a process that is necessary for strand cohesion. Roving is a coarse but soft "yarn" that can be converted into *cloth*

Row crop cultivator A *farm* implement designed to work *soil* and kill weeds between rows of plants. It may be mounted on a *tractor* or pulled separately and may be several rows wide. See *cultivator.*

Row crops Crops that are grown in rows and must be planted each year, such as *corn, soybeans,* and *sorghum.*

Royal jelly A white substance secreted by *nurse bees* and used to feed developing larvae (see *larva*).

Rumen The first large compartment of the digestive system of *ruminant* animals.

Rumensin A nonhormone (see *hormone*), *fermentation* product that helps improve the conversion of *feed* to body gain by promoting the more efficient use of *fatty acids* as energy.

Ruminant An animal, including *cattle, sheep,* goats, *bison,* deer, elk, and others, that has a stomach with four compartments (see *rumen*) and that chews a cud consisting of regurgitated, partially digested food.

Running bale See *bale (cotton).*

Runoff The surface flow of water from an area; or the total volume of surface flow during a specified time.

Runt The animal in a *litter* that is of small size and poor quality.

Rural Open country areas and towns with less than 2,500 population that are not within a *Census Bureau*–designated urbanized (see *urban*) area.

Rural Business and Cooperative Development Service (RBCDS) A *U.S. Department of Agriculture (USDA)* agency created by merging selected functions of the *Rural Development Administration,* the *Rural Electrification Administration,* and the *Agricultural Cooperative Service.* The RBCDS is responsible for USDA's business development programs.

Rural Development Administration (RDA) A *U.S. Department of Agriculture* (USDA) agency established by the *Food, Agriculture, Conservation, and Trade Act of 1990 (P.L. 101-624)* to administer some existing *Farmers Home Administration* programs, including water and waste disposal loans and grants, industrial development grants, *resource* conservation and development loans, business and industry loan guarantees, community facility loans, and the *Intermediary Relending Program.* Under the 1994 USDA reorganization, RDA's functions were transferred to the newly created *Rural Utilities Service,* the *Rural Housing and Community Development Service,* and the *Rural Business and Cooperative Development Service.*

Rural Electrification Administration (REA) The name of the former *U.S. Department of Agriculture (USDA)* agency that assisted *rural* electric and telephone utilities in obtaining financing. REA was established in 1935 and abolished in 1994. Under a 1994 USDA reorganization plan, REA's functions were transferred to the newly created *Rural Utilities Service,* the *Rural Housing and Community Development Service,* and the *Rural Business and Cooperative Development Service.*

Rural Housing and Community Development Service (RHCDS) A *U.S. Department of Agriculture (USDA)* agency created in October 1994 by merging selected functions of the *Farmers' Home Administration,* the *Rural Development Administration,* and the *Rural Electrification Administration.*

Rural Utilities Service (RUS) A *U.S. Department of Agriculture (USDA)* agency created in October 1994 by merging selected functions of the *Rural Electrification Administration (REA)* and the *Rural Development Administration (RDA).* The RUS is responsible for USDA's *rural* telephone, electric, sewer, and water programs.

Russet Brownish, roughened areas on the skin of fruit resulting from *diseases*, insects, or spray injuries.

Rust A *disease* caused by a rust *fungus*; or the fungus itself. On *grains* and *legumes*, for example, the disease causes defoliation (see *defoliate*) and reddish brown lesions. *Quarantines* are imposed on trade from affected areas.

Rutabaga A vegetable that is similar to a turnip; however rutabagas are larger, rounder, denser, and sweeter.

Rye A *cereal* grass whose *seeds* are used as *grain*. Rye is used for bread, *flour*, feed, and distilled spirits. Poland, the former Soviet Union, and Germany are the major producers.

S

Saccharide A *sugar* carbohydrate (see *carbohydrate*).

Saccharimeter See *polarization*.

Saccharin A *low-calorie sweetener* with a sweetening power of 180 to 200 times that of *sucrose*.

Safeguards Temporary measures implemented in order to protect an industry while it adjusts to increased competition by *foreign* suppliers. Safeguards can include *tariffs* or *quantitative restrictions*.

Safety shield A metal or plastic device to protect against accidental contact with the item being shielded, such as a *power-take-off (PTO)* or exhaust manifold.

Safflower seed The *seed* of a plant with orange flowers; the seeds yield an oil (see *fats and oils*) used in cooking, cosmetics, medicines, and paints. China, India, the United States, and Egypt are the major safflower producers.

Salad or cooking oils Products prepared from vegetable oils (see *fats and oils*) that are usually refined (see *refining*), bleached (see *bleaching*), deodorized (see *deodorizing*), and sometimes lightly hydrogenated (see *hydrogenation*).

Sales agent or sales agency A person or firm who finds buyers for products and, on behalf of the principal (*packer*-shipper), negotiates a selling *price* and other terms of sale for a fee paid by the principal.

Saline soil A *soil* containing enough soluble salts to impair its *productivity* for plants and not containing an excess of exchangeable sodium.

Salmonellosis A *disease* of humans frequently associated with consumption of undercooked or improperly handled *poultry*, *eggs*, beef, pork, dairy products, or occasionally seafood or raw *produce*. Symptoms range from a day or two of mild diarrhea and vomiting to hospitalization for dehydration and diarrhea, blood poisoning, or sometimes death. The severity of the symptoms depends, in part, on how many *bacteria* are consumed and how well the body can fight off the bacteria. See also *campylobacteriosis*.

Salsify A root vegetable, resembling parsnips, with a delicate, oysterlike flavor. Also called oysterplant.

Sample fiber A *cotton fiber* used because its length varies within a relatively narrow range, from about 7/8-inch to 1/34 inch.

Sanitary and phytosanitary regulations Laws imposed by government regulatory agencies to protect human, animal, and plant life and health from risks arising from *additives,* contaminants, *toxins*, *diseases*, and pests in or on agricultural products. In some cases, these regulations can be used as *nontariff trade barriers*. See *eradication, pasteurization, irradiation,* and *control*. See also *Codex Alimentarius.*

Sanitary quality (grain) *Grain* characteristics associated with cleanliness. They include the absence of *foreign material* that detracts from the overall value and appearance of the grain, including the absence of dust, broken grain, rodent excreta, insects, residues, fungal (see *fungus*) infection, and nonmillable (see *milling*) matter.

Sapote An *apple*-sized fruit, native to the highlands of Central America, with a flavor similar to a blend of *banana* and *peach.*

Satellite image data products See *imagery.*

Saturate In *soil,* to fill all the openings among the particles with water.

Saturated fat See *fatty acids (classification of).*

Scallions See *onions.*

Scarified seed A hard *seed* treated to germinate (see *germination*) more quickly.

Schedule B The official U.S. schedule of *commodity* classifications used in reporting *export* shipments from the United States. It contains approximately 4,000 seven-digit classifications, with a numbering system based on the "Tariff Schedules of the United States, Annotated," which is used for classifying goods imported into the United States.

School Breakfast Program (SBP) A federal program that provides financial and *commodity* assistance to public and nonprofit private schools agreeing to serve nourishing breakfasts according to *U.S. Department of Agriculture* (USDA) meal patterns. Public or private nonprofit, licensed, residential child care institutions are also eligible. Meals are offered free, at reduced, or full prices, depending on the child's family income. The program was authorized in 1975.

Scours An infectious *disease* of young animals that often occurs soon after birth. The disease is similar to diarrhea.

Scouting The inspection of a field for insects, weeds, or *pathogens* to determine if intervention is necessary and what is the best method of control.

Scrape gutter A gutter in the floor of a *livestock* facility through which *manure* is mechanically scraped periodically into a storage or disposal facility. Gutters may be open or under a slotted cover.

Scrapie A *disease* of the nervous system of *sheep* caused by an unidentified infectious *agent*. The disease is characterized by itching and nervous movement of the mouth.

Scratch-mix format *Production* process in which 95 percent or more of the bakery's output is made from scratch or from a commercial mix at the plant (see *plant or establishment*) or store.

Screw augur conveyor A round tube containing a continuous screw on a spiral. This is the principal means of moving *grain* on *farms* where inexpensive portable equipment is needed.

Sea Island cotton An *extra-long staple (ELS) cotton* first grown in the United States about 1786 from *seed* received from the Bahama Islands. Sea Island cotton was relatively unimportant as a commercial crop until the nineteenth century. It was produced in the coastal areas of South Carolina, Georgia, and Florida until the early 1920s, when U.S. *production* virtually ceased because of increasing competition from *foreign* production of ELS cotton, the growing *American-Egyptian cotton* industry in the Western States, and production problems associated with Sea Island cotton. Sea Island cotton was commonly about 1 1/2 inches in length but ranged up to 2 inches.

Seasonal agricultural worker A person who is employed in seasonal or other temporary agricultural work and is not required to be absent overnight from his or her permanent place of residence.

Seasonal prices *Prices* or price patterns that regularly occur at a particular season(s). Seasonal prices are usually related to a marketing (see *agricultural marketing*) period rather than to a part of the calendar year.

Secondary feed manufacturing *Processing* and *mixing* one or more ingredients with formula *feed* supplements. *Supplements* are usually used at a rate of 300 pounds or more per ton of finished feed, depending on the *protein* content of the supplement and the percentage of protein desired in the finished feed.

Secondary (resale) markets for agricultural loans See *Federal Agricultural Mortgage Corporation.*

Secondary wholesaler A *wholesaler* who buys from local *wholesale handlers* and resells to other wholesalers, such as *jobbers* and *truck jobbers.* A secondary wholesaler handles the merchandise and takes title.

Second clear flour The lower-grade portion of *clear flour* from the tail-end reductions of the *milling* system. Second clear *flour* has a higher ash content and poorer color than *first clear* flour.

Second heads Fragments of *rice* grains (see *grain*) broken during *milling*, which are at least one half as long as whole *kernels*, but less than three-fourths. This is the largest size of broken rice (see *brokens*).

Section 22 A section of the Agricultural Adjustment Act of 1933 (P.L. 73-10) (see Appendix 3) that authorizes the U.S. president to restrict *imports* by imposing *quotas* or fees if the imports interfere with federal *price support programs* or substantially reduce U.S. *production* of products processed (see *processing*) from *farm* commodities (see *commodity*).

Section 32 A section of the Agricultural Adjustment Act Amendment of 1935 (P.L. 74-320) (see Appendix 3) that authorizes use of *customs* receipts funds to encourage increased consumption of agricultural *commodities* by means of purchase, *export*, and diversion programs. Section 32 is funded by a continuing appropriation of 30 percent of the *import* duties (see *tariff*) imposed on all commodities, both agricultural and nonagricultural. Domestic acquisition and donations constitute the major use of section 32.

Section 201 Part of the U.S. Trade Act of 1974 (P.L. 93-618) (see Appendix 3) that allows the U.S. president to provide relief to industries hurt by competing *imports*. Growers or trade associations must petition the *International Trade Commission* to investigate complaints of trade practices.

Section 301 A provision of the U.S. Trade Act of 1974 (P.L. 93-618) (see Appendix 3) that allows the U.S. president to take appropriate action to persuade a *foreign* government to remove any act, policy, or practice that violates an international agreement. The provision also applies to practices of a foreign government that are unjustified, unreasonable, or discriminatory and that burden or restrict U.S. commerce.

Section 416 A section of the Agricultural Act of 1949 (P.L. 89-439) (see Appendix 3) intended to dispose of agricultural *commodities* to prevent waste. It permits donations of agricultural products to public and private nonprofit humanitarian organizations, *foreign* governments, and international organizations.

Sediment *Soil* particles deposited by water.

Sedimentation test A test that measures the quality of *protein* content in *wheat.* Ground (see *grinding*) wheat is suspended in water and treated with lactic acid. The portion that settles to the bottom of a graduated cylinder within five minutes is the sedimentary value.

Seed 1: A mature ovule consisting of an embryonic (see *embryo*) plant together with a store of food, surrounded by a protective wall.
2: To plant seeds in *soil.*

Seed bank A facility designed to preserve and disseminate *seeds*, particularly *varieties* that are not used commercially or that may be threatened with extinction.

Seed, certified *Seed* that has been determined by an official seed-certifying agency to meet standards for genetic purity and identity.

Seed cleaner A machine used to separate *chaff*, weed *seeds*, and other debris from *grain*, grass, or other seed. See also *fanning mill* and *carter mill.*

Seed cotton The raw *cotton* product that has been harvested (see *harvest*) but not ginned (see *gin*), containing the *lint*, *seed*, and *foreign material.* Within each boll of cotton, there are from five to nine seeds per lock. In *Upland cotton*, the seeds have fuzzy, short *fibers*, as well as long fibers.

Seeder A device to sow *seed.* See *drill*, *planter*, *grass seeder*, and *pasture seeder.*

Seed, pure live The percentage of *seeds* in a lot that is pure and that will germinate (see *germination*).

Seed, treated *Seed* to which a substance designed to control *disease* organisms, insects, or other pests has been applied.

Selective hedging or discretionary hedging *Hedging* that takes into account anticipated changes in the *futures price.*

Self-feeder A *feed* container consisting of a storage compartment designed to release feedstuffs gradually into an attached bunk (trough), manger, or other feed provider to which *livestock* have free access.

Self-sufficiency The ability of a nation to produce all that it consumes.

Semen The impregnating fluid of male animals that contains spermatozoa. See *artificial insemination.*

Semiworsted system A sequence of machinery similar to those employed in the *worsted system* but one in which *combing* is omitted.

Semolina A coarse separation of *endosperm* extracted from *durum wheat* used to make *pasta.*

Service cooperative See *agricultural cooperative.*

Service wholesaler A *wholesaler* that carries a complete line of grocery products (including nonfood items and some perishables) and sells mainly to independent retail *foodstores.* Service wholesalers perform at least two services for retail stores, such as suggesting retail *prices*, training personnel, handling items, and assisting with merchandising and advertising.

Sesame An East Indian *herb* grown primarily in India, China, Mexico, and Japan. Products from sesame include confectionery *seed*, edible oil (see *fats and oils*), and oilmeal (see *meal*).

Set The number of fruit or blossoms on a tree.

Set-aside A voluntary program to limit *production* by restricting the use of *land.* Introduced in 1970, set-asides may be implemented at the discretion of the secretary of agriculture but have not been offered since 1979. When a set-aside program is in effect, the total of the planted acreage of the designated crops and the set-aside acreage cannot exceed the *normal crop acreage.* Producers must comply to be eligible for *commodity* loan programs or *deficiency payments.*

Settling basin A basin designed to settle out and retain most of the solid materials in *runoff* from a *feedlot* before it passes to a *manure lagoon* or *vegetative filter.* The solids must be removed periodically.

Sewing machine See *tying machine.*

Shaddock See *pummelo.*

Shaker A device, either mounted on a *tractor* or self-propelled, that is used to shake trees to remove fruit or nuts.

Shallot See *onion.*

Shank A long steel bar extending down from the frame of a *cultivator*, *chisel plow*, or similar implement, upon which various sized and shaped shovels are mounted.

Shape A pointed steel blade mounted at the end of a *shank* designed to penetrate and loosen the *soil* and kill weeds. A shape may be different sizes from about 2 inches to a foot or more wide. It may be straight or have "wings" extending out and back from the point.

Share, plow See *plow share.*

Sharecropper A tenant who accepts a portion of crops, *livestock*, or livestock products from the landowner as payment for labor. The landowner often extends credit to and closely supervises the tenant. The sharecropper generally supplies only labor.

Shearing Removing *wool* from *sheep.*

Sheep A *ruminant* animal bred for meat and *wool.* Rambouillet is one of the major wool *breeds*, while primary meat breeds include Hampshire, Southdown, Dorset, Shropshire, and Suffolk. Australia, the former Soviet Union, New Zealand, India, and Turkey are major sheep producers. See *lamb* and *mutton.*

Sheet and rill erosion See *erosion.*

Shelf-stable foods *Vacuum-packed* foods in plastic containers that require no refrigeration or preservatives.

Sherman Antitrust Act See *price fixing.*

Shippers Export Declaration A form required by the U.S. government for almost all commercial shipments leaving the United States (except *meal* shipments of small value). The declaration is used to compile trade statistics.

Shipping holiday A provision of *federal marketing orders and agreements* that prohibits commercial shipping during periods following certain holidays, usually for 3 to 7 days after Thanksgiving and Christmas, when demand is historically low.

Shoat A young *pig* up to five months old that weighs 50 to 160 pounds.

Shorn mohair Grease *mohair* sheared from a live Angora *goat.* Shorn mohair does not include pelts or mohair removed from pelts, scoured or dyed mohair or *yarn, skeins* or forms of mohair other than its natural greasy state.

Shorn wool Grease *wool* sheared from live *sheep* or *lambs*, including black wool, tags, *crutchings,* and murrain or other wool removed from dead animals. Shorn wool does not include pelts or wool removed from pelts, scoured, carbonized, or dyed wool or *yarn*, skeins or forms of wool other than its natural greasy state.

Short A person or firm who holds a *contract* to sell a *commodity* at a specified *price.*

Shortening A semisolid, fatty food product made from edible vegetable *oils* (see *fats and oils*), a blend of vegetable oils and animal fats or pure animal fats for use in cooking, baking, and frying. There are two general classes: (1) compound or blended, classified on the basis of the raw material used as ingredients; and (2) hydrogenated (see *hydrogenation*), which may be further identified on the basis of their intended use.

Short-grain rice *Rice* that meets a government-established standard for *grain* sizes and shape types. Short-grain rice has a length/width ratio of 1.9 or below. Short-grain rice accounts for less than 1 percent of U.S. rice *production,* and almost all short-grain rice is produced in California. See also *long-grain rice* and *medium-grain rice.*

Short position The holding of a *contract* to sell a *commodity* for a specified *price.* See *open position.*

Shorts Fine particles of *bran, endosperm,* and *wheat germ* that remain after *milling* is completed. Shorts are used as *feed.*

Short staple fibers See *staple fibers.*

Short ton A measure of weight equal to 2,000 pounds or 907 kilograms. See also *metric ton* and *long ton.* See Appendix 6.

Shovel 1: A common hand tool with a blade and a long wooden handle, used to move *soil*, *manure*, and so on.
2: See *shape* and *sweep*.

Shrink (grain) The loss of weight in *grain* due to the removal of water.

Shrink (livestock) *Livestock* weight loss often associated with handling (*processing*) and transporting during the marketing (see *agricultural marketing*) process.

Sickle mower A machine that cuts standing *forage* crops.

Side-delivery rake An implement with long, thin, flexible tines mounted on off-set rotating wheels and used to push *hay* and other *forage* crops along the ground in long rows. Hay is later picked up by a *baler* or *forage harvester*.

Sideliners A person who keeps *honeybees* for profit to supplement income from other employment. The industry generally considers sideliners as maintaining 25 to 299 bee *colonies*. See also *commercial beekeeper*.

Sideweight The quantity in pounds of one-half of a beef *carcass*. The carcass is cut in half from head to tail down the back into two "sides."

Silage A fermented (see *fermentation*) *forage* from *grain* or *hay* crops. Usually *corn*, *sorghum*, or various *legumes* and grasses that have been preserved in moist, succulent condition by partial fermentation in a *silo* or other tight container above or below the ground. Silage is mainly used as cattle *feed*.

Silo A structure or trench for green *forage* that seals off air and allows forage to ferment (see *fermentation*) and form *silage*. Two common types include a tall circular structure or a trench made in the ground, sometimes concrete-lined, usually covered with plastic sheeting to keep out air.

Silviculture The management or cultivation of forest trees.

Simplesse A *protein-based fat substitute* made from *egg* whites and skim *milk* (see *milkfat content*) or whey. Simplesse could be used in manufacturing certain foods, such as ice cream, *yogurt*, *cheese*, sour cream, dips, as well as oil-based (see *fats and oils*) foods like salad dressings and *mayonnaise*. The Simplesse compound cannot be cooked.

Single-cross hybrid A first-generation *hybrid* between two selected and usually inbred (see *breed*) lines.

Single-factor productivity index See *productivity*.

Sire The male animal used in breeding.

Sisal A *perennial* crop grown in Brazil, Kenya, Tanzania, Madagascar, Haiti, and other tropical countries. Sisal is used primarily as a bailing twine. See also *hard fibers*.

Sitosterol See *stigmasterol*.

Size An adhesive applied to *warp* yarns (see *yarn*) prior to weaving (see *weave*) by the process of slashing, often with a lubricant, to increase weaving efficiency by improving abrasion resistance.

The size is generally removed in the first process of fabric (see *cloth*) finishing.

Skep A dome-shaped *beehive* made of *straw,* wood, or other material.

Skid loader A self-propelled *loader* used for the same purposes as a *front-end loader.*

Skid steer loader A highly maneuverable, four-wheel-drive machine equipped with a hydraulically operated bucket or fork, used for moving *manure, grain, hay,* or other materials.

Skins See *shorn wool.*

Skip-row planting Planting in uniform spaces one or more rows to a *commodity* (especially dryland *cotton*), then skipping one or more rows to leave them empty.

Slaughter The killing and butchering of *cattle, swine, poultry,* and other animals for food.

Slaughter, farm Animals slaughtered (see *slaughter*) on *farms* primarily for home consumption. Excludes custom slaughter for farmers at commercial establishments (see *plants or establishments*), but includes mobile slaughtering on farms.

Slaughter hog A hog (see *swine*) finished for the *slaughter* market (see *market*) usually weighing 220 to 240 pounds.

Slaughterhouse A place where animals are killed to produce meat. Also called slaughter plant.

Slaughter plant See *slaughterhouse.*

Slip A cutting, shoot, or leaf to be rooted for vegetative propagation.

Slit tillage A *plowing* method that cuts deeper and more widely spaced *furrows* and maintains crop *productivity* while decreasing *runoff* and *erosion.* The objective in slit tillage is to create slits and fracture the earth between the slits so that rainwater can infiltrate the ground and not run off.

Sliver A thick strand or rope of *fibers* without twist. In *yarn* manufacturing, a sliver is formed by the *carding* machine and in the subsequent product of the *drawframe.* It has a greater diameter than *roving.* In the worsted industry (see *semiworsted system* and *worsted system*), sliver of well-prepared combed fibers is known as "top." The equivalent in the *man-made fiber* industry is "tow."

Slope The incline of the surface of a *soil.* It is usually expressed in percentage of slope, which equals the number of feet of fall per 100 feet of horizontal distance.

Slotted floors Buildings or pens with floors partially or completely covered with slats of concrete or other material spaced to allow feces and urine to fall to a storage area below or nearby.

Slot tillage See *no-till.*

Slotting fee A practice originating in the late 1980s in which some retailers charge manufacturers a fee to cover the costs of stocking selected new food or grocery items. The fee may include payment in the form of additional product "over-order." Slotting fees are effectively a rationing device used by retailers to manage limited shelf space.

Small grain pasture 1: *Land* on which small *grains* (*wheat, oats, barley*, and/or *rye*) are grazed (see *graze*) during the early vegetative growth stage prior to subsequent *grain* or *forage* harvesting (see *harvest*) or complete utilization for grazing.

2: Plant materials grazed from immature small grains.

Small-scale farms *Farms* that generate $40,000 or less of agricultural sales each year. Small farms constitute about 75 percent of the 2.2 million U.S. farms.

Smoker A device used to blow smoke on bees (see *honeybee*) to reduce stinging.

Smoking tobacco Pipe *tobacco*, although it may also include tobacco used for roll-your-own *cigarettes*. Smoking tobacco is manufactured in several forms, such as granulated, plug cut, long cut, cube cut, and others. *Burley tobacco* and other types of tobacco are used in smoking tobacco.

Smut Any of various fungal (see *fungus*) *diseases* affecting *grain*. The disease is characterized by black powdery masses of spores on the affected parts of the plant. *Quarantines* are often imposed on trade from affected areas.

Snap-back provision A provision in an agreement that allows a signatory to withdraw *concessions* under specific circumstances, such as a surge of *imports* or *balance of payments* disequilibria. For example, the snapback provision of the *U.S.–Canada Free Trade Agreement (CFTA)* guards against imports of fruits and vegetables from the United States or Canada depressing domestic *prices* in the other country. Either country may use the snapback provision, effective for 20 years from January 1, 1989, to temporarily return *tariffs* on fresh fruits and vegetables imported from the other country to their *most-favored nation* level.

Snow pea An edible *pea* pod containing tiny peas. Also called Chinese pea pod and sugar pea.

Snuff A finely cut or pulverized *tobacco* used by placing a small quantity between the lower lip and gum. The United States produces three basic types of snuff: dry, moist, and semimoist that are either fine or coarse, flavored or toasted, and plain scented.

Soap A commercial cleansing or sudsing agent made of water-soluble salts of *fatty acids* and other natural acids.

Soapstock The *byproduct* obtained when vegetable oils (see *fats and oils*) are deacidized by various *refining* methods. Soapstock is often referred to as "foots" since it accumulates at the bottom (foot) of the refining tank.

Sodbuster A provision first authorized by the *Food Security Act of 1985 (P.L. 99-198)* (see Appendix 3) that is designed to discourage the conversion of *highly erodible land* from extensive *conserving uses* to intensive agricultural *production*. If highly erodible grassland or woodland that was not cropped in 1981–1985 is used for crop production without approved conservation measures, *producers* may lose eligibility for several *U.S. Department of Agriculture* programs.

Sod (turfgrass) All *varieties* of specialized grasses cultivated for sale as sod for golf courses, homes, parks, and businesses. Examples are fine fescue, tall fescue, Kentucky bluegrass, Bermuda grass, buffalo grass, zoysia grass, St. Augustine grass, and centipede grass.

Soft currency A currency that may not be exchanged without restrictions for another currency, or an unstable currency, the value of which is likely to decline. Also referred to as inconvertible currency.

Soft fibers Flexible *fibers* of soft texture obtained from the inner bark of dicotyledonous plants. Soft fibers are fine enough to be made into *cloth* and cordage. Examples are *flax*, *hemp*, *jute*, *kenaf*, and *ramie*. See *hard fibers*.

Soft manufactured dairy products Manufactured dairy products with a limited storage life, including ice cream, cottage *cheese*, *yogurt*, and sour cream.

Soft red winter wheat (*Triticum aestivum*) A fall-seeded *wheat* with low to medium *protein* content and a soft *endosperm*. The wheat is used primarily in the *production* of cakes and other pastries.

Soft wheat *Wheat* with a chalky *endosperm* and low amounts of *protein*.

Soft wheat flour See *flour*.

Softwood A botanical group of trees that in most cases have needlelike or scalelike leaves, known as conifers. Softwood is used to produce furniture, structural supports for homes, plywood, and veneer.

Soil A dynamic natural body on the surface of the earth composed of mineral and organic materials and living forms in which plants grow. A kind of soil is a collection of soils that are alike in specified combinations of characteristics. In the United States, about 70,000 kinds of soil are recognized in the nationwide system of classification.

Soil association A group of defined and named kinds of *soil* associated together in a characteristic geographic pattern.

Soil auger A tool for obtaining small *soil* samples for field or laboratory observations.

Soil Bank A program mandated by the Soil Bank Act of 1956 (see Agricultural Act of 1956, P.L. 84-540) (see Appendix 3) to decrease the *supply* of agricultural products by reducing the amount of *land* used in crop *production*. The program was also initiated to establish and maintain protective vegetative cover or other needed *conservation practices*. Land was retired for three, five, or 10 years to a specified type of use, such as grass, trees, or water impoundments. The official name of the "Soil Bank" was the Conservation Reserve Program. The program was voluntary, and participating farmers agreed to comply with any *acreage allotments* on the *farm* and to reduce total cropped acreage (see *acre*) by the amount of land placed in the reserve. The Soil Bank Act was repealed by the Food and Agriculture Act of 1965 (P.L. 89-321) (see Appendix 3).

Soil characteristic A feature of a *soil* that can be seen and/or measured in the field or in the laboratory on soil samples. Examples include soil *slope* and stoniness, as well as the texture, structure, color, and chemical composition of soil horizons.

Soil conservation See *conservation, soil*.

Soil Conservation Service (SCS) The name of the former *U.S. Department of Agriculture (USDA)* agency responsible for USDA's soil and water programs. Under a 1994 USDA reorganization plan, SCS became part of the newly created *Natural Resource Conservation Service*.

Soil erosion See *erosion*.

Soil erosion tolerance values (T-values) The maximum rate of *erosion* under which a high level of crop *production* can be maintained indefinitely. *Soil* maintains its crop *productivity* when erosion does not exceed the T-value. The T-value for 71 percent of U.S. cropland is 5 tons per acre per year (t/a/yr), while some *cropland's* T-value is as low as 1 t/a/yr.

Soiling crops Harvested (see *harvest*) crops fed fresh to *livestock.* Also more commonly known as *green chop.*

Soil injection wastes Injection of liquid wastes from *cattle* feeding below the surface of the *soil* to act as a *fertilizer.*

Soil management The preparation, manipulation, and treatment of *soils* for the *production* of plants, including crops, grasses, and trees.

Soil map A map that shows the distribution of *soil* types or other soil characteristics.

Soil-moisture tensiometer An instrument for measuring the tension with which water is held by *soil* in order to detect drainage problems or estimate when to irrigate (see *acres irrigated, irrigated farms,* and *irrigation methods*) land.

Soil population The group of organisms that normally live in the *soil.*

Soil quality An attribute of *soil* that cannot be seen or measured directly from the soil alone but that is inferred from *soil characteristics* and soil behavior under defined conditions. Fertility (see *fertility, soil*), *productivity,* and erodibility (see *erosion)* are examples of soil qualities (in contrast to soil characteristics).

Soil reaction A measure of the acidity or alkalinity of a *soil* expressed in *pH* value. See *acid soil, alkali soil,* and *alkaline soil.*

Soil series A principal unit of *soil* classification that groups soils that are similar in all characteristics except texture (see *texture, soil*) of the surface layer. See also *soil type.*

Soil sterilization The process of treating *soil* to kill living things in it.

Soil structure The configuration of *soil* solids and spaces resulting from the bonding of soil particles. The principal forms of soil structure are platy, prismatic, columnar, blocky, and granular.

Soil survey The systematic examination of *soils* in the field and in the laboratory; their description and classification; the mapping of kinds of soil; the interpretation of soils according to their adaptability for various crops, grasses, and trees; their behavior under use or treatment for plant *production* or other purposes; and their *productivity* under different management systems.

Soil texture A *soil's* coarseness or fineness as determined by the relative proportion of sand, silt, and clay.

Soil type A subdivision of a *soil series* that includes all *soils* of a series with similar characteristics, such as texture (see *texture, soil*).

Solasodine A compound occurring in eggplant and used in an Australian medication for treating some skin cancers.

Solid feedlot wastes *Manure* and other residues from *livestock* feeding containing insufficient water to be handled as a liquid.

Soluble fiber See *fiber.*

Solubles Liquids containing dissolved substances obtained from *processing* animal or plant materials. Solubles may contain some fine suspended solids.

Solvent extraction See *extraction, solvent.*

Sorghum A cultivated plant derived from a genus of Old World tropical grasses. Similar to Indian *corn*, sorghum is grown for food and *feed grains.* Major sorghum producers include the United States, Argentina, India, Mexico, China, and Nigeria.

Sorrel A leaf vegetable with smooth, bright green, arrow-shaped leaves with a taste similar to *spinach.*

Sous vide foods Processed (see *processing*) foods that are cooked and packaged under vacuum. This processing method is popular in Europe.

Southern Cone Common Market See *Mercosur.*

Sow A female pig (see *swine*) after production of its first *litter.*

Soybean A *legume* widely grown for its oil-rich, (see *fats and oils*) high-protein (see *protein*) *seeds* and for *forage* and *soil* improvement. The United States and Brazil are the leading producers. Products include edible *soybean oil* and *soybean meal.*

Soybean meal The product obtained by grinding (see *grind)* the cake, chips, or flakes that remain after removing most of the oil (see *fats and oils*) from *soybeans* by a mechanical (see *extraction, mechanical*) or solvent extraction (see *extraction, solvent*) process. This product contains 41–50 percent *protein*, 3–7 percent crude *fiber,* and usually about 1 percent or less fat. Soybean meal is used as *feed* for *livestock* and *poultry. Soy grits and flour* are commonly used for human consumption.

Soybean millfeed A *byproduct* resulting from the manufacture of *soy grits and flour*, composed of *soybean hulls* and the *tailings of the mill.*

Soybean oil The product resulting from crushing *soybeans* using a mechanical (see *extraction, mechanical*) or solvent extraction (see *extraction, solvent*) process. The oil (see *fats and oils*) generally amounts to about 20 percent of the weight of uncrushed (see *crushing*) soybeans. The United States is the major producer.

Soy grits and/or flour The ground (see *grind*), screened, *graded* product obtained after extracting (see *extraction, mechanical* and *extraction, solvent*) most of the oil (see *fats and oils*) from selected, sound, clean, dehulled (see *dehull*) *soybeans.* It is obtained from grinding the *defatted soy flakes. Grits* are coarser ground than soy *flour. Protein* range is 40–60 percent.

Soy protein concentrate Produced from *defatted soy flakes* or *flour* by a process that immobilizes the *protein* and removes soluble *sugars*, minerals, and so on. Soy protein concentrate has a protein content of 70 percent.

Special agricultural worker (SAW) program A provision of the Immigration Reform and Control Act of 1986 allowing illegal aliens who had performed farmwork in perishable *commodities* to become temporary resident aliens under employment and residency requirements.

Special drawing rights (SDR) International reserve assets, created by the *International Monetary Fund (IMF)* and allocated to individual member nations. Within conditions set by the IMF, a nation with a *balance of payments* deficit can use SDRs to settle debts with another nation or with the IMF.

Specialized government counties See *county type classification.*

Special Milk Program (SMP) A program that provides cash reimbursements for *milk* served to all children in public and private nonprofit schools of high school grade or under, and in nonprofit residential or nonresidential child care institutions, provided they do not also participate in other federal meal service programs. However, schools in the *National School Lunch Program (NSLP)* and/or *School Breakfast Program (SBP)* that operate split-session prekindergarten and kindergarten programs may participate in the SMP to provide milk to children in those programs who do not have access to NSLP or SBP meals.

Schools and institutions have several options in how they operate the SMP. They may (1) sell milk to all children at a locally set, sale *price*, with a partial federal reimbursement; (2) provide milk free to children who meet the income eligibility criteria, while selling it to all other children; (3) provide milk free to all children, with partial federal reimbursement.

The federal reimbursement for each half pint of milk sold to children during the 1993-1994 school year was 11 cents. The figure is adjusted each July. The federal reimbursement for milk served free to eligible participants is the net purchase price per half pint of milk.

Special policy See *marine insurance.*

Special Supplemental Food Program for Women, Infants, and Children (WIC) A program created in 1972 to provide food assistance to people determined to be at nutritional risk by local health professionals due to inadequate income and nutrition. Categories of eligibility include pregnant women, postpartum mothers (up to six months), breastfeeding mothers (up to 12 months), and infants and children up to five years old. Local WIC agencies provide participants with either vouchers redeemable for specified foods at participating retail food stores or with a food package prepared according to federal guidelines.

Specialty crops Crops with a limited number of *producers*, such as *artichokes*. Also crops with high *production* costs and per-*acre* value, such as *greenhouse* crops, ornamentals, and many fruits and vegetables.

Specialty flours A wide range of *flours* other than white bread flour, including *rye*, pumpernickel, and whole *wheat*, and used in making variety breads.

Species A group of closely related organisms potentially capable of interbreeding.

Specific tariff A *tariff* expressed as a fixed amount per unit. See also *ad valorem tariff.*

Spectrophotometer An instrument that measures the relative intensities of light in different parts of the spectrum. See *near-infrared spectroscopy.*

Speculation Holding an asset (see *asset, current* and *asset, fixed*) or a fixed-price *forward contract* to buy or sell an asset, with the objective of profiting from *price* change.

Speed sprayer A portable tank and pump mounted on wheels used to spray *orchards* and *vineyards.*

Sphagnum A group of mosses that grow in moist places. By annual increments of growth, deep layers of fibrous and highly absorbent *peat* may be built up. Sphagnum grows best in cool, humid regions.

Spider wheel cultivator A *cultivator* (see also *rotary hoe*) designed to break the *soil* crust and leave *crop residue* on the surface.

Spike See *chisel point*.

Spike tooth harrow See *harrow* and *peg harrow*.

Spinach An *annual* vegetable cultivated for its green, edible leaves. *Varieties* include savoy (wrinkled), semi-savoy, and flat-leafed. The United States is the major producer of spinach.

Spinning 1: The process of *yarn* formation that involves attenuating the assembly of *fibers* supplied, introducing improved fiber-to-fiber cohesion (typically by twisting or interlacing), and winding the product on to a bobbin or other suitable holder.
2: In the *man-made fiber* industry, the process of extruding and winding continuous filament yarn.

Spinning quality The ease with which *fibers* lend themselves to *yarn*-manufacturing processes.

Split application The application of *fertilizer* two or more times during the growing season in order to supply *nutrients* more evenly. With split applications, fertilizer can also be applied when crops can make most effective use of it.

Split-phase hog production The *production* and finishing of *pigs* in separate operations instead of combined in the same operation as in a *farrow-to-finish operation*.

Spot Available for delivery immediately or within a short time interval, typically one day, 10 days, or within the month. Most cash (non*futures contract*) trading of agricultural *commodities* is for spot delivery, but a sizeable portion is in *contracts* for *deferred delivery*. See also *spot market*.

Spot market An arrangement for selling products to be delivered to the buyer immediately or within a short time interval. See also *spot*.

Sprayer Any number of devices used to apply a mist of *pesticides* or liquid *fertilizers* on crops or *insecticides* on *livestock*. Field sprayers usually consist of a long, horizontal boom fitted with hoses, nozzles, and a tank, pulled separately, mounted on *tractor* or *farm truck*, or a self-propelled unit with tank and boom attached. See *applicator*.

Spread 1: The difference between two *prices*—for example, a *bid*-ask spread or a December-March spread in *futures contract* trading.
2: A position taken in two or more *options* or futures contracts in order to benefit from anticipated changes in their relative price.

Spring tine harrow A *harrow* consisting of flexible, curved, thin, steel rods to lightly work the *soil* and prepare the seedbed.

Spring wheat *Wheat* that is grown in the spring and harvested (see *harvest*) in the summer or fall. Spring wheat has a relatively high *protein* content and is used in bread *flours*.

Sprinkler irrigation See *irrigation methods*.

Sprouts Edible, sprouted *seeds* and *beans* used on sandwiches and salads and in Asian dishes. *Alfalfa* seeds and mung beans are most frequently used for making sprouts.

Square An unopened flower *bud* of *cotton.*

Square baler See *baler* and *large-bale baler.*

Squash A plant with long vines bearing fruits that are eaten as a vegetable. *Varieties* include summer or soft-shelled, such as zucchini and yellow squash and winter or hard-shelled, such as acorn and butternut. The United States and Mexico are the leading producers. Florida is the primary domestic supplier.

Squeeze An attempt by one or a few *futures contract* traders to profit by threatening to stand for (or make) such large deliveries that opposite traders are forced to buy (sell) futures at a distorted *price* to avoid delivery.

Squeeze chute A *cattle* restraint consisting of a *headgate* and adjustable side(s) with access doors used to immobilize cattle for detailed inspection or treatment. A squeeze chute that can be pivoted to raise the animal's feet from the ground is sometimes called a calf table or bull table.

STABEX See *Lome Convention.*

Stable A building used to house and feed *horses.*

Stacker wagon A self-propelled or *tractor*-powered machine that picks up, stacks, transports, and unloads rectangular *bales* of *hay.*

Stag A male hog (see *swine*) castrated after reaching sexual maturity.

Stalk cutting A *harvest* method in which the entire *tobacco* stalk or plant is cut. *Light air-cured, dark air-cured, fire-cured,* and *cigar* filler and binder are usually stalk cut.

Stand A row of seedlings or plants. A good stand is when all spaces are filled, while a poor stand is characterized by many skips or open spaces between plants.

Standard "C" family corporation A separately taxed entity in which equity ownership is represented by stock held by family members, and management is centralized and controlled by a board of directors.

Standard International Trade Classification (SITC) A product classification system originally developed by the United Nations in 1950 to provide a standard for reporting international trade. The SITC system includes about 400 five-digit codes for agricultural products and was developed in close coordination with the *Brussels Tariff Nomenclature.* See also *Customs Cooperation Council Nomenclature* and *Harmonized Commodity Description and Coding System.*

Standardization The process of adjusting the *butterfat* and solids, not fat (see *milk solids, not fat*) content of *milk* to meet a required standard. In California, for example, whole milk must contain at least 3.5 percent butterfat and 8.7 percent solids, not fat (see *milkfat content*).

Standard Metropolitan Statistical Area (SMSA) See *Metropolitan Area.*

Standards of Identity Standards of quality established under the *Federal Food, Drug, and Cosmetic Act of 1938 (P.L. 75-717)* to prevent foods known by a traditional name from being debased by the substitution of inferior ingredients. Standards of identity require foods that do not conform to the official recipe to be clearly labeled. For example, a product that contains less *fat* (see also *fats and oils*) or *eggs* than the official recipe for *mayonnaise* must be labeled "imitation mayonnaise."

Standing orders An agreed-upon schedule of continuing delivery of specified quantities of a product of specified quality, frequency of delivery, and delivery point, at a formula price (see *formula pricing*).

Stand oils See boiled, blown, and bodied.

Staple fibers 1: Naturally occuring *fibers* that are clearly limited in length. The term "short staple" is applied to fibers whose lengths do not exceed 2 inches. *Cotton*, whose *staple length* typically lies between 1 and 1 1/2 inches, is an example. Long staple fibers include *wool* and *mohair*. Depending on the breed of *sheep*, wool staple can vary from about 3 to more than 12 inches.

2: *Man-made fibers* that have been cut to the length of the various *natural fibers* to facilitate *blending* and further *processing* with those fibers. Polyester is generally cut to a staple length of 1 1/2 inches for blending with cotton and 3 to 4 inches for blending with wool for men's suiting.

Staple length A measure of *fiber* length, particularly in the *cotton* industry, that can be based on personal judgement. With the development and introduction of *high volume intruments* to the cotton classing system, however, the description of fiber length is on a more objective basis. For cotton, the staple length is often expressed in thirty-seconds of an inch, or in millimeters. The value is descriptive of the length of the longer fibers within the sample taken.

Starch A complex *carbohydrate* found in most plant *seeds*, *bulbs*, and *tubers*.

Starfruit See *cherimoya*.

State marketing boards Government-controlled trading agencies used by countries such as Canada, Australia, and New Zealand to receive and *market* domestic products in domestic and international markets. Many *developing countries* and countries with *centrally planned economies* also use marketing boards for all *import* purchases.

State trading The practice of conducting trade exclusively through a government agency. *Centrally planned economy* countries follow this practice for all products, while many other nations, particularly *developing countries,* use state trading for *commodities* of critical economic importance, like *grains*.

State Water Quality Coordination Program A program authorized by the *Food, Agriculture, Conservation, and Trade Act of 1990 (P.L. 101-624)* (see Appendix 3) under which the secretary of agriculture will establish a water quality coordination program in each state and designate a program leader from among the federal agencies represented. The group will coordinate water programs with state and federal agencies and prioritize research needs within the state. State ad hoc advisory panel of farmers and private groups are authorized.

Steagall commodities See *Steagall Amendment of 1941 (P.L. 77-144)* in Appendix 3.

Stearic acid See *fatty acids (classification of)*.

Stearin The solid portion of *palm oil*. See also olein.

Steepwater Water used to soak *corn* during the *wet milling* process.

Steer A castrated, male, *bovine* animal, usually intended for *slaughter.*

Stemming Removing the stem or midrib from the *tobacco* leaf at the stemmery.

Steroid hormones *Hormones* that are endogenously produced in animals and are used in beef *production* to increase growth rate and *feed* efficiency. See also *growth hormones.*

Stevedore See *longshoreman.*

Stigmasterol A natural compound that is a *byproduct* of *soybean* processing (see *processing*) used to produce steroid medications.

Stocker cattle *Cattle* (calves [see *calf*] or older animals) maintained primarily on *pasture*, *range*, or harvested (see *harvest*) *forages* to increase weight and maturity before being placed in a *feedlot.*

Stocker-feeder enterprise An enterprise in which grazing (see *graze*) or harvested (see *harvest*) *forages* are the predominate *feed* used to grow *stocker cattle* into *feedlot*-ready *feeder cattle.*

Stocks See *buffer stock*, *carryover*, and *Commodity Credit Corporation.*

Storable manufactured dairy products Manufactured dairy products, including *butter*, *nonfat dry milk*, and *cheese*, that can be stored for extended periods of time.

Storage capacity The amount of water that can be stored in the *soil* for future use by plants or evaporation.

Store label brand See *private label brand.*

Straight breed Animals produced from *sires* and dams of the same *breed.*

Straight flour All of the *flour* extracted from a given blend of *wheat* without division or addition of flour from other runs.

Strain See *variety.*

Straw Stalks of *grain* that are left to dry after *threshing*. Straw is used as food and bedding for animals.

Strawberry The sweet, red fruit of low-growing, bushy plants. The United States, Canada, Mexico, and Europe are the major producers.

Stress Factors, such as temperature and *disease,* that affect *production* and the well-being of animals.

Stress cracks Cracks in the horny *endosperm* of *corn* caused by rapidly drying *kernels* with heated air. Stress-cracking causes increased breakage during handling and reduces *flaking* grit (see *grits*) *yields.*

Strict cross-compliance See *cross-compliance.*

Strict Low Middling 1 1/16-inch cotton The *grade* and *staple length* used as the basis for loan rates by the *Commodity Credit Corporation* on *cotton*. Higher qualities receive loan premiums and generally higher *market* prices, while lower qualities receive lower loan rates and lower *prices.* See *cotton quality.*

Strike price The stipulated *price* at which a *put option* or *call option* contract (see *contract*) may be exercised.

Stripcropping Growing crops in a systematic arrangement of strips or bands to serve as vegetative barriers to wind and water *erosion.* The strips or bands are laid out approximately on the contour on erosive (see *erosion*) *soils* or at approximate right angles to the prevailing direction of the wind where soil blowing is a hazard. See *contour farming.*

Strip till A *conservation tillage* method in which the *soil* is left undisturbed prior to planting. Tillage (see *till*) in the row is done at planting. See *no-till, mulch-till,* and *ridge-till.*

Strong flour See *flour.*

Structure Industry features including the number of buyers and sellers, *product differentiation, barriers to entry,* costs, and the degree of *integration* and *diversification.* Also called *market* structure.

Structure, soil See *soil structure.*

Stubble The part of the stems of plants remaining in a field after cutting.

Stubble drill A *drill* designed to cut through the residue from a previous crop, open the ground, deposit *seed,* and cover the seed with *soil.* It may also apply *fertilizer* and *pesticides* in the same operation. A stubble drill is generally of sturdy construction to handle hard ground and heavy residues. It is designed to leave the previous *crop residue* on the surface to prevent soil *erosion.* See also *air drill, no-till drill,* and *grain drill.*

Stubble mulch A protective cover provided by leaving plant residues (see *crop residue*) of any previous crop as a *mulch* on the *soil* surface when preparing for the following crop.

Subchapter "S" family corporation Similar to a *Standard "C" family corporation* except generally not separately taxed and subject to certain restrictions to maintain subchapter "S" status. It is taxed like a partnership.

Subirrigation Irrigation (see *acres irrigated, irrigated farms,* and *irrigation methods*) through controlling the water table in order to raise it into the root zone. Water is applied in open ditches or through tile until the water table is raised enough to wet the *soil.* Some soils along streams are said to be naturally "subirrigated."

Subprimals Smaller cuts fabricated (see *fabrication*) from *primals.* A beef primal round cut, for example, may be cut into subprimals of top round, bottom round, and knuckle.

Subsidy A direct or indirect benefit granted by a government for the *production* or distribution (including *export*) of a good. Examples include any national tax rebate on exports; financial assistance on preferential terms; financial assistance for operating losses; assumption of *costs of production, processing,* or distribution; a differential *export tax* or duty (see *tariff*) exemption; domestic consumption *quota;* or other method of ensuring the availability of raw materials at artificially low *prices.* Subsidies are usually granted for activities considered to be in the public interest.

Subsistence farm A *farm* where the emphasis is on *production* for use of the *farm operator* and the operator's family rather than for sale.

Subsoiler A *tillage tool* (see also *till*) used to break up compacted subsoil (see *soil*).

Substitutes Agricultural raw materials or products that replace other agricultural products in traditional uses.

Substrate 1: A substance that is acted upon, as by an *enzyme*.
2: A culture medium, the substance, or object on which an organism lives and from which it gets its nourishment.

Subterminal elevator An *elevator* located in the *production* area that receives *grain* from *country elevators* as well as from farmers. These facilities are intermediate-sized grain assemblers, often having facilities for official weights and *grades*.

Subvention See *restitution*.

Succulence A condition of plants characterized by tenderness, juiciness, and freshness, making them appetizing to animals.

Sucrose A sweet, crystallizable, colorless substance derived from either *sugarcane* or *sugarbeets*. Refined cane and beet sugar (see *refined sugar*) is essentially 100 percent sucrose.

Sufferance The entry of goods into a country and the subsequent exportation (see *export*) without the payment of duties (see *tariff*). See also *entrepot*, *foreign trade zone*, and *reexport*.

Sugar A sweet, water-soluble, crystalline *carbohydrate*. Most sugar used for human consumption is obtained from *sugarcane* or *sugarbeets*. Major sugarcane producers include Brazil, India, Cuba, and Thailand. The major sugarbeet producers include the former Soviet Union, the *European Union*, the United States, and Poland. *Sorghum*, sugar maple, and sugar palms are also sources of sugar.

Sugar Association of America, Inc. (SAI) An organization that represents cane (see *sugarcane*) and beet (see *sugarbeet*) *sugar* processors and refiners. The SAI disseminates scientifically based information on *sucrose*.

Sugarbeets A cool-weather, common beet that is a leading raw material source for the *production* of manufactured *sugar* in the United States. In the United States, sugarbeets are grown in 14 States, with Minnesota, North Dakota, Idaho, and California leading in production. See also *sugarcane*.

Sugarbeet topper A machine that cuts the green or above ground part of *sugarbeets* from the root.

Sugarcane A tall grass native to the East Indies and the principal source of *sugar*. Sugarcane is grown in tropical and semitropical climates. Four U.S. States—Florida, Hawaii, Louisiana, and Texas—produce sugarcane.

Sugar-containing products Products containing at least 10 percent embodied *sugar*.

Sugar pea See *snow pea*.

Sulfonation A process where fatty oils (see *fats and oils*) are treated with sulfuric acid to obtain sulfonated oils used for *textile* and *leather* processing.

Summer fallow The practice of leaving *cropland* idle for a year to rebuild subsoil (see *soil*) moisture. The *land* is usually tilled (see *till*) for weed control. In drier areas of the *Great Plains*, crops are often produced in a two-year, crop-*fallow* rotation (see *crop rotation*).

Summer Food Service Program (SFSP) A program that provides meal service to children from areas where poor economic conditions exist. Sponsors must qualify by operating programs in areas where at least one-half of the children would qualify for free or reduced-price meals under the *National School Lunch Program* and the *School Breakfast Program*, or they must provide meals as part of an organized program for children enrolled in camps.

Sunchoke See *Jerusalem artichoke.*

Sun-cured A process in which plant material is dried by exposure in open air to the direct rays of the sun.

Sunflower A tall plant with large yellow flowers whose *seeds* are used as a food product and a source of oil (see *fats and oils*). See *sunflower seed.*

Sunflower Oil Assistance Program (SOAP) One of two programs under which the *Commodity Credit Corporation (CCC)* provides bonuses in vegetable oils (see *fats and oils*) to *exporters* to assist in exports to targeted *markets*. Funds for the programs were first authorized in fiscal year 1988 from *Section 32* of Agricultural Adjustment Act Amendment of 1935 (P.L. 74-320) (see Appendix 3). See also *Cottonseed Oil Assistance Program (COAP).*

Sunflower seed The *seeds* of the *sunflower* that are used as confectionery seed, edible sunflower oil, and sunflower *meal*. The major producers include the former Soviet Union, the United States, Argentina, China, and Eastern Europe.

Super An extra division of a *beehive* above the *brood* nest area in which frames of *honeycomb* are placed, usually for *honey* storage.

Supermarket See *foodstore.*

Supima The trademark of an *extra-long staple cotton*, commonly referred to as American Pima cotton. Supima is produced in Arizona, New Mexico, and west Texas. Supima Association of America is a producer association headquartered in Phoenix, Arizona.

Supima Association of America See *supima.*

Supplement A formula *feed* used with other feed ingredients to improve the nutritive balance of the diet. Supplements are intended to be fed separately in controlled amounts, or offered free-choice as a dietary supplement to other feeds, or mixed with other ingredients to formulate a *complete feed.*

Supplementary imports See *competitive imports.*

Supplementing The process of supplying additional *nutrients* to animals on poor *feed.*

Supply The quantity of an economic good available for sale in the *market,* including current *production*, *carryover*, and *imports.*

Supply control The policy of changing the amount of acreage (see *acre*) permitted to be planted to a *commodity* or the quantity of a commodity allowed to be sold by a program participant. *Supply* control is used to maintain a desired *carryover* or *price* level. See also *mandatory supply controls*, *acreage reduction program*, and *set-aside*.

Support price A legislated minimum *price* for a particular *commodity*, maintained through a variety of mechanisms, such as *minimum import prices*. In the United States, support-price mechanisms include *nonrecourse loans* and purchase programs.

Surcharge A charge levied in addition to other taxes and duties (see *tariff*). The surcharge may be levied as a percentage of the other charges or may be imposed on the original tax base.

Surimi A minced fish product used to manufacture products that simulate crab, shrimp, and other popular seafood.

Surplus *Production* in excess of *demand* for the current crop production-*marketing year*.

Surtax See *surcharge*.

Surveillance 1: Monitoring trade practices to ensure that governments implement their obligations under trade agreements.
2: A system by the Centers for Disease Control and Prevention to monitor foodborne *disease* incidence.

Susceptibility Lacking the inherent ability to resist *disease*.

Sustainable agriculture Alternative farming practices and systems that substitute higher levels of management and diversified farming practices for chemical *inputs*. See also *low-input sustainable agriculture*.

Swampbuster A provision of the *Food Security Act of 1985 (P.L. 99-198)* (see Appendix 3) that discourages the conversion of natural *wetlands* to *cropland* use. Producers converting a wetland area to cropland lose eligibility for several *U.S. Department of Agriculture (USDA)* program benefits. The exceptions include conversions that began before December 23, 1985, conversions of wetlands that had been created artificially, crop *production* on wetlands that became dry through *drought*, and conversions that the *Soil Conservation Service* has determined have minimal effect on wetland values.

Swarm A natural division of a *colony* of bees (see *honeybee*). *Worker bees*, *drones*, and *queen bees* leave the mother colony in a swarm to establish a new colony.

Swather A *farm* implement used to cut and lay crops, primarily *grain* and *hay* crops, on top of stubble in long rows to help the crop ripen.

Sweep A *shape* or *shovel* attached to *cultivator* or chisel (see *chisel*) *shanks* designed to penetrate and loosen the *soil*, kill weeds, and conserve moisture.

Sweet clover See *clover*.

Sweet Potato Council of the United States (SPCUS) An organization of sweet potato *producers*, *packers*, *processors*, and suppliers that was founded in 1962.

Swine *Pigs*, hogs, and boars. China, the former Soviet Union, the United States, Brazil, and Germany are the major swine producers.

Swine vesicular disease A highly contagious, viral *disease* of *swine*, with symptoms similar to *foot-and-mouth disease*. Swine vesicular disease can be transmitted to humans from infected hogs. Trade in infected animals and meat products is restricted.

Swinging beef, hanging beef, on the rail Sides and/or quarters of meat on hooks that move along an overhead rail in the *slaughterhouse* cooler, rail car, or store back room. Selling "on the rail" is sometimes used in place of *grade and yield* sales.

Swiss chard A type of *beet* with lush leaves rather than a fleshy root.

Switch trading See *clearing accounts*.

Symbiosis The process in which two organisms live in a close, mutually beneficial relationship.

Symptom Any reaction of a *host* to *disease*; usually refers to visible reactions.

Synthesized, fabricated, engineered The process of producing a food product from a number of elements. The purpose may be to duplicate an existing food in texture, color, nutrition, and flavor or to produce a product to certain standards that may not presently exist.

Synthetic fibers *Fibers* made from petroleum-derived chemicals. The major types are polyester, nylon, acrylic, and polypropylene.

Synthetic hormones Man-made derivatives of endogenously produced *steroid hormones* used in animal *production*.

Synthetics Products derived principally from nonagricultural sources or from agricultural sources processed so that their origin is not easily identified.

Systemic A *disease* in which an infection spreads generally throughout the body.

Systemic pesticide A *pesticide* that is absorbed by the plant's system and distributed throughout the plant.

Syrup A thick, sweet liquid obtained from concentrated (see *concentrate*) clarified *sugarcane* juice before crystallization.

T

Table eggs Unfertilized (see *fertilization*) *eggs* for consumption.

Table service restaurant An operation that has tables or booths and waiters or waitresses providing service. See also *fast food restaurant.*

Tagging See *crutching.*

Tailings of the mill Particles that are removed near the end of the *milling* process and that have passed through progressively smaller sieve sizes. The particles are smaller than those removed from the *front of the mill.*

Takeout See *carryout.*

Tall oil A *byproduct* from the manufacture of chemical wood *pulp.* Tall oil is used in making soaps and for various industrial products.

Tallow Edible and inedible rendered (see *rendering*) *bovine* and *sheep* fat and inedible rendered hog (see *swine*) *fat.* Tallow is principally used in the manufacture of soap, cosmetics, salad or cooking oils (see *fats and oils*), *margarine, feed,* paint and varnish resins, plastics, and lubricants.

Tame pasture See *improved perennial pasture.*

Tandem disk harrow A type of *harrow* with two opposing rows of offset blades that enter the *soil* at an angle and throw the soil first in one direction and then the other.

Tangelo A *tangerine*-grapefruit *hybrid.* Some *varieties* resemble tangerines, while the Orlano is more *grapefruit*-flavored.

Tangerine A *citrus fruit* that is a type of mandarin. *Varieties* include Dancy, Robinson, and Clementine. Major producers include Japan, Spain, Italy, the United States, and Chile.

Tangor A *tangerine*-sweet *hybrid* of an *orange.* Temple is the most important *variety.*

Tankage A *protein* supplement made from the residues of animal tissues, including bones, and exclusive of hair, hooves, horns, and the contents of the digestive tract.

Tanker See *bulk carrier.*

Tanning The *processing* of perishable raw *hides and skins* into permanent, durable *leather* using alum, tannin potassium, or sodium bichromate.

Tare The weight of the ties (or bands) and wrapping materials that contain the *bale* of *cotton*. The quoted net weight of a bale excludes the tare, whereas the gross weight includes the tare.

Targeted Export Assistance Program (TEA) See *Market Promotion Program.*

Targeted Option Payments (TOP) A program, implemented at the secretary of agriculture's discretion, in which *producers* of *wheat*, *feed grain*, *upland cotton*, and *rice* have the option of choosing from a schedule of *target prices* and corresponding *acreage reduction* levels. This program was first authorized by the *Food Security Act of 1985 (P.L. 99-198)* (see Appendix 3). As of 1994, *TOP* had not been implemented.

Target price In the United States, a *price* level established by law for *wheat*, *corn*, *grain sorghum*, *barley*, *oats*, *rice*, and *upland* and *extra-long staple cotton*. Farmers participating in the federal *commodity* programs receive *deficiency payments* based on the difference between the *target price* and either the *market* price during a period prescribed by law or the *loan rate*, whichever is higher.

Tariff A tax imposed on *imports* by a government. A tariff may be either a fixed charge per unit of product imported (*specific tariff*) or a fixed percentage of value (*ad valorem tariff*).

Tariff preference *Tariff* treatment accorded to a country that is more favorable than that accorded to countries outside the preferential arrangement.

Tariff quota Application of a reduced *tariff* rate, or zero duty, for a specified quantity of imported goods or for goods imported during a given period. Tariff quotas do not limit the quantity of goods that may be imported.

Tariff rate quota (TRQ) system A *tariff* system for sugar *imports* authorized by a presidential proclamation on September 14, 1990. The TRQ replaces the restrictive quota system that had regulated the amount of *sugar* entering the United States since 1982. The TRQ imposes a nominal or zero tariff for import quantities up to a certain level and a very high tariff on imports above the first-tier level.

Tariff schedule A list of articles or merchandise and the rate of duty (see *tariff*) to be paid to the government for their importation.

Tariff surcharge An *import* tax that is usually assessed at a flat rate over and above whatever duties (see *tariff*) are assessed.

Tariff union See *common market.*

Taro A brown, barrel-shaped *tuber* that is similar to a *potato*. Also called dasheen.

Taxol A compound isolated from the bark of the slow-growing *Taxus brevifolia* that has shown promise in the treatment of refractory ovarian cancer and breast cancer. The compound is also found in the needles of all *species* of yew.

Tea The dried (see *drying*) leaves of a shrub native to eastern Asia. Major producers include India, Sri Lanka, China, and Kenya.

Technical barrier to trade A specification that sets forth characteristics a product must meet in order to be imported. The characteristics include levels of quality, performance, or safety.

Technical grade refined oils Includes a wide variety of oils (see *fats and oils*) that are specially *refined and further processed* to meet requirements or specifications for a definite industrial use.

Technology transfer A strategy for ensuring access to technological details that allow the *production* of a product being commercialized (see *commercialization*).

Tedder An implement with thin, flexible tines to move, loosen, and spread *hay* for *drying*. Some may also be used to put hay in *windrows*.

Tef A natural *grain* grown in Ethiopia. Tef survives well in dry *climates* and contains unusually high levels of *protein*.

Tele-auction market Auctioning (see *auction*) by description via remote contact with buyers after *livestock* are weighed and graded (see *grading*) either at an assembly point or on the *farm*. See *electronic markets*.

Tempering Adding moisture to *wheat* and *corn* during the *dry milling* process to aid the removal of *bran* from the *endosperm*. Also called conditioning.

Temporary Emergency Food Assistance Program (TEFAP) A program established in 1983 to allow donation of *commodities* owned by the *Commodity Credit Corporation* to states in amounts relative to the number of unemployed and needy persons. The food was distributed by charitable organizations to eligible recipients. The program was replaced in 1990 by *The Emergency Food Assistance Program (TEFAP)*.

Temporary pasture A *pasture* grazed (see *graze*) one crop season only.

Tender A written offer to buy or sell goods or services at a specified *price*.

Tensiometer A measuring device used in irrigation (see *acres irrigated* and *irrigated farms*) management to determine the amount of *soil* moisture and provide an indicator of how much and when to irrigate.

Terminal elevator An *elevator* located at a point of accumulation and distribution in the movement of *grain*. A terminal elevator ordinarily receives grain by carload rather than truck loads (see *country elevator*) and stores grain for hire for others. The elevator is usually operated by a wholesale grain dealer as opposed to a country grain dealer.

Terminal market 1: A *market* facility generally located in or near a metropolitan area that handles many agricultural *commodities*. The San Francisco Wholesale Produce and Fruit Market is one example.

2: Facilities that receive *livestock* from the seller and sell them to *packers* through commission firms that represent the seller for a fee. The volume at livestock terminal markets, however, is now generally low.

Terms of sale Items, such as quantity, quality, *price*, terms of delivery, shipment or delivery period, insurance, payment, commission, and controllers, that are negotiated at the time of sale.

Terms of trade The relationship over time between the *price* of a country's *exports* to the price of its *imports*. Terms of trade become more favorable as export prices received rise compared with import prices.

Terrace An embankment or ridge constructed across sloping (see *slope*) *soils* on the contour or at a right angle to the contour. The terrace intercepts surplus *runoff* in order to retard it so that it will infiltrate the soil and so that any excess will flow slowly to a prepared outlet.

Terracer A machine designed to form large, level ridges across hilly fields to control water *runoff* and *erosion* of *soil*. Some use large *moldboard plows* or *disk plows*.

Territorial price discrimination *Price discrimination* between buyers in different *market* territories where the *price* differences exceed the differences in *production*, packaging, sales, or distribution costs to buyers in the different market territories.

Test weight A quality test used to determine the weight per *bushel*. For example, the standard bushel of *wheat* is 60 pounds. See Appendix 7.

Tex The unit of the *International Organization for Standardization (ISO)* system (direct) for the linear density of *textile* materials from *fiber* to *yarns*. It is the weight of 1,000 meters, expressed in grams. It is normal to quote linear densities for yarn in units of tex, fiber in millitex (mtex) or decitex (dtex), and *sliver* in kilotex (ktex).

Textile Any product made from *fibers*, including *yarns*, *cloth*, and *end-use* products, such as apparel, home furnishings, and industrial applications.

Texture The number of *warp* threads (ends) and *filling* yarn (picks) (see *yarn*) per square inch in a woven *cloth*. For example, 88 by 72 means there are 88 ends and 72 picks per square inch in the fabric.

Textured vegetable protein *Soybean* or other vegetable *protein* that has texture imparted either by spinning a *fiber* and combining the fiber in layers to achieve the desired texture or by a thermoplastic extrusion process.

Texture, soil The relative proportions of the various size groups of individual *soil* grains in a mass of soil. Specifically, it refers to the proportions of sand, silt, and clay.

The Emergency Food Assistance Program (TEFAP) A program, established in 1990, that replaced the *Temporary Emergency Food Assistance Program (TEFAP)*, established in 1983. The Emergency Food Assistance Program allows donation of *commodities* owned by the *Commodity Credit Corporation (CCC)* to states in amounts relative to the number of unemployed and needy persons. The food is distributed by charitable organizations to eligible recipients.

Thin market A *market* characterized by few potential traders and few or infrequent trades. In general, the amount of open or negotiated trading is small relative to the total volume, making the accuracy of established or reported *prices* suspect.

Thinner A mechanical device used to thin plants in rows. Units either operate hydraulically or electrically.

Three-point hitch A method of mounting implements such as *cultivators* on the back of a *tractor*. The method allows positive depth settings.

Threshing The mechanical process of separating *grain* from the *hull* of plants.

Threshold price A mimimum *import* price set by the *European Union (EU)* under the *Common Agricultural Policy (CAP)* for certain *commodities*, equal to the target price less transportation costs to the EU region identified as the deficit area. Certain imports from nonmember countries are subject to a *variable levy* that is equal to the difference between the threshold *price* and the minimum *world price* at EU ports.

Thrifty Food Plan (TFP) The least costly of four food plans (thrifty, low-cost, moderate-cost, and liberal cost) developed by the *U.S. Department of Agriculture*. The plan suggests the amounts of food that could be consumed by males and females of different ages to meet dietary standards. The TFP for a family of four (man and woman, 20–54 years old; children, ages 6–8 and 9–11 years) by law constitutes the basis for allotments to households participating in the *Food Stamp Program (FSP)*. The cost of the TFP for the family of four is the maximum benefit level payable to any household of four.

Till The process of preparing *soil* for planting by *plowing*.

Tillage equipment Any number of different types of *farm* implements designed to lift, turn, stir, and work the *soil*, such as *cultivators*.

Tillage tool Any number of different *shovels* or *shapes* that can be mounted at the end of a *shank* and used to work the *soil*.

Tilth, soil The physical condition of a *soil* with respect to its fitness for the growth of a specified plant or sequence of plants. Ideal soil tilth is not the same for each kind of crop nor is it uniform for the same kind of crop growing on contrasting kinds of soil.

Timothy A grass cultivated for *hay*.

Tipping The process of removing the top one-third of the *tobacco* leaf that does not contain objectionable stem; the remaining two-thirds of the leaf is *threshed*.

Tissue culture The process of extracting small bits of tissue from a single plant to grow in test tubes in order to regenerate the individual *cells* into many whole plants.

Tobacco A plant of the Solanaceae family, native to tropical America, whose leaves are used to primarily to produce smoking products, such as *cigarettes*, *cigars*, and *smoking tobacco*, as well as *chewing tobacco*, cut tobacco, and snuff. The types of tobacco are *flue-cured* (Virginia), *air-cured* (*burley tobacco*, *Maryland tobacco*, and *Oriental tobacco*), *fire-cured*, and *sun-cured*. China, the United States, India, Brazil, the former Soviet Union, and Turkey are the major tobacco producers.

Tobacco Institute (TI) An organization, founded in 1958, to promote the *tobacco* industry and compile information on tobacco. TI members include manufacturers of *cigarettes*, *smoking tobacco*, *chewing tobacco*, and *snuff*.

Tokyo Round The *General Agreement on Tariffs and Trade (GATT)* negotiations formally initiated by the 1973 Tokyo Declaration and completed in 1979. More countries were involved in the Tokyo Round than previously (including many *developing countries* and several Eastern European countries), and discussions were expanded to include *nontariff trade barriers*, especially as they related to agricultural trade.

Tolerance The degree of endurance of a plant to the effects of adverse conditions, chemicals, or *parasites*. A tolerant plant is capable of sustaining a *disease* without serious injury or crop loss.

Tolerance (T) value See *soil erosion tolerance value.*

Toll milling The *milling* of *feed* by a commercial feed mill or mobile operator for other feed manufacturers.

Tom A male turkey (see *poultry*).

Tomatillo A vegetable resembling a tiny, green *cherry tomato* with clinging husks. The sticky-skinned fruit has an acidic flavor similar to green *tomatoes*. Tomatillos are eaten raw or cooked and are a major ingredient in salsa. Also called a Mexican husk tomato.

Tomato A plant native to South America but grown in the United States, Mexico, Taiwan, Spain, and Italy and cultivated (see *cultivate*) for its fruit.

Tomato repacker A firm that buys green *tomatoes* and sells ripened, packaged tomatoes sorted by size, color, and *variety* to wholesalers, retailers, institutions, and *foodservice* establishments. Tomato repackers are generally located near metropolitan areas.

Ton A measure of weight (see *long ton*, *metric ton*, and *short ton*) and volume (see *measurement ton*). See Appendix 6.

Tool bar A steel frame with horizontally mounted bars on wheels or mounted on a *tractor* and used to carry various *shanks* and *sweeps* for *soil* tillage (see *till*).

Top See *sliver.*

Top dressing A *lime*, *fertilizer*, or *manure* surface applied either to a prepared seedbed or after the plants are up.

Topography The shape of the ground surface, such as hills, mountains, or plains. Steep topography indicates steep slopes or hilly *land;* flat topography indicates flat land with minor undulations and gentle slopes.

Topper A machine that cuts the above ground part of the plant from the root. See *sugarbeet topper.*

Topping Removing blossoms and sometimes top leaves of *tobacco* plants. Topping tends to increase the size, thickness, body, and nicotine content of tobacco leaves.

Topsoil 1: A presumed fertile *soil* or soil material, usually rich in organic matter, used to topdress roadbanks, lawns, and gardens.

2: The surface *plow* layer of a soil and thus a synonym for surface soil.

3: The original, or current, dark-colored, upper soil, which ranges from a mere fraction of an inch to 2 to 3 feet on different kinds of soil.

Total Class I price See *Class I effective price.*

Total confinement *Livestock* and *poultry* kept within buildings during all stages of *production.*

Total digestible nutrients (TDN) The sum of all *nutrients* in a *feed* that are digestible by an animal.

Total farm output per unit of input See *productivity*.

Tow See *sliver*.

Toxicology The science dealing with poisons in food and their effects.

Toxin A poisonous substance formed by an organism.

Toxoplasmosis Protozoa that can cause a range of mild to severe animal and human illnesses. Toxoplasmosis is most often associated with undercooked or improperly handled pork and *lamb*.

Trace element A chemical substance used in minute amounts by organisms and essential to their physiology (magnesium, iron, copper, etc.)

Tractor An off-road power unit used mostly to pull *farm* implements. While some tractors have steel or rubber tracks (see *crawler tractor*), most have two large rubber-tired drive wheels on the rear and two smaller ones on the front, some of which are power-assisted. Larger tractors may have four or more large drive wheel (four-wheel drive), and some may be articulated in the middle for steering. Most tractors sold in the first half of the 1990s are equipped with *power-take-off (PTO)*, hydraulic systems, and cabs (many with air conditioning, radios, and other amenities). Most have *three-point hitches*, and some have front-mounting hitches for *cultivators* and other implements. Many have some form of automatic or easy-to-shift transmission. Most have at least six gear ranges, some 20 or more. Tractor size is rated by *horsepower* ranges. Farm tractors may have from 40 to as much as 300 or more horsepower. Some go as high as 670 horsepower. Many tractors are designed for specific purposes. Some have tall wheels or raised frames for high clearance in row crops. Others may have pointed fenders and shields for use in *orchards*. See *farm tractor*.

Tractor front-end stacker An implement mounted on the front of a *tractor* used to pick up and stack *hay bales*, stacks, and loose *hay*.

Trade Act of 1974 (P.L. 93-618) A U.S. law that provided the president with *tariff* and *nontariff trade barrier* negotiating authority for the *Tokyo Round* of *multilateral trade negotiations*. See Appendix 3 for more details.

Trade Agreements Act of 1979 A U.S. law that approves and implements the trade agreements negotiated by the United States in the *Tokyo Round* of the *multilateral trade negotiations*. See Appendix 3 for more details.

Trade and Tariff Act of 1984 A U.S. law that clarified the conditions under which unfair trade cases (see *unfair trade practices*) can be pursued under *Section 301* of the *Trade Act of 1974 (P.L. 93-618)*. See Appendix 3 for more details.

Trade balance The value of merchandise exported (see *exports*) by a country minus the value of its *imports*. The measure is distinct from the *balance of payments*, which includes the net account of both goods and services.

Trade barriers Regulations used by governments to restrict *imports* from, and *exports* to, other countries. Examples include *tariffs*, *nontariff barriers*, *embargoes*, and *import quotas*.

Trade deficit See *balance of trade*.

Trade liberalization The complete or partial elimination of government policies or *subsidies* that adversely affect trade. The removal of trade-distorting policies may be done by one country (unilaterally) or by many (multilaterally).

Trade negotiations Discussions of trade issues involving two or more countries. See *bilateral trade negotiations* and *multilateral trade negotiations.*

Trade policy committee (TPC) The senior U.S. government interagency trade committee established to provide broad guidance on trade issues. It is chaired by the *U.S. Trade Representative (USTR)* and is comprised of secretary-level individuals. The Trade Policy Review Group, which reports to the TPC, is chaired by a Deputy USTR and is comprised of assistant secretary–level persons. The Trade Policy Staff Committee, the level at which position papers are initiated, is chaired by a deputy assistant USTR and is comprised of office/director level individuals.

Trade preference Any of several forms of preferential trade treatments (see *preferential trade agreement*). Often used interchangeably with *tariff preference.*

Trade servicing Activities to help develop *export* markets (see *market*) include hosting trade conferences; assisting with displays at trade fairs; issuing trade press announcements; and hosting visiting study teams from *foreign* countries to learn about U.S. *production* capacity and the reliability of U.S. agricultural *supplies.*

Trade surplus See *balance of trade.*

Trade-weighted exchange rate An index-weighted average of bilateral *exchange rates* using trade volumes as weights. It measures the extent of appreciation or depreciation against the trade-weighted average of bilateral exchange rates that dominate trade in a particular *commodity.*

Trailer A towed, wheeled vehicle used for transport. A trailer can be a flatbed or have boxed sides. It may have hydraulic cylinders to lift the box and dump the contents, or it may use an *auger* or conveyor. *Farm* trailers may be classified into two categories: highway use and off-road use. Trailers for highway use include flatbed units used to haul farm machines, such as *combines, tractors,* and other implements. The front of a trailer is supported by the towing vehicle, as opposed to a *wagon,* which is supported on its own wheels.

Transpiration The loss of water vapor to the atmosphere from plant leaves and stems.

Trap nest A device that traps a *hen* after she enters in order to count *eggs* laid during the year or to identify the parent in breeding egg *production.*

Tray-ready Meat that is trimmed and cut to retail cuts but kept together in *subprimal* units, *vacuum-packed,* and placed in cartons at the packing plant (see *plant or establishment*). This meat is ready for traying, wrapping, weighing, and pricing at the retail store.

Treaties Written agreements between governments, usually ratified by the lawmaking authority of each country. In the United States, treaties are negotiated by the executive branch and must be ratified by a two-thirds majority in the Senate.

Trench silo See *horizontal silo.*

Trichinellosis A *disease* caused by an intestinal roundworm and frequently associated with undercooked or improperly handled pork and game. The symptoms can range from asymptomatic infection to death, depending on the number of *larvae* ingested.

Trichinosis See *trichinellosis.*

Tricot See *knitting.*

Trigger An antidumping (see *dumping*) mechanism designed to protect domestic U.S. industries from underpriced *imports*. First introduced in the United States in 1978 to protect the steel industry, it is the *price* of the lowest-cost *foreign* producer. Imports priced below the trigger price are assessed a duty (see *tariff*) equal to the difference between their price and the trigger price.

Trigger level Under the *U.S. Meat Import Law*, the *import* level at which import restraints are imposed for the meats subject to the law. The limit is calculated from a formula based on domestic *quota* meat *production* and cow-beef production.

Trigger price level See *Farmer-owned Reserve Program.*

Triglycerides See *fats and oils.*

Tripe The stomach lining of *ruminants* used as human food.

Triple base That portion of a producer's *crop acreage base* on which the *producer* is permitted to plant alternative crops but would not receive *deficiency payments*. Under current programs, triple base is often used to describe *normal flex acres.*

Triticale A *hybrid* between *wheat* and *rye* with a high yield and rich *protein* content but low palatability. Triticale is used for animal *feed.*

Tropical products *Commodities* grown in tropical areas, including *coffee*, *tea,* and *cocoa*; spices, gums, *essential oils* and cut flowers; certain *oilseeds* (*castor oil*, *palm oil*, *coconut*, *peanut*, etc.); *tobacco*; *rice*; cassava, and other tropical roots; tropical fruits and nuts (*bananas, pineapples*, and *citrus fruit*); tropical wood and rubber; and *jute* and other *hard fibers.*

Truck See *farm truck.*

Truck jobber A *jobber* who conducts business from a truck. Although truck jobbers do not sell from a store, they typically have a regular customer route and deliver on a fixed schedule.

Tryptophan An *amino acid* found in plants and animals. Tryptophan has been used as a sedative. Better sources of "natural tryptophan" include crops such as bean sprouts, fenugreek, pumpkin, lablab *bean*, *sesame*, *spinach*, watercress, *winged bean*, and evening primrose.

Tubers An often edible, underground stem from which new plants develop. Examples include the *potato* and sweet potato.

Tuna See *prickly pear.*

Tung A tree fruit produced in China, Argentina, and Paraguay. The inedible oil (see *fats and oils*) is used for industrial purposes.

Turkey See *poultry.*

Turkey red oil The product resulting from sulfonated (see *sulfonation*) *castoroil*. Turkey red oil is used in dying and emulsification (see *emulsifier*).

Turnip A root vegetable with reddish purple skin and crisp, white flesh.

T-values See *soil erosion tolerance values.*

Twist The number of turns per unit of length of the *fiber,* strand, *roving,* or *yarn.* In the United States, twist is measured in the number of turns per inch.

Twist (tobacco) See *chewing tobacco.*

Two-price plan *Price discrimination* between the domestic and *export* markets that occurs when *commodities* for export are sold at a different *price* than in the domestic *market.* Governments or firms may adopt a two-price plan in order to expand markets, dispose of surpluses (see *surplus*), or increase returns.

Two-tiered pricing Any *farm* program under which *commodities* for domestic use would be supported at a higher level than those grown for *export* markets (see *market*).

Tying machine A device that stitches *tobacco* leaves to sticks for *curing.* Also called a sewing machine.

Type of farm (or ranch) A classification of farming types based on a farming operation's *production* specialty. In the *Census of Agriculture,* this designation is based on the product or group of products that accounts for 50 percent or more of the total value of agricultural products sold during the reporting year. If no single *commodity* or group of commodities accounts for 50 percent or more of a *farm's* value of production, that farm is classified as either a general farm, primarily crops, or a general farm, primarily *livestock.*

Farming may also be classified by type based on the commodity or commodity grouping that represents the largest portion of *gross farm income* during the reporting year. Typically, 10 to 14 major farm types are reported. However, the following broad groupings are sometimes used to summarize detailed farm types:

Field crops: A farm primarily engaged in the production of: *wheat, rice, corn, soybeans, barley,* dry *beans, rye, sorghum, cotton,* popcorn, *tobacco, potatoes, sugar* crops, *hay, peanuts, hops,* mint, or other such crops.

Other crops: A farm primarily engaged in the production of: vegetables, melons, berry crops, *grapes,* tree nuts, *citrus fruits, deciduous tree* fruits, *avocados,* dates, figs, *olives, nursery,* or *greenhouse* crops.

Livestock or poultry: An operation primarily engaged in the production of: *cattle,* hogs (see *swine*), *sheep, goats, milk,* chickens (see *poultry*), *eggs,* turkeys (see *poultry*), or animal specialties such as furs, fish, *honey,* and so on.

U

Udder The sack-like organ of female *cows*, *sheep*, and *goats* containing the *mammary glands*. The milk is obtained from the udder's two or more nipples.

Ultra-high temperature milk (UHT) *Homogenized milk* heated to 135–150° Centigrade (275–302° Fahrenheit), then aseptically (see *aseptic processing [packaging]*) canned or packaged.

Uncertainty Lack of predictability due to randomness or incomplete information.

Underdeveloped country See *developing country*.

Understory Vegetation growing in the shade of taller plants.

Undulant fever See *brucellosis*.

Unfair trade practices Actions by a government or firms that result in competitive advantages in international trade. Such actions include *export subsidies*, *dumping*, boycotts, or discriminatory shipping arrangements. Under *Section 301* of the U.S. Trade Act of 1974 (P.L. 93-618) (see Appendix 3), the president is required to take appropriate action, including *retaliation*, to obtain removal of policies or actions by a *foreign* government that violate an international agreement or are unjustifiable, unreasonable, or discriminatory, and burden or restrict U.S. commerce.

Unfinished plant materials Plant materials that include cuttings, "lining-out" stock, plug seedlings, tissue-cultured plantlets, and other unfinished plants and propagative materials grown mostly for resale to other growers. Growers purchase rooted cuttings or unpotted plants from a specialist propagator for either "growing on" to a finished state or *market*-ready size or for further propagation of a certain *specie* or *variety*.

Uniform Grain Storage Agreement (UGSA) An agreement between commercial *elevators*, warehouses, and mills (see *milling*) and the *Commodity Credit Corporation (CCC)* to store *wheat* and other *commodities* in approved storage facilities at specified rates.

United Egg Association (UEA) An organization of individuals involved in all aspects of the *egg* industry. The UEA promotes egg consumption and *exports* and is involved in political action, nutrition, and education.

United Egg Producers (UEP) An organization of regional *egg* marketing *cooperatives* whose members are independent egg *producers*. UEP provides technical assistance in egg *production*, distribution, and marketing (see *agricultural marketing*), promotes favorable legislation, and distributes relevant industry information.

United Farm Workers of America (UFW) An organization founded in 1962 to achieve collective bargaining rights for U.S. *farm* workers. The UFW provides education and training for agricultural laborers (see *agricultural employment*).

United Nations Conference on Trade and Development (UNCTAD) A group that focuses special attention on international economic relations and on measures that might be taken by developed countries to accelerate economic development in *developing countries.*

United Nations Food and Agriculture Organization See *Food and Agriculture Organization (FAO).*

Universal density bale A *bale* of *cotton* compressed to a density of 28 pounds per cubic foot.

Unsaturated fat See *fatty acids (classification of).*

Unshorn lambs *Lambs* that have never been shorn.

Upland cotton The predominant type of *cotton* grown in the United States and in most major cotton-producing countries of the world. The *staple length* of these *fibers* ranges from about 3/4 inch to 1 1/4 inches, averaging nearly 1 3/32 inches.

Up sales, down sales The direction of a *price* from the previous day or period.

Upstream subsidization A *subsidy* provided by a government to a *producer* who sells the subsidized product to an unrelated company that performs further *processing* and ships the product to a *foreign* country.

Urban Towns with a population of 2,500 or more and all territory lying within *Census Bureau*–designated urbanized areas.

Urbanized area A central city or cities and surrounding, closely settled territory (without regard to political boundaries) with a total population of at least 50,000.

Urea A compound containing not less than 45 percent *nitrogen* produced synthetically by combining *carbon dioxide* and *ammonia* or from calcium cyanamide under heat pressure. Urea is used mainly as a *fertilizer* but also as a nonprotein (see *protein*) compound in *ruminant* diets.

Uruguay Round The most recent *round* of *multilateral trade negotiations* conducted under the *General Agreement on Tariffs and Trade (GATT).* The talks were launched in September 1986 in Punta del Este, Uruguay, although actual negotiations were conducted in Geneva, Switzerland. *Agriculture* was included, and negotiators focused on reducing the use of agricultural domestic and *export subsidies*, providing for greater *market access*, harmonizing sanitary and *phytosanitary regulations*, and strengthening the role of GATT in agricultural trade.

U.S. Beet Sugar Association (USBSA) An organization of beet (see *beet*) *sugar* processing (see *processing*) companies that was founded in 1911. See *Sugar Association of America, Inc.*

U.S.–Canada Free Trade Agreement (USCFTA) A comprehensive trade agreement between the United States and Canada enacted on January 1, 1989. The CFTA reduced or removed *tariffs* and mitigated the effects of certain *nontariff trade barriers* between the United States and Canada. Tariffs will be phased out such that all agricultural products will be duty-free by January 1, 1998. Each country also agreed to work toward elimination of technical regulations that arbitrarily restrain agricultural, food, and beverage trade. See Appendix 4.

U.S. Cane Sugar Refiners Association (USCSRA) An organization of cane (see *sugarcane*) *sugar* refiners (see *refining*) that was founded in 1934. See *Sugar Association of America, Inc.*

U.S. Customs Service An agency of the *U.S. Treasury* that collects the revenue from *imports* and enforces *customs* and related laws. The Customs Service is the principal border enforcement agency and assists in the administration and enforcement of over 400 provisions of law on behalf of more than 40 government agencies.

U.S. Department of Agriculture (USDA) A federal department with broad responsibilities related to food and *fiber* production, *rural* development, *export* expansion and promotion (see *export promotion program*), credit, conservation, food safety, and domestic and international food assistance. USDA is one of 14 cabinet-level departments in the executive branch of the U.S. government.

U.S. Department of Commerce (DOC) A federal department responsible for encouraging, serving, and promoting U.S. international trade, economic growth, and technological advancement. DOC offers assistance and information to increase U.S. competitiveness in the world economy; administers programs to prevent unfair *foreign* trade competition; provides social and economic statistics and analyses for business and government planners; provides research and support for the increased use of scientific, engineering, and technological development; works to improve our understanding of and benefits from the earth's physical *environment* and oceanic *resources*; grants patents and registers trademarks; develops policies and conducts research on telecommunications; provides assistance to promote domestic economic development; and assists in the growth of minority businesses. DOC is one of 14 cabinet-level departments in the executive branch of the U.S. government.

U.S. Department of Defense (DOD) A federal department responsible for providing the military forces needed to deter war and protect national security. DOD oversees the Army, Navy, Marine Corps, and Air Force, as well as the civilian reserves. DOD is one of 14 cabinet-level departments of the executive branch of the U.S. government.

U.S. Department of Education A federal department that establishes policy for, administers, and coordinates most federal assistance to education. The Department of Education is one of 14 cabinet-level departments of the executive branch of the U.S. government.

U.S. Department of Energy (DOE) The federal department responsible for developing a comprehensive and balanced, national energy plan through coordination and administration of the energy functions of the federal government. DOE is responsible for long-term, high-risk research and development of energy technology; the marketing of federal power; energy conservation; the nuclear weapons program; energy regulatory programs; and a central, energy data collection and analysis program. DOE is one of 14 cabinet-level departments of the executive branch of the U.S. government.

U.S. Department of Health and Human Services (HHS) A federal department with a broad mandate covering health, welfare, and income security plans. HHS houses the Public Health Service, the Administration on Aging, the *Food and Drug Administration*, the Centers for *Disease* Control, and the National Institutes of Health. HHS also administers the Social Security system. HHS is one of 14 cabinet-level departments in the executive branch of the U.S. government.

U.S. Department of Housing and Urban Development (HUD) The principal federal agency responsible for programs concerned with U.S. housing needs, fair housing opportunities, and improvement and development of communities. HUD is one of 14 cabinet-level departments of the executive branch of the U.S. government.

U.S. Department of Interior The principal conservation agency in the United States, with responsibility for most of the nationally owned public *lands* and natural *resources*. The department is responsible for, among other things, the National Park Service and the American Indian reservation communities. The Department of Interior is one of 14 cabinet-level departments of the executive branch of the U.S. government.

U.S. Department of Justice A federal department that serves as counsel for all U.S. citizens, representing them in enforcing federal laws, preventing and detecting crime, and prosecuting offenders. The department represents the government in legal matters generally and conducts all suits in the Supreme Court in which the United States is concerned. The Department of Justice is one of 14 cabinet-level departments of the executive branch of the U.S. government.

U.S. Department of Labor (DOL) A federal department that fosters, promotes, and develops the welfare of U.S. wage earners. DOL administers a variety of federal labor laws guaranteeing workers' rights to safe and healthful working conditions, a minimum hourly wage and overtime pay, freedom from employment discrimination, unemployment insurance, and workers' compensation. The department also monitors employment, *prices*, and other national economic measurements (see also *Bureau of Labor Statistics*). DOL is one of 14 cabinet-level departments of the executive branch of the U.S. government.

U.S. Department of State The federal department that advises the president in the formulation and execution of *foreign* policy. The department determines and analyzes American overseas interests, makes recommendations on policy and future action, and takes the necessary steps to carry out established policy. The Department of State is one of 14 cabinet-level departments of the executive branch of the U.S. government.

U.S. Department of Transportation (DOT) The federal agency that establishes U.S. overall transportation policy. DOT has nine administrations whose jurisdictions include highway planning, development, and construction; urban mass transit; railroads; aviation; and the safety of waterways, ports, highways, and oil and gas pipelines. DOT is one of 14 cabinet-level departments of the executive branch of the U.S. government.

U.S. Department of Treasury A federal agency that formulates and recommends economic, financial, tax, and *fiscal policies*; serves as financial agent for the U.S. government; enforces the law; and manufactures coins and currency. The Department of Treasury is one of 14 cabinet-level departments of the executive branch of the U.S. government.

U.S. Department of Veterans Affairs (VA) A federal department that operates programs to benefit veterans and members of their families, including providing compensation payments for disabilities or death related to military service; pensions, education, and rehabilitation; home loan guaranty; burial; and a medical care program incorporating nursing homes, clinics, and medical centers. The VA is one of 14 cabinet-level departments of the executive branch of the U.S. government.

U.S. flag vessel A merchant ship under U.S. registry.

U.S. Grain Standards Act An act, administered by the *Grain Inspection, Packers and Stockyards Administration*, requiring that uniform standards be developed and used when marketing (see *agricultural marketing*) *grain.* Testing is provided for, but no requirement exists as to what tests should be performed on grain moving domestically within the United States. Mandatory testing of grain for *export* is required.

U.S. International Trade Commission (USITC) An independent U.S. government agency responsible for reviewing and making recommendations concerning *countervailing duty* and antidumping (see *dumping*) petitions submitted by U.S. industries seeking relief from *imports* that benefit from *unfair trade practices.* The USITC was known as the U.S. Tariff Commission before its mandate was broadened by the *Trade Act of 1974.*

U.S. Meat Import Law A U.S. law, enacted in 1964 and amended in 1979, that provides for the imposition of import *quotas* if *imports* of certain meat products exceed the *trigger level.* The law applies to fresh, chilled, and frozen meat of *cattle, sheep* (except *lamb*), and *goats,* as well as certain prepared and preserved beef and veal products.

U.S. Tariff Commission See *U.S. International Trade Commission (USITC).*

U.S. Trade Representative (USTR) A cabinet-level head of the Office of the U.S. Trade Representative, the principal trade policy agency of the U.S. government. The U.S. Trade Representative is also the chief U.S. delegate and negotiator at all major trade talks and negotiations.

U.S. Treasury See *U.S. Department of the Treasury.*

U.S. Warehouse Act A law that provides for a voluntary warehouse licensing system and a program of periodic examinations of licensed warehouses and their contents to help prevent deterioration and loss of stored products. The program is operated by the *U.S. Department of Agriculture*'s (USDA) *Agricultural Marketing Service.*

U.S. Wheat Associates A nonprofit organization representing the U.S. wheat *producer.* The organization promotes *wheat* produced in the United States and encourages world consumption of U.S. wheat and wheat products.

Utility A subjective measure of the value of place, time, form, and possession in the marketing (see *agricultural marketing*) process. For many products, value is increased by transporting, storing, and *processing* plus the profits of ownership.

Utility, Cutter, and Canner *U.S. Department of Agriculture* (USDA) quality *grades* for beef that represent the lowest *carcass* grades. These grades are used mainly for *cow* beef but may also be applied to *bull* and stag beef. A few cuts may be sold from Utility grade animals, but most of this beef is boned and used for grinding and *processing.*

Utility trailer An all-purpose towed vehicle, often with removable sides, used to haul fuel, implements, welders, and other equipment used on a *farm.*

Utility vehicle A vehicle used for any number of general purposes on a *farm*: to go to and from fields, haul fuel to *tractors,* check *livestock,* fix fences, and so forth. It is usually a pickup truck, but it may be a small three- or four-wheeled, all-terrain vehicle, an old car, or a small tractor.

Utilization The use of a food, *feed, seed,* or industrial product. Used interchangeably with "use, disposition, and disappearance."

V

Vaccine A suspension of attenuated, live, or killed *microorganism*, or parts thereof, administered for the prevention or treatment of infectious *diseases*.

Vacuum-packed The process of placing a product (beef, for example) in a multilayered plastic bag and removing the air by creating a vacuum to shrink the bag around the product. This process is designed to extend the shelf life of a product by reducing shrinkage and deterioration.

Value-added Increasing the value of a good by further *processing*.

Value-added tax (VAT) Taxes collected at each stage of *production*. Raw material costs used from earlier stages are subtracted from each subsequent selling *price*, and the tax is applied only to the "value added." For example, the cost of *wheat* from a farmer is subtracted from the miller's (see *milling*) selling price in calculating the miller's value-added tax.

In the *European Union (EU)*, for example, VATs are set by individual member states. The tax is levied on the amount by which *processors* or merchants increase the value of items they purchase.

Value of production less cash expenses A measure of the shortrun return to *production* and the potential cashflow position of *producers*.

Variable costs The portion of total cash *production* costs used for *farm inputs* required to produce a specific amount of a crop. Variable costs include *seed*, *fertilizer*, fuel, and animal *feed*.

Variable levy A variable tax on *imports*, equal to the difference between the *price* of a *foreign* product at the port and the official "*gate price*" at which competitive imports can be sold. Variable levies are used by the *European Union (EU)*, Austria, Sweden, and Switzerland.

Variety A rank or minor unit within a *species* composed of individuals that differ slightly from the others. In plant breeding, a strain may be adopted or licensed for commercial *production*. See *cultivar*.

Variety meat Includes *offals* and head meat (tongue, cheekmeat, etc.) of animals.

Vealer Calves (see *calf*) that are raised usually on *milk* only and slaughtered (see *slaughter*) usually at less than four months old and less than 350 pounds.

Vegetable oils See *fats and oils*.

Vegetable pear See *chayote*.

Vegetable stearin The filtered-out portion of winterized vegetable oils (see *fats and oils*). These crystallized glyceride compounds usually contain appreciable quantities of stearic acid (see *fatty acids, classification of*).

Vegetative cover Trees, *perennial* grasses, *legumes*, or shrubs with an expected lifespan of five years or more.

Vegetative filter An area of close-growing crops downslope from a *feedlot* designed to absorb pollutants in *runoff* after it has first moved through a *settling basin* to remove solid materials.

Vernonia A plant native to tropical and subtropical Africa that has *seeds* containing an epoxy *fatty acid*, vernolic acid. Scientists are currently studying the feasibility of making vernonia a new *industrial crop*. One possible use for vernonia oil is in paints and coatings.

Vernolic acid See *vernonia.*

Vertical integration Control by a firm of two or more stages of operation either through *contracts* or ownership. Under contract integration, firms obtain legal commitments binding *producers* to certain *production* practices, such as selling the product or purchasing inputs from specified sources. Ownership integration exists when a firm owns two or more levels of the production and management system.

Vertical joint activity or vertical collusion Joint activity among firms operating at successive stages of the marketing (see *agricultural marketing*) channel for a product.

Vertically integrated firm A firm that performs successive stages of *production* and/or marketing (see *agricultural marketing*) of a product. See *vertical integration.*

Vertical storage The storage of *grain* in upright concrete *silos* or metal *bins* that can range in size from 3,000 bushel *farm* bins to 500,000 *bushel* commercial bins. They are easy to load, unload, aerate, and fumigate (see *fumigant*).

Vesicular stomatitis A viral *disease* of *cattle*, *swine*, *sheep*, *goats*, and *horses*, with symptoms similar to *foot-and-mouth disease*. Trade in infected animals is restricted.

Vineyard A field planted with grapevines.

Virgin wool *Wool* yarn (see *yarn*) or *cloth* that has not been through a manufacturing process.

Virulence The capacity of a pathogenic (see *pathogen*) organism to cause *disease*, and the severity of that disease. For example, *E. coli* O157:H7 is a particularly virulent foodborne pathogen.

Vital wheat gluten A *wheat* product containing 75 to 80 percent *protein*, used as a *flour* fortifier, the product of new advances in wheat *processing* technologies.

Vitamin An organic compound that functions as part of *enzyme* systems essential for the transmission of energy and regulation of metabolism in the body.

Viticulture The cultivation of vines and *grapes*.

Vitis labrusca A family of *grape* varieties (see *variety*) used in making *wine*. Most U.S. wine is derived from *Vitis vinifera* and *Vitis labrusca.*

Vitis vinifera A family of *grape* varieties (see *variety*) used in making *wine*. Most U.S. wine is derived from *Vitis vinifera and Vitis labrusca.*

Vitreous A *wheat* kernel characteristic in which the *endosperm* is flinty or glassy. Vitreous *kernels* are very hard and appear translucent and bright in strong light.

Volume control Regulations that limit sales to *markets* and set minimum levels of the *stocks* carried from one year to the next. Volume controls are used as part of the *federal marketing orders*.

Voluntary export agreement An agreement between trading partners in which the exporting nation, in order to reduce trade friction, agrees to limit its *exports* of a particular good. These agreements are generally undertaken to avoid action by the importing country against *imports* that may injure or in some way threaten the positions of domestic firms in the industry in question. Also called voluntary restraint agreement.

Voluntary restraint agreement See *voluntary export agreement.*

W

Wagon Any number of different types of towed, four-wheeled vehicles used to haul *grain*, *feed*, and other *farm* items. Wagons are often equipped with unloading mechanisms, such as *augers* or hydraulic dumps. See *auger wagon*, *gravity wagon*, *bale wagon*, and *grain cart*.

Wagon jobber A *wholesaler* who performs many of the functions of a full-service wholesaler, though operating out of a truck. Wagon jobbers provide supplies, advice and assistance, and credit. They offer only a limited line of products, and the driver/salespeople sell and deliver at the same time.

Wale A lengthwise series of loops in a knitted *cloth*.

Walnut The nut of the walnut tree. English walnut *varieties* include Payne, Hartley, and Franquette, while Eastern Black, Thomas, and Ohio are the principle varieties of black walnuts. China, France, India, and the United States are the major walnut producers.

Warehouse club See *wholesale club store*.

Warehouse (tobacco) Large buildings with skylights used for displaying *tobacco* for *auction* sales.

Warm confinement barn A fully enclosed building in which *livestock* are confined.

Warp The *yarns* that run lengthwise in a woven or warp-knit *cloth*.

Warp knitting See *knitting*.

Wash and wear See *durable press*.

Waste management The use or disposal of *manure*, used bedding, and waste water resulting from animal feeding.

Water Bank Program A *U.S. Department of Agriculture (USDA)* program available to farmers or ranchers having specified types of *wetlands* along major migratory waterfowl flyways. The program is designed to preserve and improve migratory waterfowl and other wildlife *habitat*; preserve and improve wetlands; conserve surface waters; reduce *runoff*, *erosion*, and stream sedimentation; contribute to flood control, better water quality, and improved subsurface moisture; and accomplish related conservation and environmental (see *environment*) objectives.

Water depletion allowance A provision of tax law that provides a deduction based on the depletion of certain *aquifers* used for agricultural irrigation (see *acres irrigated* and *irrigated farms*).

Watermelon A fruit that is a member of the gourd family. The United States and Mexico are the major watermelon producers.

Water rights The rights of a landowner to the banks, bed, or waters of a body of water contained or bordering on their property. Also known as riparian rights.

Watershed The total *land* area, regardless of size, above a given point on a *waterway* that contributes *runoff* water to the flow at that point. It is a major subdivision of a drainage basin. The United States is generally divided into 18 major drainage areas and 160 principal river drainage basins containing about 12,700 smaller watersheds.

Water table The upper limit in *soil* or underlying rock material that is *saturated* with water.

Water Quality Incentive Program (WQIP) The *Agricultural Water Quality Protection Program.*

Water Quality Initiative An interagency, interdepartmental program to develop, test, and deliver to farmers information on crop and *livestock* management systems that reduce the risk of *agricultural chemicals* reaching water supplies, particularly *groundwater.*

Waterway A naturally or artificially constructed course for the concentrated flow of water.

Wax glands The eight glands on the underside of the bee (see *honeybee*) abdomen from which wax is secreted after the bee has been gorged with food.

Wax moth An insect whose *larvae* (see *larva*) destroy wax combs (see *honeycomb*) by boring through the wax in search of food.

Weak flour See *flour.*

Weaning Stopping young animals from nursing or suckling their mothers.

Weather The observed, daily, atmospheric events that, when averaged for a period of time, describe *climate.*

Weave To make *cloth* by interlacing the threads of the *weft* and the *warp* on a *loom.* See also *woven fabric.*

Weeder Hand, towed, or power-driven implements used to pull, cut, or dig undesirable plants from *farm* crops.

Weft The *filling* yarns (see *yarn*) that run crosswise in woven *cloth* or weft-knit (see *knitting*) fabric.

Weft knitting See *knitting.*

Weight of fabric Three methods are used to measure *cloth* weight: (1) linear *yards* per pound, (2) ounces per linear yard, (3) ounces per square yard.

West African Economic Community An organization established in 1972 to promote regional economic development within the seven West African member nations. Members are Benin, Burkina, Côte d'Ivoire, Mali, Mauritania, Niger, and Senegal. Also called Communaute Economique de l'Afrique de l'Ouest. See *common market.*

Wetlands *Land* that is characterized by an abundance of moisture and that is inundated by surface or *groundwater*, often enough to support a prevalence of vegetation typically adapted for life in saturated *soil* conditions.

Wetlands Reserve Program (WRP) A program authorized by the *Food, Agriculture, Conservation, and Trade Act of 1990 (P.L. 101-624)* (see Appendix 3). The WRP has a goal of enrolling 1 million *acres*, including farmed and converted *wetlands*. *Producers* enrolling in the program must agree to implement an approved wetlands restoration and protection plan and provide either a permanent easement or one of 30 years or more. In return, participating producers would receive payments over a 5- to 20-year period—or in one lump sum if they grant a permanent easement.

Wet-milling The manufacturing process that separates the corn *germ*, *hull*, *gluten*, and *starch* from the corn *kernel* with water as the suspension medium. *Corn starch* is the main product. The corn germ is the raw material processed (see *processing*) to obtain corn oil (see *fats and oils*).

Wet-rendered A process in which material is cooked with steam under pressure in closed tanks.

Wheat Any of various grasses (belonging to the genus *Triticum*) high in *gluten* that are cultivated (see *cultivate*) in various temperate areas for the *grain* that they yield, which is used in a vast array of products, including animal *feed*, *flour*, *bran*, *germ*, *gluten*, and *starch*. In the United States, the main wheat-producing states are Kansas, Oklahoma, Texas, Nebraska, Colorado, and North Dakota. *Hard red winter wheat* is the main *variety* grown in the United States. See *white wheat*, *soft red winter wheat*, *hard red spring wheat*, *hard red winter wheat*, and *durum wheat*.

Wheat and Feed Grain Export Certificate Programs Two discretionary programs for the 1986-1990 crops designed to encourage *exports* of *wheat* and *feed grains* from private *stocks*. Under the Cash Export Certificate Program and the Export Marketing Certificate Program, the secretary of agriculture would have issued wheat and feed grain export certificates to all eligible *producers*. The programs were not implemented.

Wheat starch The portion of the *wheat* kernel (see *kernel*) remaining after the *gluten* has been extracted.

Wheat Trade Convention See *International Wheat Agreement*.

Wheel tractor A *tractor* with wheels as opposed to one with steel or rubber tracks. See *crawler tractor*.

Whey The liquid part of the *milk* remaining after separation of the *curd* during cheesemaking (see *cheese*). Whey is used in foods and animal *feed*.

White clover See *clover*.

White pan bread A commercially baked loaf of white bread that is distinguished from variety breads, such as French and Italian breads.

White wheat A low-*protein*, common *wheat* (*Triticum aestivum*) or *club wheat* (*Triticum compactum*) used principally for pastry *flours* and shredded or puffed breakfast foods. White wheat is grown in the fall or spring.

Whole-herd buyout program See *Milk Production Termination Program*.

Wholesale club store A hybrid wholesale-retail establishment selling food, appliances, hardware, office supplies, and similar products to its individual and small-business members at prices slightly above wholesale. See also *cash and carry wholesaler.*

Wholesale Market Development Program A federal program that conducts research to find new ways of improving the efficiency of handling and storing food products moving between the farmers and retail outlets. The program is administered by the *U.S. Department of Agriculture*'s (USDA) *Agricultural Marketing Service.*

Wholesale meat cuts *Primal, subprimal,* or *processor*-packaged retail cuts of meat that are traded at the wholesale level.

Wholesale price index A measure of average changes in the *prices* of *commodities* sold in primary U.S. markets. "Wholesale" refers to sale in large quantities by *producers*, not to prices received by *wholesalers, jobbers,* or distributors. In *agriculture*, it is the average price received by farmers for their *farm* commodities at first point of sale when the commodity leaves the farm. The principal such index compiled by the *Bureau of Labor Statistics* is called the *Producer Price Index.*

Wholesaler A person or firm who sells products, usually in large quantities, to retailers, restaurants, and so on who then resell the items at a *price* that includes the retailer's markup.

Wholesaling The operations of firms engaged in buying, assembling, transporting, storing, and distributing groceries and food products for resale by retailers, institutions, businesses, industrial users, and commercial firms. See *packer sales office*, *direct sales*, and *purveyor.*

Wholesome Meat Act Legislation that specifies that all meat produced for sale in the United States must be inspected; and further, that all meat transported interstate must be inspected in compliance with federal *(U.S. Department of Agriculture*) standards. See also *inspection, federal* and *meat and poultry inspection.*

Whole wheat flour See *flour.*

WIC **Program** See *Special Supplemental Food Program for Women, Infants, and Children.*

Wigging Shearing *wool* from the eyes and face of *sheep.*

Wild marjoram See *oregano.*

Wilt Loss of freshness and drooping of leaves, stems, or shoots of plants due to inadequate water supply or to excessive *transpiration,* from either a vascular *disease* that interferes with uptake of water by a plant or a *toxin* produced by an organism.

Wilting point (or permanent wilting point) The moisture content of *soil* at which plants wilt and fail to recover their turgidity when placed in a dark humid atmosphere. The percentage of water at the wilting point approximates the minimum moisture content in soils under plants in the field at depths below the effects of surface evaporation.

Windbreak Any protective shelter from the wind or to rows of trees or shrubs reducing the force of wind.

Wind erosion *Soil* depletion resulting from strong winds blowing across inadequately protected soil. Wind *erosion* is usually worse during *drought* conditions.

Windrow Cut *hay* or *forage* that is raked into rows and left to cure (see *curing*).

Windrow pickup An apparatus used to lift and carry *grain, hay*, or other *forage* materials from a field into a *combine, baler*, or other machine. It may have a set of chains or belts with short, flexible tines that go under a previously cut and swathed or windrowed (see *windrow*) crop to lift it onto a platform with chains or belts that move the materials into the machine.

Wine An alcoholic beverage produced from a number of different *grape* varieties. Major wine types include (Red) Zinfandel, Cabernet Sauvignon, and Pinor Noir; and (White) French Colombard, Chenin Blanc, and Chardonnay. France, Italy, the former Soviet Union, Spain, the United States, and Argentina are the leading wine producers. Most U.S. wine is derived from two basic families of grape *varieties—Vitis vinifera* and *Vitis labrusca.*

Winged bean A *legume* cultivated as a *market* and garden crop in southern India, Indonesia, Malaysia, Myanmar, Papua New Guinea, the Philippines, Thailand, and Vietnam. All parts of the plant are edible: leaves, flowers, pods, *seeds*, and *tubers*.

Winterizing Chilling refined (see *refining*) oil (see *fats and oils*), holding it at proper temperatures to separate higher-melting *glycerides* by crystallization, and filtering them out. The "wintered oil" remains clear and bright and is used as salad or cooking oil. Winterization is carried out after refining and *bleaching* but before deodorization (see *deodorizing).*

Winter wheat A *wheat* that is sown in the fall, lies dormant in the winter, and is harvested (see *harvest*) the following spring or summer.

Witches'-broom A symptom of a plant *disease* characterized by an abnormal tufted or brushlike growth/development of many shoots. Witches'-broom is caused by fungi (see *fungus*), viruses, or mycoplasma-like organisms.

Withdrawal period The number of days, after a *drug* has been administered, that it takes for levels in the tissue to go below the violative levels prescribed by the *Food and Drug Administration (FDA).*

With particular average (WPA) A form of *marine insurance* that includes all the coverage provided under *free of particular average (FPA)* insurance, as well as seawater damage.

Women Involved in Farm Economics (WIFE) A national organization of wives of *farm operators*.

Woodland pasture 1: Forested *land* used for grazing (see *graze*).
2: Plant materials grazed from forested land.

Wool The thick, soft hair shorn from *sheep* that is used for clothing and other *textiles*. In the United States, the major *varieties* of wool are Territory, Fleece, Texas, and California. Australia, China, Argentina, New Zealand, the former Soviet Union, and South Africa are the major wool producers.

Wool bag A standard 41-inch by 89-inch burlap bag for marketing *wool.*

Wool Bureau The U.S. branch of the *International Wool Secretariat (IWS).* The Wool Bureau educates the public on the merits of *wool* and assists wool manufacturers and retailers through wool product research and development. The Wool Bureau is headquartered in New York City.

Woolen System A sequence of machinery used for the conversion into yarn of coarser, shorter, more heavily contaminated *wools*, in which *yarn* is spun directly from the product of the woolen card (see *carding*), called "roping." Such yarns are used in applications where bulkiness is required, such as blankets.

Worker bee A female bee (see *honeybee*) with undeveloped reproductive organs. These bees do all the work of the *colony* except laying *eggs*.

Worker egg A fertilized (see *fertilization*) *egg* laid by a *queen bee*. The egg may produce a *worker bee* or queen bee.

World Agricultural Outlook Board (WAOB) A *U.S. Department of Agriculture* (USDA) agency created in 1977 to coordinate the Department's *supply* and use *commodity* forecasts and agricultural situation and outlook analyses, including monthly U.S. and world supply and *demand* estimates. The WAOB heads the departmental *weather*, *climate*, and *remote sensing* activities and assesses the impact of weather on *agriculture*. The WAOB also coordinates the Department's annual Outlook Conference.

World Bank See *International Bank for Reconstruction and Development*.

World Food Council (WFC) The highest level, political institution dealing with world food problems in the United Nations. The WFC meets yearly at the *ministerial* level to discuss coordinating international efforts to solve food and agricultural problems.

World Food Program (WFP) A program operated by the United Nations' *Food and Agriculture Organization (FAO)* to supply food *resources* for economic development projects in *developing countries*. Examples include child feeding and school lunch programs and food-for-work *infrastructure* projects.

World Health Organization (WHO) An agency of the United Nations designed to further international cooperation for improved health conditions. WHO was established in 1948.

World price A *price* determined by *supply* and *demand* conditions in all parts of the world. Technically, it is generally the *cost, insurance, and freight (c.i.f.)* price of an imported agricultural *commodity* at the principal port of a major importing (see *import*) country or area.

Worsted system A sequence of machinery used to process finer or longer *wool* fibers that are relatively clean, involving the process of *combing* to produce *yarns* that are leaner and smoother. Worsted yarns are used in men's suiting and in fine knitwear, for example.

Wort The liquid portion of malted *grain*. It is a solution of *malt* sugar and other soluble extracts from malted mash.

Woven fabric *Cloth* made by interlacing two sets of *yarn* at right angles. The *warp* yarns run lengthwise in the cloth; the *filling* (*weft*) yarns are passed over and under the warp yarns.

Wrapper tobacco The class of *tobacco* grown for the outside cover of *cigars*. This is the most difficult and expensive tobacco to grow. The leaves must be elastic, uniform, free of injury, uniform in color, and have good burning qualities. To produce leaves of such quality, it is necessary to protect them from the sun and extremes of weather. Therefore, many fields of tobacco are covered with cheese cloth to filter the sun and create the artificial *environment* favorable to the specialized product desired.

Writer or grantor A person who sells an *option*.

Y

Yarn 1: A continuous strand of twisted (spun) *fibers* of any kind and of varying *staple length*, usually used in the *weaving* or *knitting* of *cloth*.
2: A discrete assembly of continuous *filaments*, which may be twisted or interlaced together.

Yarn size 1: A description of *yarn* in terms of weight per unit of length (direct systems) or length per unit of weight (indirect systems). Examples of direct systems are those that use *denier* or *tex* as units of linear density. Indirect systems, of which there are many, express yarn sizes as "*counts*" or "numbers" that increase as yarns become finer. Yarn count (number) is usually the number of hanks per pound, and it is the length of the hank that varies between systems. In the U.S. *textile* industry, the length is 840 yards for *cotton* and 560 yards for *wool*. A "1s" cotton yarn has 840 yards per pound; a "30s" cotton yarn has 25,200 yards per pound. A "30/2" is a two-ply yarn containing two strands of 30s and is effectively a 15s yarn. See *cotton count*.
2: See *size*.

Yearling loss rate The number of *yearlings* lost due to death, theft, and other causes during a calendar year per 100 calves (see *calf*) born alive during that calendar year.

Yeast A *fungus* used to ferment *starches* and *sugars*. Yeast is important for brewing and baking.

Yellows A plant *disease* caused by a *fungus*, mycoplasma-like organism, or virus and characterized by yellowing and stunting of the affected plants.

Yield The quantity of a *commodity* resulting from *production* or *harvest*.

Yield, economic maximum The most that can be produced by full, efficient application of technology presently known by all farmers. Assumes there are no limitations on management, materials, equipment, capital, and experience.

Yield grades (1-5) A *U.S. Department of Agriculture (USDA)* system that provides a nationally uniform method of identifying "quantity" or "cutability" (lean *yield*) differences among beef carcasses (see *carcass*). Yield grade 1 has the least *fat* and waste, and yield grade 5 has the most.

Yogurt A product produced by curdling *milk* with certain *bacteria*.

Young chicken See broiler.

Yolk 1: The yellow part of an *egg*, consisting chiefly of *protein* and *fat*.
2: The *lanolin* and other fats and waxes secreted by *sheep* that show up as grease in the *wool*. See *grease weight*, *grease mohair*, and *grease wool*.

Yucca A popular root crop used in Asian, African, and Latin cooking. It is cooked in a variety of ways and is widely known in its processed (see *processing*) form as tapioca.

Z

Zero cultivation See *no-till.*

0/85/92 A provision, originally known as 0/92, that allows *wheat* and *feed grain* producers to devote all or a portion of their *maximum payment acreage (MPA)* to *conserving uses*, minor *oilseeds* (*sunflowers, safflowers, canola, rapeseed, flaxseed,* and *mustard seed*) and receive *deficiency payments* on a maximum of 92 percent of a *farm's* MPA. Under the *Food, Agriculture, Conservation, and Trade Act of 1990 (P.L. 101-624)* (see Appendix 3), *producers* planting minor oilseeds on 0/92 acres must choose either the 0/92 payment or the oilseed *marketing loan.* If producers choose the 0/92 payment, they lose their marketing loan eligibility for that oilseed. See also *50/92.* The Omnibus Budget Reconciliation Act of 1993 (see Appendix 1) reduced the percentage of deficiency payments producers may receive from 92 to 85 percent of *MPA*. However, producers who plant approved industrial or experimental crops, or who suffer prevented planting or failed acres may earn deficiency payments on up to 92 percent of the farm's *MPA*.

Zoonosis Any infectious *disease* that can be transmitted between vertebrate animals and humans. The disease is not transmitted from person to person.

Appendix 1

Commonly Used Acronyms

AAM	American Agricultural Movement
AAMP	American Association of Meat Processors
AARC	Alternative Agricultural Research and Commercialization Center
ACA	Accession compensatory amount
ACA	Agricultural Credit Association
ace-k	Acesulfame-k
ACP	Agricultural Conservation Program
ACP States	African, Caribbean, and Pacific States
ACS	Agricultural Cooperative Service
ADC Program	Animal Damage Control Program
AFDB	African Development Bank
AID	Agency for International Development
AIMS	Agricultural Information and Marketing Service
AMAA	Agricultural Marketing Agreement Act of 1937
AMF	Anhydrous milk fat
AMI	American Meat Institute
AMS	Agricultural Marketing Service
ANEC	Association Nacional dos Exportadores de Cereais
APHIS	Animal and Plant Health Inspection Service
APM	Aspartame
ARCP	Agricultural Resource Conservation Program
ARP	Acreage Reduction Program
ARS	Agricultural Research Service
ASC	Agricultural Stabilization Committee
ASCS	Agricultural Stabilization and Conservation Service
ASFMRA	American Society of Farm Managers and Rural Appraisers
ASGA	American Sugarbeet Growers Association
ASTA	American Seed Trade Association
ATO	Agricultural trade office
AVE	Ad valorem equivalent
AWB	Australian Wheat Board
AWQPP	Agricultural Water Quality Protection Program
BC	Bank for Cooperatives
BCFM	Broken corn and foreign material
BDEAC	Banque de Developpement des Etats de l'Afrique Centrale
BENELUX	Belgium-Luxembourg Economic Union
BIFAD	Board for International Food and Agricultural Development
BLS	Bureau of Labor Statistics
BPSY	Bleachable prime summer yellow

bST	Bovine Somatotropin
B.t.	*Bacillus thuringiensis*
BTN	Brussels Tariff Nomenclature
CACFP	Child and Adult Care Food Program
CACM	Central American Common Market
CAP	Common Agricultural Policy
CARICOM	Caribbean Community and Common Market
CARIFTA	*See* Caribbean Community and Common Market
CAT	Computer-assisted trading
CBERA	Caribbean Basin Economic Recovery Act
CBI	Caribbean Basin Initiative
CCC	Commodity Credit Corporation
CCCN	Customs Cooperation Council Nomenclature
CCI	Cotton Council International
CDB	Caribbean Development Bank
CEEAC	Economic Community of Central African States
CES	Consumer Expenditure Survey
CFTA	U.S.–Canada Free Trade Agreement
CFTC	Commodity Futures Trading Commission
CGIAR	Consultative Group on International Agricultural Research
CI	Cotton Incorporated
c.i.f	Cost, insurance, and freight
CIPS	Commodity Import Programs
CMSA	Consolidated Metropolitan Statistical Area
CNI	Community Nutrition Institute
COAP	Cottonseed Oil Assistance Program
CONCEX	Conselho Nacional do Comercio Exterior
CP	Contracting party
CPI	Consumer Price Index
CRIS	Current Research Information System
CRP	Conservation Reserve Program
CSD	Committee on Surplus Disposal
CSE	Consumer subsidy equivalent
CSFA	Consolated Farm Service Agency
CSFII	Continuing Survey of Food Intakes by Individuals
CSFP	Commodity Supplemental Food Program
CSPI	Center for Science in the Public Interest
CSREES	Cooperative State Research, Education, and Extension Service
CSRS	Cooperative State Research Service
CVD	Countervailing duty
CWB	Canadian Wheat Board
CWE	Carcass weight equivalent
cwt	Hundredweight
DEIP	Dairy Export Incentive Program
DNA	Deoxyribonucleic acid
DOC	U.S. Department of Commerce
DOD	U.S. Department of Defense
DOE	U.S. Department of Energy
DOL	U.S. Department of Labor
DOT	U.S. Department of Transportation
EAGGF	European Agricultural Guidance and Guarantee Fund
EAI	Enterprise for the Americas Initiative

EBT	Electronic Benefits Transfer
EC	European Community
ECARP	Environmental Conservation Acreage Reserve Program
ECOWAS	Economic Community of West African States
ECP	Emergency Conservation Program
ECU	European currency unit
EEC	European Economic Community
EEP	Environmental Easement Program
EEP	Export Enhancement Program
EFAP	Emergency Feed Assistance Program
EFP	Emergency Feed Program
EFTA	European Free Trade Association
EIP	Export Incentive Program
ELISA	Enzyme-linked immunosorbent assay
EMS	European monetary system
EPA	Environmental Protection Agency
ERS	Economic Research Service
EU	European Union
FAO	Food and Agriculture Organization
FAS	Foreign Agricultural Service
FCA	Farm Credit Administration
FCB	Farm Credit Bank
FCC	Farm Credit Council
FCIA	Foreign Credit Insurance Association
FCIC	Federal Crop Insurance Corporation
FCRS	Farm Costs and Returns Survey
FCS	Farm Credit System
FCS	Food and Consumer Service
FDA	Food and Drug Administration
FEOGA	French abbreviation for the EAGGF (European Agricultural Guidance and Guarantee Fund)
FFDCA	Federal Food, Drug, and Cosmetic Act
FGIS	Federal Grain Inspection Service
FI	Free in
FICB	Federal Intermediate Credit Bank
FIFRA	Federal Insecticide, Fungicide, and Rodenticide Act of 1947
FIO	Free in and out
FIP	Forestry Incentives Program
FLBAs	Federal Land Bank Associations
FLCAS	Federal Land Credit Associations
FMD	Foot-and-mouth disease
FmHA	Farmers Home Administration
FNS	Food and Nutrition Service
FO	Free out
F.o.b.	Free-on-board
FOGS	Functioning of the GATT System
FOPS	Falling object protective structure
FOR	Farmer-owned Reserve Program
FPA	Free of particular average
FRAC	Food Research and Action Center
FS	Forest Service
FSMIP	Federal-State Marketing Improvement Program

FSIS	Food Safety and Inspection Service
FSP	Food Stamp Program
FSW	Farm sale weight
GATT	General Agreement on Tariffs and Trade
GCAU	Grain-consuming animal unit
GDP	Gross domestic product
GIPSA	Grain Inspection, Packers, and Stockyards Administration
GMA	Grocery Manufacturers of America
GMP	Guaranteed minimum price
GNP	Gross national product
GPCP	Great Plains Conservation Program
GRAS	Generally Recognized as Safe
GRP	Group Risk Plan
GSM	General Sales Manager
GSM-5	*See* Direct (export) credit.
GSM-102	Export Credit Guarantee Program
GSM-103	Intermediate Export Credit Guarantee Program
GSP	Generalized System of Preferences
HACCP	Hazard Analysis/Critical Control Points
HFCS	High fructose corn syrup
HHS	U.S. Department of Health and Human Services
HNIS	Human Nutrition Information Service
HQB	High-quality beef
HS	Harmonized system
HUD	U.S. Department of Housing and Urban Development
HVI	High volume instrument
HVP	High-value products
IADB	Inter-American Development Bank
IAMP	International Association of Meat Processors
IBRD	International Bank for Reconstruction and Development
ICA	International Coffee Agreement
ICA(s)	International Commodity Agreement(s)
ICAC	International Cotton Advisory Committee
ICCA	International Cocoa Agreement
IDA	International Dairy Arrangement
IFAD	International Fund for Agricultural Development
IFC	International Finance Corporation
IFPRI	International Food Policy Research Institute
IIC	International Institute for Cotton
IICA	Inter-American Institute for Cooperation in Agriculture
IGR	Insect growth regulators
IMF	International Monetary Fund
IOE	International Office of Epizootics
IOOA	International Olive Oil Agreement
IP	Initial payment
IP	Intervention price
IPM	Integrated Pest Management
IPPC	International Plant Protection Convention
IPR	Intellectual Property Rights
ISA	International Sugar Agreement
ISEO	Institute of Shortening and Edible Oils
ISO	International Organization for Standardization

ITC	International Trade Commission
IWA	International Wheat Agreement
JNG	Junta Nacional de Granos
LAFTA	Latin American Free Trade Association
LDC	Less developed country
LISA	Low-input sustainable agriculture
MA	Metropolitan area
MARR	Maximum Acceptable Rental Rates
MASDA	Midwestern Association of State Departments of Agriculture
MCA	Monetary compensatory account
MFA	Multifiber Arrangement
MFN	Most-favored nation
MIP	Minimum import price
MIU	Moisture, insoluble impurities, and unsaponifiable matter
MLRA	Major Land Resource Area
MPA	Maximun payment acreage
MPP	Market Promotion Program
MSA	Metropolitan Statistical Area
MSG	Monosodium glutamate
MSPA	Migrant and Seasonal Agricultural Worker Protection Act
MTN	Multilateral trade negotiations
M-W price	Minnesota-Wisconsin price
NAC	National Advisory Council on International Monetary and Financial Policies
NACD	National Association of Conservation Districts
NAEGA	North American Export Grain Association
NAL	National Agricultural Library
NAMP	National Association of Meat Purveyors
NAP	Noninsured Assistance Program
NASDA	National Association of State Departments of Agriculture
NASS	National Agricultural Statistics Service
NAWG	National Association of Wheat Growers
NAWGA	National American Wholesale Grocers Association
NBC	National Broiler Council
NCA	National Cattlemen's Association
NCA	Normal crop acreage
NCC	National Cotton Council of America
NCFAP	National Center for Food and Agricultural Policy
NCGA	National Corn Growers Association
NCPA	National Cottonseed Products Association
NDBC	National Dry Bean Council
NDM	Nonfat dry milk
NEASDA	Northeastern Association of State Departments of Agriculture
NEPA	National Environmental Policy Act
NFBA	National Food Brokers Association
NFFA	National Frozen Food Association
NFI	National Fisheries Institute
NFO	National Farmers Organization
NFPA	National Food Processors Association
NFU	National Farmers Union
NGA	National Grocers Association
NGFA	National Grain and Feed Association
NGTC	National Grain Trade Council

NIC	Newly industrialized country
NIOP	National Institute of Oilseed Products
NIRS	Near-infrared reflectance spectroscopy
NMPF	National Milk Producers Federation
NOA	National Onion Association
NPA	National Program Acreage
NPC	National Peanut Council
NPC	National Potato Council
NPCW	National Pork Council Women
NPMC	National Pecan Marketing Council
NPPC	National Pork Producers Council
NRA	National Restaurant Association
NRCS	National Resource Conservation Service
NRECA	National Rural Electric Cooperative Association
NRHC	National Rural Housing Coalition
NRTC	National Rural Telecommunications Cooperative
NSCIC	National Soybean Crop Improvement Council
NSDA	National Soft Drink Association
NSGC	National Swine Growers Council
NSLP	National School Lunch Program
NSPA	National Soybean Processors Association
NTF	National Turkey Federation
NWGA	National Wool Growers Association
OAS	Organization of American States
OAU	Organization of African Unity
OE	Office of Energy
OECD	Organization for Economic Cooperation and Development
OICD	Office of International Cooperation and Development
OMA	Orderly marketing agreement
OMB	Office of Management and Budget
OPA	Office of Public Affairs
OPIC	Overseas Private Investment Corporation
OT	Office of Transportation
OTA	Office of Technology Assessment
PA	Purchase authorization
PCAs	Production Credit Associations
PCP	Posted county price
PIK	Payment-in-kind
PPI	Producer Price Index
PSA	Packers and Stockyards Administration
PSE	Producer Subsidy Equivalents
PTO	Power-take-off
PVO	Private voluntary organization
QR	Quantitative restriction
RAW	Replenishment agricultural worker
RBCDS	Rural Business and Cooperative Development Service
RCAU	Roughage consuming animal unit
RDA	Rural Development Administration
REA	Rural Electrification Administration
RHCDS	Rural Housing and Community Development Service
RMA	Rice Millers Association
RNA	Ribonucleic acid

ROPS	Roll over protective structure
RUS	Rural Utilities Service
SAI	Sugar Association of America, Inc.
SAW	Special agricultural worker
SBP	School Breakfast Program
SCS	Soil Conservation Service
SDR	Special drawing rights
SFSP	Summer Food Service Program
SITC	Standard International Trade Classification
SMP	Special Milk Program
SMSA	Standard Metropolitan Statistical Area
SOAP	Sunflower Oil Assistance Program
SPCUS	Sweet Potato Council of the United States
STABEX	*See* Lome Convention.
TDN	Total digestible nutrients
TEA	Targeted Export Assistance Program
TEFAP	Temporary Emergency Food Assistance Program
TEFAP	The Emergency Food Assistance Program
TFP	Thrifty Food Plan
TI	Tobacco Institute
TOP	Targeted Option Payments
TPC	Trade policy committee
TRQ	Tariff rate quota
UDEAC	*See* Central African Customs and Economic Union.
UEA	United Egg Association
UEP	United Egg Producers
UFW	United Farm Workers of America
UGSA	Uniform Grain Storage Agreement
UHT	Ultra-high temperature milk
UNCTAD	United Nations Conference on Trade and Development
USBSA	U.S. Beet Sugar Association
USCFTA	U.S.–Canada Free Trade Agreement
USCSRA	U.S. Cane Sugar Refiners Association
USDA	U.S. Department of Agriculture
USITC	U.S. International Trade Commission
USTR	U.S Trade Representative
VA	U.S. Department of Veterans Affairs
VAT	Value-added tax
WAOB	World Agricultural Outlook Board
WFC	World Food Council
WFP	World Food Program
WHO	World Health Organization
WIC	Special Supplemental Food Program for Women, Infants, and Children
WIFE	Women Involved in Farm Economics
WPA	With particular average
WQIP	Water Quality Incentive Program
WRP	Wetlands Reserve Program

Appendix 2

Dictionary Terms by Subject Area

Subject Areas Included in This Appendix

Agribusiness, commodity, farm, and trade associations
Agricultural development
Agricultural policies and programs, foreign
Agricultural policies and programs, U.S.
Agricultural production
Animal and plant science and research
Bees and beekeeping
Climate
Commodity agreements
Conservation and environmental protection
Corn
Cotton
Credit and banking
Crop insurance and disaster assistance
Dairy
Diseases, crop
Diseases, livestock
Economics (micro and macro)
Environmental horticulture and floriculture
European Union
Farm and agriculture-related labor
Farm and food prices and expenditures
Farm equipment and buildings
Farm income (costs and returns)
Farm structure and organization
Fats and oils
Feeds and fodder
Feedlot
Fertilizers
Fibers and textiles
Food and feed grain
Food assistance, domestic and international
Food away from home
Food expenditures
Food grading and inspection
Food manufacturing and processing
Food marketing, wholesaling, and retailing
Food prices and expenditures
Food safety, nutrition, and health
Foodservice
Forestry and forestry products
Fruits and tree nuts
Futures and options markets
General Agreement on Tariffs and Trade (GATT)
Grain drying, milling, and storage
Herbs and spices
Industrial crops and products
Insecticides, pesticides, fungicides, and fumigants
Labor
Legislation
Legumes
Livestock
Marketing orders
Market performance and structure
Meat and poultry packing and processing
Medicinal products from agriculture
Milk
Natural resources (soil and water)
Oilseeds
Poultry
Public Law 480
Rice
Rural and urban
Science
Seeds
Shipping and transportation
Soil
Soybeans
Sugar and sweeteners
Sunflowers
Supply and demand
Tariffs and trade
Tobacco
Trade
Tropical, root, tuber, and other crops and products
U.S. Department of Agriculture
U.S. government
Vegetables
Water
Weights and measures
Wheat

Dictionary Terms by Subject Area

Agribusiness, Commodity, Farm, and Trade Associations

Agribusiness
Agriculture Council of

America
American Agricultural Movement (AAM)
American Association of Meat Processors (AAMP)
American Bankers Association
American Farm Bureau Federation
American Frozen Food Institute
American Meat Institute (AMI)
American Seed Trade Association (ASTA)
American Society of Farm Managers and Rural Appraisers (ASFMRA)
American Soybean Association
American Sugarbeet Growers Association (ASGA)
American Textile Manufacturers Institute
Association of American Feed Control Officials
Association of American Pesticide Control Officials
Association of American Seed Control Officials
Association Nacional dos Exportadores de Cereais (ANEC)
Bargaining association
Catfish Farmers of America
Farm Bureau
Fertilizer Institute
Food Marketing Institute
Grocery Manufacturers of America
Institute of Shortening and Edible Oils (ISEO)
International Association of Meat Processors (IAMP)
International Institute for Cotton (IIC)
International Meat Council
Latin American Free Trade Association
Midwestern Association of State Departments of Agriculture (MASDA)
National Association of Conservation Districts
National Association of Margarine Manufacturers (NAMM)
National Association of Meat Purveyors
National Association of State Departments of Agriculture (NASDA)
National Association of Swine Growers (NASG)
National Association of Wheat Growers (NAWG)
National Broiler Council (NBC)
National Cattlemen's Association (NCA)
National Corn Growers Association (NCGA)
National Cotton Council of America (NCC)
National Cottonseed Products Association (NCPA)
National Council of Farmer Cooperatives
National Dry Bean Council (NDBC)
National Farmers Organization (NFO)
National Farmers Union (NFU)
National Fisheries Institute (NFI)
National Food Brokers Association (NFBA)
National Food Processors Association (NFPA)
National Frozen Food Association (NFFA)
National Grain and Feed Association (NGFA)
National Grain Trade Council (NGTC)
National Grange
National Grocers Association (NGA)
National Institute of Oilseed Products (NIOP)
National Milk Producers Federation (NMPF)
National Onion Association (NOA)
National Peanut Council (NPC)
National Pecan Marketing Council (NPMC)
National Pork Council Women (NPCW)
National Pork Producers Council (NPPC)
National Potato Council (NPC)
National Restaurant Association (NRA)
National Rural Electric Cooperative Association (NRECA)
National Rural Housing Coalition (NRHC)
National Rural Telecommunications Cooperative (NRTC)
National Soft Drink Association (NSDA)
National Soybean Crop Improvement Council (NSCIC)
National Soybean Processors Association (NSPA)
National Swine Growers Council (NSGC)
National Turkey Federation (NTF)
National Wildlife Federation
National Wool Growers Association (NWGA)
North American Export Grain Association (NAEGA)
Northeast Association of State Departments of Agriculture (NEASDA)
Overseas Development Council (ODC)
Patrons of Husbandry
Poultry and Egg Export Council
Poultry and Egg Institute of America
Rice Millers' Association (RMA)
Southern Association of State Departments of Agriculture (SASDA)
Sugar Association, Inc. (SAI)
Sweet Potato Council of the United States (SPCUS)
Tobacco Institute (TI)
United Egg Association (UEA)
United Egg Producers (UEP)

United Farm Workers of America (UFW)
U.S. Beet Sugar Association (USBSA)
U.S. Cane Sugar Refiners' Association (USCSRA)
Western Association of State Departments of Agriculture (WASDA)
Women Involved in Farm Economics (WIFE)

Agricultural Development

African Development Bank (AFDB)
Agency for International Development (AID)
Board for International Food and Agricultural Development (BIFAD)
Bread for the World Education Fund
Bread for the World Institute on Hunger and Development
Developing country
Food and Agriculture Organization (FAO)
Inter-American Development Bank (IDB)
Inter-American Institute for Cooperation in Agriculture (IICA)
International Bank for Reconstruction and Development (IBRD)
International Development Association
International Finance Corporation (IFC)
International Food Policy Research Institute (IFPRI)
International Fund for Agricultural Development (IFAD)
International Monetary Fund (IMF)
Less developed country (LDC)
Newly industrialized country (NIC)
Overseas Private Investment Corporation (OPIC)
Underdeveloped nations
World Bank
World Food Program (WFP)

Agricultural Policies and Programs, Foreign

A-Fix-A
African, Caribbean, and Pacific States (ACP States)
Andean Group
Association Nacional dos Exportadores de Cereais (ANEC)
Australian Wheat Board
Banque de Developpement des Etats de l'Afrique Centrale (BDEAC)
Belgium-Luxembourg Economic Union (BENELUX)
Canadian Wheat Board (CWB)
Caribbean Basin Economic Recovery Act (CBERA)
Caribbean Basin Initiative (CBI)
Caribbean Community and Common Market (CARICOM)
Caribbean Development Bank (CDB)
Central African Customs and Economic Union
Central African States Development Bank (BDEAC)
Central American Common Market (CACM)
Centrally planned economy
Codex Alimentarius Commission
Common market
Communaute Economique des Etats de l'Afrique Centrale (CEEAC)
Conasupo
Customs union
East African Community
Economic Community of Central African States (CEEAC)
Economic Community of West African States (ECOWAS)
Economic development
Foreign
General Agreement on Tariffs and Trade (GATT)
Guaranteed Minimum Price (GMP)
Junta Nacional de Granos (JNG)
Marketing board
Mercosur
Ministerial
North American Free Trade Agreement (NAFTA)
Organization of African Unity (OAU)
Organization of American States (OAS)
Receival standards
Single European Act
Union Douaniere et Economique de l'Afrique Central (UDEAC)
West African Economic Community

See also ***European Union***

Agricultural Policies and Programs, U.S.

Acreage allotment
Acreage base
Acreage diversion
Acreage reduction program (ARP)
Acreage reserve
Acreage slippage
Advance deficiency payments
Advance recourse loans
Agricultural Conservation Program (ACP)
Agricultural Resource Conservation Program (ARCP)
Agricultural Stabilization and Conservation Service (ASCS)
Agricultural Stabilization Committee (ASC)

Agricultural Trade Development and Assistance Act of 1954
Allocation procedure
Assessment programs
Balanced Budget and Emergency Deficit Control Act of 1985
Base acreage
Basic commodities
Brannan plan
Cash-out option for generic certificates
CCC commercial credit
Certs
Check-off programs
Commodity Credit Corporation
Commodity Import Programs (CIPs)
Conservation Reserve Corn Bonus Program
Conservation Reserve Program (CRP)
Cost-sharing program
Cottonseed Oil Assistance Program (COAP)
County committee
County loan rate
County office
Credit guarantees
Crop acreage base
Cropland set-aside
Cross-compliance
Dairy Diversion Program
Dairy Export Incentive Program
Dairy Indemnity Payment Program
Dairy Termination Program
Decoupling
Deficiency payment
Designated nonbasic commodities
Direct (export) credit
Direct government subsidies
Direct payments
Disaster payments
Diversion payments
Emergency Compensation Payments
Emergency Conservation Program (ECP)
Emergency Feed Assistance Program (EFAP)
Emergency Feed Program (EFP)
Enrollment (signup) period
Environmental Conservation Acreage Reserve Program (ECARP)
Environmental Easement Program (EEP)
Export Credit Guarantee Program (GSM-102)
Export Credit Revolving Fund
Export Enhancement Program (EEP)
Export Incentive Program (EIP)
Export Marketing Certificate Program
Export promotion programs
Farm acreage base
Farm bill
Farmer Mac
Farmer-Owned Reserve (FOR) Program
Farm Storage Facility Loan Program
Farm program payment yield
Federal Crop Insurance Corporation Reinsurance
Federal Crop Insurance Program
Federal Food, Drug, and Cosmetic Act (FFDCA)
Federal Insecticide, Fungicide, and Rodenticide Act of 1947 (FIFRA)
Federal marketing orders and agreements
Federal milk marketing orders
Feed Cost-sharing Program
50/92
Findley loan rates
Findley payments
Flex acres
Food, Agriculture, Conservation, and Trade Act of 1990 (P.L. 101-624)
Food for Development Program
Food for Peace Program
Food for Progress
Food Security Act of 1985 (P.L. 99-198)
Food Security Wheat Reserve
Food Stamp Program (FSP)
Generic commodity certificates
Government payments
Grain Reserve Program
Gramm-Rudman-Hollings Deficit Reduction Act
Great Plains Conservation Program (GPCP)
GSM
GSM-5
GSM-102
GSM-103
Guaranteed Export Credit
Homestead Protection
Incentive payments
Income insurance
Income support payment
Income support programs
Integrated Farm Management Program
Interest Payment Certificates
Intermediary Relending Program
Intermediate Export Credit Guarantee Program (GSM-103)
Inventory Reduction Program
Land grant
Loan deficiency payments
Loan forfeiture
Loan rate
Make allowance
Mandatory supply controls
Market certificate
Market development programs
Marketing certificate
Marketing loan
Marketing loan program
Marketing order
Marketing quota(s)
Market Promotion Program (MPP)
Maximum Acceptable Rental Rates (MARR)
Maximun payment acreage (MPA)
Meat Products Export Incentive Program payments

Milk Diversion Program
Milk Production Termination Program
Minnesota-Wisconsin (M-W) price
National Organic Standards Board
National Program Acreage (NPA)
National Wool Act
No net cost
No-Net-Cost Act of 1982
Nonprogram crops
Nonrecourse loans
Normal crop acreage (NCA)
Normal flex acres
Offsetting compliance
Optional flex acres
Packers and Stockyards Act
Paid land diversion
Parity
Parity index
Parity ratio
Payment-in-kind (PIK)
Payment limitation
Permanent legislation
Permitted acreage
PIK and roll
Posted county price (PCP)
Poundage quotas
Prevented planting acreage
Prevented planting disaster payments
Price support level
Price support programs
Producer option payments
Production controls
Program (agricultural)
Program costs
Program crop
Program slippage
Program yield
Promotion, research, and package regulations
Public Law 480 (P.L. 480)
Referendum
Release price
Reopening signup payments
Replenishment agricultural worker (RAW) program
Reseal
Revenue insurance
School Breakfast Program (SBP)
Secondary (resale) markets for agricultural loans
Section 22
Section 32
Section 201
Section 301
Section 416
Set-aside
Sodbuster
Soil Bank
Special agricultural worker (SAW) program
Special Milk Program (SMP)
Special Supplemental Food Program for Women, Infants, and Children (WIC)
State Water Quality Coordination Program
Steagall commodities
Summer Food Service Program (SFSP)
Sunflower Oil Assistance Program (SOAP)
Supply control
Support price
Swampbuster
Targeted Export Assistance Program (TEA)
Targeted Option Payments (TOP)
Target price
Temporary Emergency Food Assistance Program (TEFAP)
The Emergency Food Assistance Program (TEFAP)
Triple base
Two-tiered pricing
U.S. Grain Standards Act
Wetlands Reserve Program (WRP)
Wheat and Feed Grain Export Certificate Programs
0/92

See also ***Conservation***

Agricultural Production

Agricultural inputs
Agricultural resource base
Agriculture
Alternative farming
Aquaculture
Attainable yield
Biodiversity
Biological control of pests
Cash crop
Cash grain farm
Catch crop
Chemigation
Close-grown crops
Companion crop
Continuous crop
Contour farrow
Contract farming
Conventional tillage
Cover crop
Crop failure
Crop forecast
Cropland
Cropland harvested
Cropland pasture
Cropland used for crops
Crop report
Crop residue
Crop residue pasture
Crop rotation
Crop year
Crop yield
Cultivate
Cultivated summer fallow
Custom work
Denitrification
Double cropping
Dryland farming
Failed acreage
Farm
Farm management
Farm operator
Full bloom
Gleaning
Harvest
Harvested acres
Idled acres
Imagery
Integrated crop management
Integrated resource management
Intercropping
Land capability
Living mulch

Low-input sustainable agriculture (LISA)
Low-intensity animal production
Minimum tillage
Monoculture
Multifactor productivity index
No-till
Once-over harvester
One-person baling
Operator (farm)
Organic farming
Output
Pasture
Plowing
Polyculture
Prime farmland
Producer
Production flexibility
Productive capacity
Resources
Rotation, crop
Row crops
Scouting
Slot tillage
Specialty crops
Stripcropping
Stubble mulch
Summer fallow
Sustainable agriculture
Technology transfer
Terrace
Till
Total farm output per unit of input
Variable costs
Zero cultivation

See also ***Conservation and Environmental Protection; Farm and Agriculture-Related Labor; Farm Equipment and Buildings; Fertilizers; Insecticides, Pesticides, Fungicides, and Fumigants***

Animal and Plant Science and Research

Aeroponics
Agricultural economics
Agricultural research
Agricultural Research Service (ARS)
Agriculture
Agronomy
Annual
Assay
Autotrophic
Backcross
Biennial
Biochemical
Biochemistry
Biodegradation
Biomass
Biotechnology
Buffer
Callus
Carbohydrate
Carbon
Carbon dioxide
Carotene
Catalyst
Cell
Cell culture
Cellulose
Chlorophyll
Chloroplasts
Chlorosis
Chromosome
Clone
Consultative Group on International Agricultural Research (CGIAR)
Cooperative State Research Service (CSRS)
Cotyledons
Cross
Crossbreeding
Cross-fertilization
Cross-pollinated crops
Cross-pollination
Cultivar
Current Research Information System (CRIS)
DNA
E. coli
Ecology
Ecosystems
Embryo
Embryo transfer
Endosperm
Enzyme
Enzyme-linked immunosorbent assay (ELISA)
Epidermis
Equilibrium
Escherichia coli
Essential amino acid
Floriculture
Full bloom
Fungus
Gene
Genetic engineering
Genetics
Genome
Genotype
Genotypic variability
Germ
Germination
Germplasm
Growth inhibitor
Growth regulator
Growth stimulant
Herbaceous
Herbivore
Heterozygous
Homozygous
Hormone
Hormone-type materials
Hybrid
Hybrid vigor
Hydrolysis
Hydroponics
Imagery
Joint Council on Food and Agricultural Sciences
Luxury consumption
Metabolism
Microingredients
Microorganisms
National Agricultural Advisory Board
Nitrogen
Nitrogen cycle
Nitrogen fixation
Nutrient
Nutrient, plant
Ovum
Parasite
Pathogen
Pathology
Perennial
Pheromone

Photosynthesis
Physiology
Plant breeding
Pollination
Pomology
Protein
Radicle
Recombinant DNA
Remote sensing
Rhizobium
RNA
Satellite image data products
Silviculture
Single-cross hybrid
Slip
Species
Substrate
Technology transfer
Tissue culture
Toxin
Transpiration
Variety
Viticulture
Wholesale Market Development Program
Yeast

Bees and Beekeeping

Apiarist
Apiary
Apiculture
Apis
Apis mellifera
Bee glue
Beehive
Beekeeper
Beekeeping associations
Bee pasture
Beeswax
Brood
Brood chamber
Colony
Comb
Combless package
Commercial beekeeper
Cross-pollination
Drone
Drone egg
Extractor
Field bees
Food chamber
Frame
Hive
Hobbyist beekeeper
Honey
Honeybee
Honeycomb
Honey flow
Honey stomach
Larva
Nectar
Nectaries
Nurse bees
Package bees
Pollen
Pollen basket
Pollination
Propolis
Pupa
Queen bee
Queen cell
Royal jelly
Sideliners
Skep
Smoker
Super
Swarm
Wax glands
Wax moth
Worker bee
Worker egg

Climate

Acid rain
Arid climate
Arid region
Climate
Continental climate
Drought
Environment
Normal climate
Weather

Commodity Agreements

International Cocoa Agreement (ICCA)
International Coffee Agreement (ICA)
International Commodity Agreement(s) (ICA)
International Dairy Arrangement (IDA)
International Olive Oil Agreement (IOOA)
International Sugar Agreement (ISA)
International Wheat Agreement (IWA)
Multifiber Arrangement (MFA)

Conservation and Environmental Protection

Acid rain
Agricultural Conservation Program (ACP)
Agricultural Environmental Quality Council
Agricultural Resource Conservation Program (ARCP)
Agricultural Water Quality Protection Program (AWQPP)
Bench terraces
Best management practices
Broad-base terrace
Buffer strip
Climate
Colorado River Salinity Control Program
Conservation compliance provision
Conservation district
Conservation easement
Conservation plan
Conservation practices
Conservation Reserve Corn Bonus Program
Conservation Reserve Program (CRP)
Conservation, soil
Conservation tillage
Conserving uses
Contour farming
Ecology
Ecosystems
Emergency Conservation Program (ECP)
Endangered species
Environment
Environmental Conservation Acreage Reserve Program (ECARP)

Environmental Defense Fund
Environmental Easement Program (EEP)
Environmental Protection Agency (EPA)
Ephemeral erosion
Erodibility index
Erodible (soil)
Erosion
Eutrophication
Fallow
Farmland protection
Filter strip
Level terrace
Mulch-till
National Association of Conservation Districts
National Audubon Society
National Wildlife Federation
No-till
Organic farming
Ridge-till
Slit tillage
Soil Bank
Soil conservation
Soil Conservation Service (SCS)
Soil erosion tolerance values (T-values)
Soil management
Stripcropping
Strip-till
Stubble mulch
Summer fallow
Sustainable agriculture
T-values
Terrace
Water Bank Program
Water Quality Incentive Projects (WQIP)
Wetlands Reserve Program
Windbreak
Wind erosion
Zero cultivation

Corn

Adulterated grain
Aflatoxin
Amylopectin
Basic commodities
Broken corn and foreign material (BCFM)
Broken kernel
Cash grains
Cereals
Coarse breaks
Coarse grains
Conditioning
Corn
Corn Belt
Corn gluten
Corn-hog ratio
Corn sweeteners
Corn syrup
Deficiency payment
Degermination
Dehull
Dry milling
Endosperm
Ethanol
Ethyl alcohol
Flaking grits
Gasohol
Grain
Grain screenings
Heat-damaged kernels
Kernel
National Corn Growers Association (NCGA)
National farm program
Nonrecourse loans
On-farm dryers
Prevented planting disaster payments
Program crop
Steepwater
Stress cracks
Tempering
Wet-milling

See also ***Food and feed grains***

Cotton

Acreage reduction program (ARP)
American-Egyptian cotton
America-Pima cotton
Bale
Bale (cotton)
Basic commodities
Boll
BPSY (bleachable prime summer yellow)
Cellulosic fibers
Chicago Rice and Cotton Exchange
Cotton
Cotton Board
Cotton compress
Cotton Council
Cotton Council International
Cotton count
Cotton exchange
Cotton Incorporated
Cotton lint
Cotton quality
Cottonseed
Cottonseed hulls
Cottonseed meal
Cottonseed oil
Cottonseed Oil Assistance Program (COAP)
Cotton system
Count
Delinting
Denier
Denim
End-use
Extra-long staple cotton
Fiber crop
Fiber maturity
Fiber strength
Finishing
Gin
Gossypol
Group "B" mill price
Hand
High density
Honeydew
HVI (high volume, instrument testing)
International Cotton Advisory Committee (ICAC)
International Institute for Cotton (IIC)
Lint
Linters
Long staple cotton
Man-made fibers
Micronaire
Micronaire reading
Middling
Moduled seed cotton
Motes

Naps
National Cotton Council of America (NCC)
National Cottonseed Products Association (NCPA)
National Program Acreage (NPA)
Natural fibers
Neps
New York Cotton Exchange
Nonrecourse loans
Oilseed
Oilseed crops
Oilseed hulls
Oilseed meal
Oilseed processing
Point
Price
Program crop
Running bale
Sample fiber
Sea Island cotton
Seed cotton
Square
Staple fibers
Staple length
Strict Low Middling 1 1/16 inch cotton
Supima
Supima Association of America
Tare
Universal density bale
Upland cotton

Credit and Banking

Agricultural Credit Associations (ACAs)
Agricultural Credit Banks (ACBs)
Bank for Cooperatives
Board of Governors
CCC commercial credit
Credit, supervised
Discount rate
District Banks for Cooperatives
Eurodollar market
Farm Credit Administration (FCA)
Farm Credit Assistance Board
Farm Credit Banks (FCBs)
Farm Credit Council
Farm Credit Leasing Services Corporation
Farm Credit System (FCS)
Farm Credit System Financial Assistance Corporation
Farm Credit System Insurance Corporation
Farmer Mac
Farmers Home Administration (FmHA)
Farm Facility Loan Program
Federal Agricultural Mortgage Corporation (Farmer Mac)
Federal Farm Credit Banks Funding Corporation
Federal Intermediate Credit Bank (FICB)
Federal Land Bank Associations (FLBAs)
Federal Land Credit Associations (FLCAs)
Federal Reserve
Federal Reserve System
Letter of credit
National Bank for Cooperatives (NBC)
Production Credit Associations (PCAs)
Special drawing rights (SDR)

Crop Insurance and Disaster Assistance

Ad Hoc Disaster Assistance
Crop failure
Crop insurance
Crop insurance indemnity
Crop insurance premium
Federal Crop Insurance Corporation (FCIC)
Federal Crop Insurance Program
Group Risk Plan (GRP)
Indemnity
Master marketers
Noninsured Assistance Program (NAP)
Reinsurance company

Dairy

Allocation procedure
Anhydrous milk fat
Animal protein
Babcock test
Balancing
Balancing plant
Blend price
Bovine growth hormone (bGH)
bovine Somatotropin (bST)
Butter
Butterfat
Buttermilk
Butteroil
Casein
Cheese
Class I differential
Class I effective price
Class I minimum price
Classes of milk
Classified pricing
Cold wall storage tank
Colustrum
Compensatory payment
Condensery
Creamline milk
Curd
Dairy Diversion Program
Dairy Export Incentive Program (DEIP)
Dairy Indemnity Payment Program
Dairy Termination Program
Dry cow
Fluid differential
Fluid (milk) products
Fluid utilization
Foot-and-mouth disease (FMD)
Freshen
Ghee
Give-up-charge
Grade A dairy
Grade B dairy
Grade A milk
Grade B milk
Handler
Handler price
Hard manufactured dairy products
Homogenized
International Dairy Arrangement (IDA)

Lactase
Lactation period
Lactose
Lactose intolerance
Manufacturing grade milk
Milk
Milk contract
Milk Diversion Program
Milk equivalent
Milkfat
Milk fat content
Milk-feed price ratio
Milkfever
Milk marketing order
Milk Production Termination Program
Milk solids, not fat
Minnesota-Wisconsin (M-W) price
National Milk Producers Federation (NMPF)
Nonfat dry milk (NDM)
Over-order payment
Pasteurization
Perishable manufactured dairy products
Reconstituted milk
Revenue pool
Reverse osmosis filtration
Soft manufactured dairy products
Special Milk Program (SMP)
Standardization
Storable manufactured dairy products
Ultra-high temperature milk (UHT)

Diseases, Crop

Aspergillus flavus
Bacteria
Bacterial ring rot
Bactericide
Blight
Chlorosis
Citrus canker
Damping off
Downy mildew
Fungicide
Fungus
Host
Host range
Immunity
Karnal bunt
Microorganism
Mildew
Mold
Necrosis
Pathogen
Pathology
Phylloxera
Phytosanitary
Rot
Russet
Rust
Sanitary and phytosanitary regulations
Smut
Tolerance
Toxin
Wilt
Witches'-broom
Yellows

Diseases, Livestock

African swine fever
Aftosa
Anthelmintic
Antibiotic
Avian influenza
Bang's disease
Bloat
Bluebag
Bluetongue
Brucellosis
E.coli
Escherichia coli
Foot-and-mouth disease
Foot rot
Founder
Hardware disease
Hog cholera
Hoof-and-mouth disease
Host
Immunity
Leukosis
Liver fluke
Mastitis
Microorganism
Milk fever
Newcastle disease
Pseudorabies
Rinderpest
Sanitary and phytosanitary regulations
Scours
Scrapie
Susceptibility
Swine vesicular disease
Symptom
Systemic
Toxin
Vaccine
Vesicular stomatitis

Economics

Capital account
Capital account surplus
Capital replacement
Centrally planned economy
Consumer subsidy equivalent (CSE)
Convertible currency
Cost-benefit analysis
Cost-saving input
Council of Economic Advisors
Discounting
Discount rate
Economies of scale
Economies of scope
Economies of size
Economies of speed
Externality
Fiscal policy
Fixed costs
Free rider
Gross Domestic Product (GDP)
Gross National Product (GNP)
Hard currency
Imperfect competition
Inconvertible currency
Increasing economies of scale
Macroeconomic policies
Microeconomics
Monetary policy
Nominal dollars
Nonmarket economy
Plant-specific economies
Producer subsidy equivalent (PSE)
Productivity
Product-specific economies
Real dollars

Single-factor productivity index
Soft currency
Utility
Variable costs

See also ***Supply and Demand***

Environmental Horticulture and Floriculture

Annual
Aquatic plant
Auxins
Bed
Bedding and garden plants
Biennial
Bolt
Bud
Cut Christmas trees
Cut cultivated greens
Cut flowers
Deciduous fruit
Deciduous trees
Defoliate
Dormancy
Environmental horticulture
Floriculture
Flower and vegetable seed
Foliage plants
Grafting
Greenhouse
Green industry
Little leaf
Nursery plants
Perennial
Potted flowering plants
Pruning
Set
Sod (turfgrass)
Stand
Unfinished plant materials

European Union

Accession compensatory amount (ACA)
African, Caribbean, and Pacific States (ACP States)
CAP
Coefficient of equivalence
Common Agricultural Policy (CAP)
Common external tariff
Common market
Compensatory tax
Customs union
EC Commission
EC Council
EC Court of Justice
EC 92
EC Parliament
Entry price
European Agricultural Guidance and Guarantee Fund (EAGGF)
European Community (EC)
European currency unit (ECU)
European Economic Community (EEC)
European Monetary Cooperation Fund
European monetary system (EMS)
European Union (EU)
FEOGA
Green rate
Green rate of exchange
Intervention
Intervention price (IP)
Minimum Import Price (MIP)
Monetary compensatory account (MCA)
Norm Price
Reference price
Restitution
Threshold price
Value-added tax (VAT)
Variable levy

Farm and Agriculture-Related Labor

AgriAbility Project
Agricultural association
Agricultural commissioner
Agricultural employment
Agricultural inputs
Agricultural service
Agricultural work
Census of Agriculture
Custom work
Dockworker
Farm inputs
Farm labor contractor
Farm operator
Farmworker
Field work
Field worker
H-2A temporary foreign workers
Hired worker
Illegal alien
Inputs
Labor force participant
Longshoreman
Lumpers
Migrant and Seasonal Agricultural Worker Protection Act (MSPA)
Migrant farmworker
Payment-in-kind (PIK)
Piece work
Seasonal agricultural worker
Sharecropper
United Farm Workers of America (UFW)

Farm and Food Prices and Expenditures

Base period price
Consumer Price Index (CPI)
Continuing Consumer Expenditures Survey (CCES)
Dual pricing
Ex (point of origin)
Farm-to-retail price spread
Farm value
Farm value share
Federal-state market news service
Gross margin
Grower price
Handler price
Index of Prices Paid
Index of Prices Paid by Farmers for Commodities, Services, Interest, Taxes, and Farm Wage Rates
Index of Prices Received
Index of Prices Received by Farmers
Inflation rate

Mandatory reporting
Market basket
Market basket of farm foods
Market-clearing prices
Marketing bill
Marketing margin (or marketing spread)
Market Stabilization Price (MSP)
Merit pricing
Negotiated pricing
Net market price
Parity
Parity index
Parity ratio
Price
Price competition
Price coordination
Price discrimination
Price fixing
Price index
Price leadership
Prices-paid index
Prices-received index
Price support level
Price support programs
Price umbrella
Producer Price Index (PPI)
Product differentiation
Real prices
Retail price
Thin market
Up sales, down sales
Wholesale price index
World price

See also ***Futures and Options Markets***
See also ***Marketing Orders***

Farm Equipment and Buildings

Air drill
Air planter
Air seeder
Anhydrous ammonia tank wagon
Applicator
Auger
Auger wagon
Back hoe
Bale pickup loader
Baler
Bale shredder
Bale stacker
Bale twine
Bale wagon
Bale wrapper
Bank out wagon
Barn cleaner
Bed leveler
Bed shaper
Belt horsepower
Bin, grain
Blower, feed and grain
Broadcasting
Broadcast seeder
Bulk milk tank
Bulldozer
Bunker silo
Bunk feeder system
Carter mill
Catching frame
Cat skinner
Cattle feeder
Center pivot irrigation
Chisel
Chisel plow
Chisel point
Chopper, flail
Chopper, rotary
Chute
Clod buster
Cold confinement barn
Cold wall storage tank
Combine
Conservation tillage equipment
Controlled lighting
Corn harvester
Corn planter
Coulter
Coulter chisel plow
Crawler tractor
Crib
Crop dryer
Crop duster
Cultipacker
Cultivator
Cultivator disk hiller
Cultivator fenders
Cultivator shanks
Cultivator shapes
Cultivator standards
Cutterbar
Disk
Disk blade
Disk harrow
Disk hiller
Disk marker
Disk plow
Disk ripper
Ditcher
Double-disk opener
Drawbar
Drawbar horsepower
Drill
Drilling
Dryer
Dump rake
Duster
Egg washer
Electronic control
Electronic GPS (Global Positioning System)
Electronic milking monitor
Electronic monitor
Electronic yield measurer
Elevator
End-gate seeder (broadcast seeder)
Fanning mill
Farm inputs
Farm tractor
Farm truck
Farrowing crate
Farrowing house
Feed bunk
Feed grinder
Feed mixer
Fence charger
Fertilizer side dresser
Fertilizer spreader
Field cultivator
Flail cutter
Flame cultivator
Float
Floating elevator
Flush gutter
Foam marker
FOPS (falling object protective structures)
Forage blower
Forage harvester
French plow
Fresno (Fresno scraper)
Front-end loader
Gin
Grain bin unloader
Grain blower
Grain cart

Grain cleaner
Grain conveyor
Grain drill
Grain elevator
Grain harvester
Grain truck
Grass drill
Grass seeder
Gravity-drain gutter
Gravity wagon
Greenhouse
Grinder
Hammer grinder
Hammer mill
Harrow
Harvester
Hay baler
Hay conditioner
Hay mower
Hay rake
Header
Headgate
Herringbone milking parlor
High-clearance tractor
Homestead
Horsepower
Horsepower hour
Implement carrier
Individual cage
Inputs
Irrigation furrow opener
Irrigation levee plow
Irrigation pipe
Irrigation pump
Irrigation sprinkler head
Irrigation system, center pivot
Irrigation system, drip
Irrigation system, stationary gun
Land plane
Large bale baler
Large bale mover
Leveler
Lister
Livestock scale
Livestock truck
Loader
Manure spreader
Mechanical bulk feeder
Mechanical nests
Mechanical stack mover
Middlebreaker
Middlebuster
Milking machine
Mist blower
Modular
Moldboard plow
Mower
Mower-conditioner
Mulcher
Mulch tillage
Mulch tillage equipment
Multiple hatcher
No-till
No-till drill
One-way disk
Open-front shed
Open upright silo
Packer
Pasture seeder
Peanut harvester
Peg harrow
Pellet machine
Pellet mill
Pen
Picker
Planter
Planter, no-till
Plow
Plow point
Plow share
Portable auger
Portable bunk feeder
Portable housing
Post-hole digger
Potato digger
Potato harvester
Potato planter
Power-take-off (PTO)
Rake
Ridge-till cultivator
Ripper
Rock picker
Rodweeder
Roll-away nest
Roll bar
Roller mill
Roll-over frame
ROPS (rollover protective structures)
Rotary hoe
Rotary tiller
Round baler
Row crop cultivator
Safety shield
Scrape gutter
Seed cleaner
Shank
Shape
Share, plow
Shovel
Sickle mower
Side delivery rake
Silo
Skid loader
Skid steer loader
Slotted floors
Speed sprayer
Spider wheel cultivator
Spike
Spike tooth harrow
Sprayer
Spring tine harrow
Square baler
Squeeze chute
Stable
Stacker wagon
Stubble drill
Subsoiler
Sugar beet topper
Swather
Sweep
Tandem disk harrow
Tedder
Terracer
Three-point hitch
Tillage equipment
Tillage tool
Tool bar
Topper
Tractor
Tractor front-end stacker
Trailer
Trap nests
Trench silo
Truck
Utility trailer
Utility vehicle
Wagon
Warm confinement barn
Weeder
Wheel tractor
Windrow pickup

Farm Income (Costs and Returns)

Capital expenditures
Cash receipts
Cost of production
Economic costs
Equity level

Farm Costs and Returns Survey (FCRS)
Farm income
Farm income and balance sheet
Farm-related income
Government payments
Gross cash income
Gross farm income
Gross value of production
Gross value of sales
Net cash farm income
Net cash flow
Net cash household income
Net cash income
Net farm income
Nonmoney farm income
Off-farm income
Per capita income
Production expenses
Value of production less cash expenses

Farm Structure and Organization

Corporate farm
Family farm
Farm
Farm size
Standard "C" family corporation
Subchapter "S" family corporation
Subsistence farm
Type of farm (or ranch)
  Field crops
  Livestock or poultry
  Other crops

Fats and Oils

American Soybean Association
Animal fats
Baking or frying fats (shortening)
Bleaching
Blown oils
Bodied oil
Boiled, blown, bodied oils
Boiled oils
BPSY (bleachable prime summer yellow)
Butteroil
Carbohydrate-based fat substitute
Castormeal
Castoroil
Castorseed
Cottonseed Oil Assistance Program (COAP)
Crude vegetable oil
Degummed oil
Deodorizing
Diglyceride
Double-zero rapeseed
Dry rendering
Drying oils
Emulsifier
Essential oils
Extraction, mechanical
Extraction, solvent
Fat
Fats and oils
Fat splitting (hydrolysis of fat)
Fatty-acid based fat substitute
Fatty acids (classification of)
Flaking
Flaxseed
Foots
Free fatty acids
Glycerine
Grease
Hydraulic pressing
Hydrogenation
Institute of Shortening and Edible Oils (ISEO)
International Olive Oil Agreement (IOOA)
Jojoba
Jojoba oil
Lard
Lecithin
Linoleic acid
Linolenic acid
Linseed
Linseed meal
Linseed oil
Margarine
Mayonnaise
Mellorine
MIU
Monoglycerides
Monounsaturated fats
Mustard seed
National Association of Margarine Manufacturers (NAMM)
National Cottonseed Products Association (NCPA)
National Institute of Oilseed Products (NIOP)
National Soybean Crop Improvement Council (NSCIC)
National Soybean Processors Association (NSPA)
N-oil
Oil cake
Oil palm
Oils
Oilseed
Oilseed crops
Oilseed hulls
Oilseed meal
Oilseed processing
Olein
Olive
Once-refined oil
Owala
Palmetic acid
Palm kernel
Palm oil
Peppermint
Pod seed
Polyglycerol esters
Protein-based fat substitute
Refined and further processed
Refined, not further processed
Refining
Render
Rendering
Salad or cooking oils
Saturated fat
Shortening
Simplesse
Soap
Soapstock
Solvent extraction
Stand oils
Stearic acid
Stearin
Sulfonation
Sunflower Oil Assistance Program (SOAP)

Tall oil
Tallow
Technical grade refined oils
Triglycerides
Tung
Turkey red oil
Unsaturated fat
Vegetable oils
Vegetable stearin
Winterizing

See also ***Oilseeds***
See also ***Soybeans***

Feeds and Fodder

Additive
Adulterated grain
Aeration, grain
Aflatoxin
Agricultural cooperative
Agricultural inputs
Air classification
Airtight upright silo
Alfalfa
Alsike clover
Animal unit
Animal unit month
Annual pasture
Antioxidant
Aspiration
Association of American Feed Control Officials
Backgrounding
Balanced ration
Biomass
Blocks
Blood meal
Buckwheat
Carriers
Carrying capacity
Clover
Commercial supplement
Complete feed
Concentrate
Corn gluten
Crambe
Crimped
Crop residue pasture
Custom feeding
Custom grinding and mixing
Emergency Feed Assistance Program (EFAP)
Emergency Feed Program (EFP)
Extrusion
Feed
Feed bunk
Feed conversion ratio (feed efficiency)
Feed Cost-sharing Program
Feeder calf
Feeder cattle
Feed grain
Feeding margin
Feedlot
Feedmilling establishment
Feedstuff
Fenceline bunk feeder
Fish meal
Forage
Formula feeds
Free choice feeding
Grain-consuming animal unit (GCAU)
Grain lupins
Grass silage
Graze
Green chop
Hay
Haying and grazing
Haylage
Hay silage
High-protein animal unit
Improved perennial pasture
Ladino clover
Least-cost feed formulation
Lespedeza
Meat meal and meat and bone meal
Mechanical bulk feeder
Milk-feed price ratio
Millet
Millfeed
Mixing concentrate (pre-mix)
Mobile grinder mixer
National grassland
Native pasture
Nonairtight upright silo
Nonfed cattle
Nontillable pasture
Palatability
Pasture
Pasture production system
Pellet binders
Pellet machine
Pellet mill
Pellets
Permanent pasture
Portable bunk feeder
Premix
Primary feed manufacturing
Prime farmland
Processing
Processor
Range
Rangeland
Red clover
Roller mill
Rolling
Roughage
Roughage consuming animal units
Rumensin
Secondary feed manufacturing
Self-feeder
Shorts
Silage
Small grain pasture
Soil injection wastes
Soiling crops
Sorghum
Soybean millfeed
Supplement
Supplementing
Sweet clover
Tame pasture
Toll milling
Utilization
Vegetative filter
Waste management
Whey
White clover
Windrow
Woodland pasture

Feedlot

Capacity of feedlot
Cattle and calves on feed
Cattle disposition
Cattle feeding
Commercial feedlot
Diversion terrace
Farmer feedlot
Farm lot feeding

Fed cattle
Feeder pig finishing operation
Feeder pig production
Feeders
Feeder yearling
Feedlot
Feedlot placements
Liquid feedlot wastes
Open-lot system
Settling basin
Solid feedlot wastes
Stocker-feeder enterprise

See also ***Feeds and Fodder***

Fertilizers

Additive
Agricultural chemicals
Ammonia
Anhydrous
Ash
Band application
Banding (of fertilizer)
Chemigation
Complete fertilizer
Conditioner (of fertilizer)
Fertilizer
Fertilizer analysis
Fertilizer formula
Fertilizer grade
Fertilizer material or carrier
Fertilizer side dresser
Fertilizer spreader
Green manure
Lime, agricultural
Lime requirement
Marl
Nitrogen
Nitrogen cycle
Nutrient, plant
Phosphate
Potash
Rhizobium
Top dressing
Urea

Fibers and Textiles

Abaca
Acrylic
Air-jet spinning
American Textile Manufactures Institute
Blowroom
Carding
Cellulosic fibers
Cloth
Coir
Combing
Corduroy
Course
Crutching
Decitex (dtex)
Denier
Denim
Drafting
Drawframe
Drawing
Durable press
End
Fabric
Face
Fiber
Fiber crop
Filament
Filling
Filling pick
Finishing
Flax
Grease mohair
Grease weight
Grease wool
Hard fibers
Hemp
Henequen
Industrial fabrics
Jute
Kilotex (Ktex)
Knitting
Lanolin
Linen
Loom
Man-made fibers
Man-made fiber fabric
Mill consumption
Millitex (mtex)
Mill (textile)
Mill run
Mohair
Mouton
Multifiber Arrangement (MFA)
Murrain
National Wool Act
National Wool Growers Association
Natural fibers
Noncellulosic fibers
Nonwoven fabrics
Nylon
Open-end spinning
Opening line
Permanent press
Pick
Pile
Ply
Polyester
Polypropylene
Ramie
Raw fibers
Ring spinning
Roving
Semi-worsted System
Sewing machine
Shearing
Short staple fibers
Sisal
Size
Sliver
Soft fibers
Spinning
Spinning quality
Staple fibers
Staple length
Synthetic fibers
Tagging
Tex
Textile
Texture
Top
Tow
Twist
Virgin wool
Wale
Warp
Warp knitting
Wash and wear
Weave
Weft
Weft knitting
Weight of fabric
Wigging
Wool
Wool bag
Woolen system
Worsted system
Woven fabric
Yarn
Yarn size

See also ***Cotton***

Food and Feed Grain

Acreage reduction program

(ARP)
Adulterated grain
Aeration
Aflatoxin
Air classification
Alfalfa
Aspiration
Barley
Basic commodities
Bin dryers
Blending
Bran
Break system
Broken corn and foreign material (BCFM)
Broken kernel
Bunker silo
Cereals
Chaff
Chess
Cleaning (grain)
Club wheat
Coarse breaks
Coarse grains
Combination dryers
Combine
Concurrent-flow dryers
Conditioning
Country elevator
Crimped
Crossflow dryers
Cultivated summer fallow
Dark northern wheat
Deficiency payment
Dehull
Dehydrate
Digestible protein
Dockage
Drying
Dry-milling
Durum wheat
Dust
Electrophoresis
Elevator
Elevator leg
Endosperm
Ensiled
Ethanol
Ethyl alcohol
Extraction rate
Farina
Farmer-owned Reserve Program
Federal Grain Inspection Service (FGIS)
Feed conversion ratio
Fine breaks
Flakes
Flaking grits
Floating elevator
Fobbing
Food grains
Foreign material
Free choice feeding
Front of the mill particles
Gasohol
Gliadin
Glume
Grade-determining factors
Grading
Grain
Grain breakage
Grain-consuming animal unit (GCAU)
Grain crop silage
Grain quality
Grain reserve
Grain Reserve Program
Grain screenings
Grain storage
Grind
Grits
Groats
Handling technologies
Hard red spring wheat
Hard red winter wheat
Hard wheat
Heat-damaged kernels
High-moisture grain
Hops
Horizontal storage
Hull
Husk
Incline belt
International Wheat Agreement (IWA)
Intrinsic quality
Isoglucose
Karnal bunt
Kernel
Lupulin
Maize
Malt
Meal
Meslin
Microorganisms
Middling
Middlings rolls
Mill byproduct
Millfeed
Milling
Mill mix
Milo
Mixed-flow dryers
Mold
National farm program acreage
National Grain and Feed Association (NGFA)
National Grain Trade Council (NGTC)
National Program Acreage (NPA)
Near-infrared reflectance spectroscopy (NIRS)
Nonbin dryers
Nongrade-determining factors
Nonrecourse loans
North American Export Grain Association (NAEGA)
Northern wheat
Oats
Off-farm dryers
On-farm dryers
On-ground pile storage
Pasta
Permitted acreage
Physical quality
Plant breeding
Precleaning
Prevented planting disaster payments
Price pooling
Price-support programs
Program crop
Proso millet
Protein premium
Quinoa
Radicle
Receival standards
Red wheat
Release price
Rye
Sanitary quality (grain)
Screw augur conveyor
Sedimentation test
Semolina
Shorts
Shrink (grain)
Silo
Small grain pasture
Smut
Soft red winter wheat (*Triticum aesturim*)
Soft wheat
Sorghum

Spring wheat
Steepwater
Strategic Grain Reserve
Stress-cracks
Subterminal elevator
Tailings of the mill
Tempering
Terminal elevator
Test weight
Triticale
Uniform Grain Storage Agreement (UGSA)
U.S. Grain Standards Act
Vertical storage
Vital wheat gluten
Vitreous
Wet-milling
Wheat
Wheat and Feed Grain Export Certificate Programs
Wheat starch
Wheat Trade Convention
White wheat
Winter wheat
Wort
Yield

See also ***Agricultural Policies and Programs, Foreign; Agricultural Policies and Programs, U.S.; Corn; Flour; Rice; Wheat***

Food Assistance, Domestic and International

Bonus commodities
Bread for the World
Bread for the World Education Fund
Bread for the World Institute on Hunger and Development
Cashing out
Child and Adult Care Food Program (CACFP)
Child nutrition programs
Committee on Surplus Disposal
Commodity distribution
Commodity Supplemental Food Program
Electronic Benefits Transfer (EBT)
Food Aid Convention
Food and Nutrition Service
Food Research and Action Center (FRAC)
Food Stamp Program (FSP)
Hunger Prevention Act of 1988 (P.L. 100-435)
International Wheat Agreement
National School Lunch Program (NSLP)
School Breakfast Program (SBP)
Special Milk Program (SMP)
Special Supplemental Food Program for Women, Infants, and Children (WIC)
Summer Food Service Program (SFSP)
Temporary Emergency Food Assistance Program (TEFAP)
The Emergency Food Assistance Program (TEFAP)
WIC Program

Food Away from Home

See ***Foodservice***

Food Expenditures

See ***Farm and Food Prices and Expenditures***

Food Grading and Inspection

Antemortem condemnation
Carcass trim (Y4 to Y3)
Certified ready-to-cook
Commercial meat production
Dressing percentage and/or yield
Federally inspected slaughter
Grade
Grade and yield
Grades
Grading
Inspection (federal)
Inspection (meat)
In the beef
No-roll
Meat grading and inspection
Merit pricing
Postmortem condemnations
Quality and yield
Quality grades for steer and heifer slaughter
Rolling
Unbranded
Wholesome Meat Act
Yield grades (1-5)

Food Manufacturing and Processing

Agribusiness
Alternative agricultural product
American Association of Meat Processors (AAMP)
Aseptic processing (packaging)
Assembler's margin
Carbohydrate-based fat substitute
Color additives
Condensed
Condensery
Cutback juice
Direct marketing
Fabrication
Fatty-acid based fat substitute
Fermentation
Flakes
Food additives
Food and fiber system
Food, farm-produced
Food manufacturing
Fortified foods
Further processor
Further processed
High-value products (HVP)
Industrial uses
International Association of Meat Processors (IAMP)
Joint products

Manufacturer deals
Mixing
Monosodium glutamate (MSG)
National Food Processors Association (NFPA)
National Frozen Food Association (NFFA)
National Soybean Processors Association (NSPA)
Nonfarm foods
Perishable commodities
Protein-based fat substitute
Ready-to-eat
Reciprocal baking
Shelf-stable foods
Sous vide foods
Sun-cured
Surimi
Synthesized, fabricated, engineered
Vacuum-packed
Value-added
Wet-milling
Wet-rendered
White pan bread
Yeast

Food Marketing, Wholesaling, and Retailing

Agent
Agricultural marketing
Bake-off format
Buying broker
Captive bakery
Carryout
Cash and carry wholesaler
Chain
Chain warehouse
Combination bakery
Commission merchant
Computer assisted trading (CAT)
Consign
Consumer
Controlled brand
Cooperative-owned wholesaler
Creamline milk
Delivered sale
Direct sales
Distressed sales
Electronic market
Federal-State Marketing Improvement Program
F.o.b. acceptance final
F.o.b. shipping point
F.o.b. shipping price
Food broker
Foodservice
Foodstore
Generic advertising
Grocery Manufacturers of America (GMA)
Grocery store
Home delivery route
Independent grocer
Institutional wholesaler
Instore bakery
Jobber
Labor coverage
Leveraged buyout
Lumpers
Manufacturers' deals
Marketing contract
National Food Brokers Association (NFBA)
National Grocers Association (NGA)
National Restaurant Association (NRA)
Off-beat trades
Open contract
Packer sales offices
Packer-shipper
Portion-controlled product
Private label brand
Private trade
Production contract
Purveyor
Rack jobber
Refining loss
Regional brand
Retailer-owned wholesaler
Retail price
Sales agent or sales agency
Scratch-mix format
Secondary wholesaler
Slotting fee
Spot market
Standing orders
Store label brand
Supermarket
Takeout
Tele-auction market
Teletype auctions
Truck jobber
Vertically integrated firm
Wagon jobber
Warehouse club
Wholesale club store
Wholesale Market Development Program
Wholesaler
Wholesaling

Food Prices and Expenditures

See ***Farm and food prices and expenditures***

Food safety, nutrition, and health

Absorption
Acute
Additive
Adulterated grain
Aflatoxin
Allergy
Animal and Plant Health Inspection Service
Animal drug residue
Antemortem condemnations
Anthelmintic
Antibiotic
Antivitamins
Aspergillus flavus
Avian influenza
Bacteria
Bang's disease
Bovine growth hormone (bGH)
Bovine Somatotropin (bST)
Brucellosis
Campylobacteriosis
Carcinogen
Center for Science in the Public Interest (CSPI)
Cholesterol
Chronic
Codex Alimentarius
Codex Alimentarius Commis-

sion
Color additives
Community Nutrition Institute (CNI)
Contamination
Continuing Survey of Individual Intakes (CSFII)
Control
Cross-contamination
Danger zone
Decontamination
Delaney Clause
Dietary Guidelines for Americans
Disease
Drug
Drug residue
E. coli
E. coli O157:H7
Egg Products Inspection Act
Elimination
Eradication
Escherichia coli
Fats and oils
Federal Food, Drug, and Cosmetic Act (FFDCA)
Federal Grain Inspection Service (FGIS)
Federal Insecticide, Fungicide, and Rodenticide Act (FIFRA)
Federal Meat Inspection Act
Fiber
Food and Drug Administration (FDA)
Food and Nutrition Service (FNS)
Food disappearance data
Food Safety and Inspection Service (FSIS)
Foot and mouth disease
Free range
Fumigant
Fumigation
Fungicide
Fungus
General Agreement on Tariffs and Trade (GATT)
Generally Recognized as Safe (GRAS)
Grade and size (produce)
Grading
Growth hormones
Hazard Analysis/Critical Control Points (HACCP)
Hormone
Human Nutrition Information Service (HNIS)
Incidence (of foodborne disease)
Insecticide residue
Insoluble fiber
International Office of Epizootics (IOE)
International Plant Protection Convention (IPPC)
Irradiation
International Organization for Standardization (ISO)
ISO 9000
Kwashiorkor
Lactose intolerance
Latency period
Listeriosis
Leukosis
Lignan
Limonene
Liver fluke
Mastitis
Meat and poultry inspection
Metabolism
Metabolite
Methyl bromide
Microbial contaminant or pathogen
Microbiology
Microorganisms
Mildew
Milk fever
Mold
Natural Resources Defense Council (NRDC)
Organic farming
Parasite
Pasteurization
Pathogen
Pesticide
Pesticide residue
Poultry Products Inspection Act
Private or store label brand
Protein quality
Quarantine
Quarantine, sanitary and health laws and regulations
Risk assessment
Salmonellosis
Sanitary and phytosanitary regulations
Soluble fiber
Surveillance
Susceptibility
Symptom
Systemic pesticide
Synthetic hormone
Toxicology
Toxin
Toxoplasmosis
Trace element
Undulant fever
Vaccine
Virulence
Vitamin
World Health Organization (WHO)
Zoonosis

See also ***Food Assistance, Domestic and International***

Foodservice

Fast food restaurant
Foodservice
Foodservice industry
National Association of Meat Purveyors (NAMP)
Table-service restaurant

Forestry and Forestry Products

Conifers
Fiberboard
Forestry Incentives Program (FIP)
Forest Service
Green wood
Hardwood
Lumber
Methanol
Methyl alcohol
National forest
National grassland

Nonindustrial private forest lands
Pulp
Pulpwood
Softwood
Wood alcohol

Fruits and Tree Nuts

Almond
Apple
Apricot
Asian pear
Avocado
Banana
Blackberry
Blueberry
Cactus pear
Cantaloupe
Carambola
Cherimoya
Cherry
Chinese gooseberry
Citrus canker
Citrus fruit
Coconut
Copra
Deciduous fruit
Dewberry
Feijoa
Fig
Filbert
Fresh weight equivalent
Grade and size (produce)
Grape
Grapefruit
Guava
Hazelnut
Honeydew melon
Indian fig
Kiwano
Kiwifruit
Kumquat
Lemon
Lichees
Lime
Litchi
Longan
Lychee
Macadamia nut
Mango
Muskmelon
Nectarine
Orange
Papain
Papaya
Passion fruit
Peach
Pear
Pecan
Pectin
Pepino
Pineapple
Pineapple guava
Pistachio
Plum
Pomace
Pomegranate
Precooling
Prickly pear
Produce
Produce grower
Produce wholesaler
Prune
Pulp
Pummelo
Quince
Raisin
Rambutan
Raspberry
Sapote
Shaddock
Snapback provision
Starfruit
Strawberry
Tangelo
Tangerine
Tangor
Terminal market
Tuna
Tung nut
Vitis labrusca
Vitis vinfera
Walnut
Watermelon
Wine

Futures and Options Markets

Basis
Basis risk
Bearish and bullish
Bear market
Bias
Bid
Booking the basis
Broker
Call option
Call pricing
Cash basis
Cash commodity
Cash contract
Cash forward contract
Cash future spread
Cash market
Chicago Board of Trade
Commodity futures exchange
Commodity Futures Trading Commission (CFTC)
Commodity options
Contract
Deferred payment
Deferred pricing
Delayed pricing
Delivery month
Delivery point
Distant futures
Exercise
Fixed-price contract
Formula pricing
Forward buying or selling
Forward contract
Forward market
Forward pricing
Futures
Futures contract
Futures market
Futures-options program
Futures trading
Hedging
In-the-money option
Incomplete risk markets
Initial margin
Intrinsic value
Invitation for bids
Long position
Lumpiness
Maintenance margin
Margin
Margin call
Maturing future
Near futures
Offer
Open position
Option grantor (writer)

Option premium
Options
Options contract
Out-of-the money option
Premium
Price-later contracts
Put option
Risk averse
Risk premium
Risky
Routine hedging
Selective hedging or discretionary hedging
Short position
Speculation
Spot
Spot delivery
Spot market
Squeeze
Strike price
Tender
Thin markets
Uncertainty
Writer or grantor

Grain Drying, Milling, and Storage

Air classification
Aspiration
Baking quality
Break flour
Break system
Clear flour
Combination dryer
Concurrent-flow dryer
Crossflow dryer
Defatted soy flour
Dehull
Electrophoresis
Elevator
Elevator leg
Family flour
Fancy clear
Fancy patent flour
Fanning mill
Farina
First clear flour
Floating elevator
Flour
Flour extraction rate
Flour stream
Fobbing
Front of the mill particles
Hard wheat flour
High-fat soy flour
Isolated soy protein
Low-fat soy flour
Lupulin
Malt
Meal
Middling
Middlings rolls
Mill byproduct
Millfeed
Milling
Mill mix
Near-infrared reflectance spectroscopy (NIRS)
Nonbin dryer
Patent flour
Prepared flour mixes
Reduction rolls
Second clear flour
Sedimentation test
Soft wheat flour
Specialty flours
Straight flour
Strong flour
Weak flour
Whole wheat flour

Herbs and Spices

Artichoke
Buckwheat
Celeriac
Celery
Celery knob
Celery root
Cilantro
Coriander
Eggplant
Garbanzo bean
Garlic
Oregano
Organy
Papaya
Pepper, black
Rape
Sesame
Wild marjoram

Industrial Crops and Products

Abaca
Alternative Agricultural Research and Commercialization (AARC) Center
Biodegradation
Biodiesel
Biofuels
Biopulping
Black liquor
Carbon disulfide
Chitin
Coir
Commercialization
Crambe
Dimethyl sulfoxide
Guayule
Hevea tree
Industrial crops
Industrial fabrics
Industrial rapeseed
Jojoba
Jojoba oil
Jute
Kenaf
Lesquerella
Lignin
Milkweed
Sisal
Vernonia
Vernolic acid

Insecticides, Pesticides, Fungicides, and Fumigants

Bacillus thuringiensis
Bactericide
Biocide
Biological control of pests
B.t.
Disinfectant
Fumigant
Fumigation
Fungicide
Germicide
Herbicide
Hydrogen phosphide
Insect growth regulators (IGRs)

Insecticides
Integrated management system
Integrated Pest Management (IPM)
Irradiation
Methyl bromide
Nematocide
Nematodes
Parasite
Pathogen
Pesticide
Pheromone
Pyrethrum
Systemic pesticide

Labor

See ***Farm and Agriculture-Related Labor***

Legislation

Act of April 14, 1971
Agricultural Act of 1948
Agricultural Act of 1949
Agricultural Act of 1954
Agricultural Act of 1956
Agricultural Act of 1964
Agricultural Act of 1970
Agricultural Adjustment Act Amendment of 1935
Agricultural Adjustment Act of 1933
Agricultural Adjustment Act of 1938
Agricultural Credit Act of 1987
Agricultural Credit Technical Corrections Act
Agricultural Fair Practices Act
Agricultural Programs Adjustment Act of 1984
Agricultural Trade Development and Assistance Act of 1954 (Food for Peace)
Agriculture and Consumer Protection Act of 1973
Agriculture and Food Act of 1981
Animal Welfare Act
Appropriations bill
Authorization bill
Balanced Budget and Emergency Deficit Control Act of 1985
Capper-Volstead Act
Caribbean Basin Economic Recovery Act
Consolidated Farm and Rural Development Act
Consolidated Omnibus Budget Reconciliation Act of 1985
Dairy and Tobacco Adjustment Act of 1983
Desert Land Act of 1877
Disaster Assistance Act of 1988
Disaster Assistance Act of 1989
Egg Products Inspection Act
Extra-Long Staple Cotton Act of 1983
Farm Credit Restructuring and Regulatory Reform Act of 1985
Farm Disaster Assistance Act of 1987
Federal Crop Insurance Act of 1980
Federal Crop Insurance Reform and Department of Agriculture Reorganization Act of 1994
Federal Meat Inspection Act
Federal Seed Act
Food, Agriculture, Conservation, and Trade Act of 1990
Food and Agricultural Act of 1962
Food and Agricultural Act of 1965
Food and Agriculture Act of 1977
Food Security Act of 1985
Food Security Improvements Act of 1986
Food Stamp Act of 1964
Food Stamp Act of 1977
Futures Trading Act of 1986
Grain Quality Improvement Act of 1986
Granger legislation
Homestead Act of 1862
Horse Protection Act
Hunger Prevention Act of 1988
Land Reclamation Act of 1902
Making Continuing Appropriations for the Fiscal Year 1987, and for Other Purposes
Morrill Acts of 1862 and 1890
National Agricultural Research, Extension, and Teaching Policy Act
National Environmental Policy Act
National Wood Act of 1954
Omnibus Budget Reconciliation Act of 1982
Omnibus Budget Reconciliation Act of 1986
Omnibus Budget Reconciliation Act of 1987
Omnibus Budget Reconciliation Act of 1989
Omnibus Budget Reconciliation Act of 1990
Omnibus Trade and Competitiveness Act of 1988
Perishable Agricultural Commodities Act
Plant Variety Protection Act
Poultry Products Inspection Act
Processed Products Inspection Improvement Act of 1986
Sherman Antitrust Act
Soil Conservation and Domestic Allotment Act of 1936
Steagall Amendment of 1941
Stewart B. McKinney Homeless Assistance Act
Technical Corrections to Food Security Act of 1985 Amendments
Temporary Emergency Food

Assistance Act of 1983
Trade Act of 1974
Section 201
Section 301
Trade Agreements Act of 1979
Trade and Tariff Act of 1984
Uniform Cotton Classing Fees Act of 1987
United States–Canada Free Trade Agreement Implementation Act of 1988
U.S. Warehouse Act

Legumes

Bambara groundnut
Bean
Black bean
Blackeye pea (bean)
Chick pea
Clover
Earth pea
Garbanzo bean
Great Northern bean
Groundnut
Haylage
Jack bean
Legumes
Lentil
Lespedeza
Licorice
Mesquite bean
National Dry Bean Council (NDBC)
National Peanut Council (NPC)
Navy bean
Peanut
Peanut meal
Pinto bean
Pulses
Red kidney bean
Round bean
Soybean
Winged bean

Livestock

American Association of Meat Processors (AAMP)
American Meat Institute (AMI)
Animal protein
Animal unit
Animal unit month
Artificial insemination
Auction
Auction market
Band
Barrow
Beefalo
Beef cattle
Beef cows
Beef cow-calf production
Bison
Blood meal
Boar
Bovine
Bovine growth hormone (bGH)
Bovine Somatotropin (bST)
Brahma
Breed
Breeding herd
Breeding unit index
Brood cows
Buffalo
Bull
Bull table
Bypass protein
Byproduct
Calf
Calf loss rate
Calf table
Calving
Calving rate
Capacity of feedlot
Capacity of housing
Carcass
Carcass weight, dressed
Carcass weight equivalent (CWE)
Cash expenses
Catalo
Cattle
Cattle and calves on feed
Cattle cycle
Cattle disposition
Cattle feeding
Central farrowing house
Classification of cattle for type
Cold confinement barn
Colt
Commercial feedlot
Commercial slaughter
Computer assisted trading (CAT)
Confinement
Confinement production system
Contract production
Corn-hog ratio
Corral
Country commission firm
Cow
Cow and replacement heifer loss rate
Crossbreeding
Crutching
Cull
Cull breeding animals
Custom feeding
Draft horse
Dressed animal
Dressing percentage
Dry lot
Dry lot cattle feeding
Equine
Ewe
Farrow
Farrowing crate
Farrowing house
Farrow-to-finish operation
Feeder calf
Feeder cattle
Feeder pig finishing operation
Feeder pig production
Feeders
Feeder yearling
Feeding margin
Feedlot
Feedlot placements
Fermentable protein
Filly
Finishing
Finishing house (barn)
First litter gilt
Flush gutter
Foal
Foot-and-mouth disease (FMD)
Foundation herd
Gaits
Gelding
Gestation period
Gilt
Grade and yield (weight)
Grain-consuming animal unit (GCAU)

Grain-fed animals
Grass-fed animal slaughter
Grease mohair
Grease weight
Grease wool
Growing-finishing house
Growth hormones
Hand
Headgate
Heifer
Hide
Hides and skins
High-quality beef (HQB)
Hog
Hog cholera
Hog-corn price ratio
Hog cycle
Horse
International Wool Secretariat (IWS)
Kobe beef
Lamb
Lard
Light horse
Litter
Livestock
Livestock work
Liveweight
Lot
Manure
Manure lagoon
Marbling
Market hog
Meat and poultry inspection
Meat Import Law
Monogastric
Mutton
National Cattlemen's Association (NCA)
National Pork Council Women (NPCW)
National Pork Producers Council (NPPC)
Net energy
Nonruminant
Offal
Omnivore
Open-front shed
Overfinished
Ovine
Packer
Pedigree
Peewee
Pen
Pig
Porcine
Porcine Somatotropin
Pulled wool
Ram
Ranch
Regular auction market
Replacement heifers
Rumen
Ruminant
Runt
Scrape gutter
Scrapie
Semen
Sheep
Shoat
Shorn mohair
Shorn wool
Sire
Skins
Shrink (livestock)
Slaughter
Slaughter hog
Sow
Split-phase hog production
Squeeze chute
Steer
Stocker cattle
Stocker-feeder enterprise
Straight breed
Swine
Swine vesicular disease
Tele-auction market
Terminal market
Total confinement
Tripe
Type of farm (or ranch)
U.S. Meat Import Law
Vealer
Waste management
Weaning
Wool
Wool Bureau
Yearling
Yearling loss rate
Yield grades (1-5)

See also ***Crop and livestock diseases***
See also ***Feed, feedlots, and fodder***
See also ***Meat and poultry packing and processing***
See also ***Poultry***

Marketing Orders

Administrative committee
Allocation procedure
Allotments
Balancing
Basing point
Blend price
Classified Pricing
Class I differential
Class I minimum price
Class I, II, and III milk
Compensatory payment
Flow-to-market
Fluid differential
Free rider
Generic advertising
Marketing agreement
Market allocation
Marketing order
Marketing order policy
Market support tools
Over-order payment
Producer allotments
Prorate
Revenue pool
Shipping holiday

Market Performance and Structure

Acquisition
Aggregate concentration
Agricultural cooperative
Backward integration
Bargaining association
Bargaining power
Barrier to entry
Barrier to exit
Bilateral oligopoly (monopoly)
Capital budgeting
Cartel
Company
Competitive advantage
Competitive market
Concentration
Concentration ratio

Conditions of entry
Conditions of exit
Conduct
Cooperative
Cooperative arrangement
Cooperative federation
Corporation
Countervailing power
Divestiture
Duopoly
Economies of scale
Economies of scope
Economies of size
Economies of speed
Excess capacity
Farmer cooperative
Foreign investment
Free rider
General partnership
Horizontal integration
Horizontal merger
Imperfect information
Individual activity
Individual profit maximization
Individual proprietorship
Industry
Integration
Intermediate handler
Joint activity or collusion
Limited partnership
Market
Market concentration
Market conduct
Market entry
Market exit
Market-extension merger
Market institution
Market performance
Market power
Market power or bargaining power
Market shares
Market structure
Merger
Minimum optimum size
Monopoly
Monopsony
Multinational firms
Multiplant economies
Nonprice competition
Oligopoly
Oligopsony
Partnership
Performance
Plant or establishment
Price competition
Price coordination
Price discrimination
Price fixing
Price leader
Price leadership
Product competition
Product differentiation
Product-specific economies
Pure conglomerate merger
Risk averse
Service cooperative
Structure
Vertically integrated firm
Vertical integration
Vertical joint activity or vertical collusion

Meat and Poultry Packing and Processing

American Association of Meat Processors (AAMP)
Antemortem condemnation
Bark
Bloom
Boned out
Boners
Boning
Boxed beef
Box industry
Branded processed meat operation
Breakers and/or breaking house
Breaking (beef)
Carcass
Carcass cutout
Carcass proportion
Carcass trim (Y4 to Y3)
Carcass weight, dressed
Carcass weight, equivalent
Central breaking
Central cutting
Commercial slaughter
Competing brand
Country commission firm
Cutability
Cutout
Cutup
Dressed animal
Dressed weight
Dressing percentage and/or yield
Dry-rendering
Eviscerated
Fabrication
Fabricators
Formed steaks
Grain-fed animals
Grass-fed animals
High-quality beef (HQB)
International Association of Meat Processors (IAMP)
In-the-beef
Kobe beef
Leather
Marbling
Meat analogs
Meat and bone meal
Meat and poultry inspection
Meat distribution
Meat extenders
Meat meal and meat and bone meal
Meat packer
Meat packing
Meat price list
Meat Products Exports Incentive Program payments
Multispecie slaughter plant
National Association of Meat Purveyors (NAMP)
New York dressed weight
No-roll
Offal
Packer
Packer brand
Packer sales offices
Packers and Stockyards Act
Packers and Stockyards Administration (PSA)
Packer-to-packer sales
Primal cuts
Prime, Choice, Select Pullout
Quality grades for steer and heifer slaughter
Quarters, fore and hind
Ready-to-cook poultry
Red meat
Retail beef or retail cuts

Rolled beef
Rolled out
Rollout, switchout, or pullout
Shrink
Sideweight
Slaughter
Slaughterhouse
Subprimals
Swinging beef, hanging beef, on the rail
Switchout
Tankage
Tanning
Terminal market
Tray-ready beef
Variety meat
Wholesale meat cuts
Wholesome Meat Act

Medicinal Products from Agriculture

Capsaicin
Diosgenin
L-dopa
Lignan
Limonene
Sitosterol
Solasodine
Stigmasterol
Taxol
Tryptophan

Milk

See ***Dairy***

Natural Resources (Soil and Water)

Acid soil
Acres irrigated
Aeration, soil
Aggregate (of soil)
Agricultural resource base
Alkaline soil
Alkali soil
Amendment
Anthropic soil
Aquifers
Arable soils
Available nutrient in soils
Available water in soils
Basin irrigation (or level borders)
Bedding soil
Blowout
Bog soil
Border irrigation
Buffer
Bureau of Reclamation
Carbon-nitrogen ratio
Center pivot irrigation
Compost
Converted wetlands
Creep, soil
Crust
Dehydration
Drainage
Drainage, soil
Drip irrigation
Ecosystems
Effluent
Environmental Protection Agency (EPA)
Erodibility index
Erodible (soil)
Erosion
Fertility, soil
Field, capacity
Fixation (in soil)
Flood plain
Friable soil
Furrow
Galled spots
Gravitational water
Groundwater
Gully erosion
Gumbo soil
Habitat
Hardpan
Highly erodible cropland
Humus
Hygroscopic water
Irrigable farms
Irrigable land
Irrigation
Irrigation methods
Irrigation water management
Land
Land capability
Land capability classification
Land classification
Land tenure
Land-use planning
Leaching
Light soil
Loam
Major Land Resource Area (MLRA)
Marginal land
Marl
Mulch
National Environmental Policy Act (NEPA)
Nonpoint-source pollution
Normal soil
Organic soil
Peat
Percolation (soil water)
Permanent wilting point
Permeability, soil
pH
Plow pan
Productive soil
Productivity (of soil)
Renewable natural resources
Riparian areas
Riparian rights
Runoff
Saline soil
Saturate
Sediment
Sheet and rill erosion
Slope
Soil
Soil association
Soil auger
Soil Bank
Soil characteristic
Soil conservation
Soil Conservation Service (SCS)
Soil erosion
Soil erosion tolerance values (T-values)
Soil injection wastes
Soil management
Soil map
Soil moisture tensiometer
Soil population
Soil quality
Soil reaction
Soil series
Soil sterilization
Soil structure

Soil survey
Soil texture
Soil type
Sphagnum
Storage capacity
Structure, soil
Subirrigation
Tensiometer
Texture, soil
Tillage
Tilth, soil
Topography
Topsoil
Water Bank Program
Water depletion allowance
Water Quality Incentives Program (WQIP)
Water rights
Watershed
Water table
Waterway
Wetlands
Wetlands Reserve Program
Wilting point (or permanent wilting point)
Wind erosion

Oilseeds

Canola
Crush
Crusher
Crushing
Industrial rapeseed
Lesquerella
Meadowfoam
Mustard seed
National Cottonseed Products Association (NCPA)
National Institute of Oilseed Products (NIOP)
National Soybean Processors Association (NSPA)
Nonindustrial rapeseed
Oilseed
Oilseed crops
Oilseed hulls
Oilseed meal
Oilseed processing
Peanut
Peanut meal
Rapeseed
Safflower seed
Soybean
Sunflower
Sunflower seed

See also ***Legumes; Soybeans; Sunflowers***

Poultry

Animal protein
Avian
Avian influenza
Blood meal
Broiler
Broiler complex
Brooder house
Brooding
Candling
Capacity of housing
Capon
Case of eggs
Checks
Chicken
Cockerel
Confinement rearing
Contract grower (poultry)
Controlled lighting
Duck
Egg
Egg dealer
Egg grader
Egg layers
Egg size
Egg washer
Flash candling
Free range
Fryer-roaster turkey
Fryers
Further processed
Further processor
Gobbler
Goose
Gradeout
Grain-consuming animal unit (GCAU)
Hatching eggs
Hatching flock
Hen
Incubation
Layer
Mature chickens
Mechanical nests
Molting
Monogastric
Multiple hatcher
National Broiler Council (NBC)
National Turkey Federation (NTF)
New York dressed weight
Nonruminant
Old turkeys (breeders)
Partial house brooding
Poult
Poultry
Poultry and Egg Export Council
Poultry and Egg Institute of America
Poultry Products Inspection Act
Pullet
Roaster
Roll-away nests
Rooster
Stress
Tom
Total confinement
Trap nest
Type of farm (or ranch)
United Egg Association (UEA)
United Egg Producers (UEP)
Young chicken
Young turkey (mature)

See also ***Crop and Livestock Diseases***
See also ***Meat and Poultry Packing***

Public Law 480

Additionality
Agency for International Development (AID)
Agricultural Trade Development and Assistance Act of 1954
Cargo Preference Act
Convertible local currency credit
Currency use payment
Excess foreign currencies
Export limitation

Fair share
Food for Development Program
Food for Peace Program
Food for Progress
Food Security Wheat Reserve
Initial payment
Letter of commitment
Local currency sale
Ocean freight differential
P.L. 480
Private voluntary organization (PVO)
Public Law 480
Purchase authorization

Rice

Acreage reduction program (ARP)
Aromatic rice
Basic commodities
Basmati rice
Brewers' rice
Brokens
Brown rice
Cereals
Deficiency payment
Food grain
Grain
Grain sizes
Head rice
Long-grain rice
Medium-grain rice
Milling
National Farm Program Acreage
Nonrecourse loans
Parboiled rice
Precooked rice
Prevented planting disaster payments
Program crop
Rice bran
Rice Millers Association (RMA)
Rice oil
Rough rice
Second heads
Short-grain rice

See also ***Food and Feed Grains***

Rural and Urban

Consolidated Metropolitan Statistical Area (CMSA)
County-type classification
Farming-dependent county
Government-dependent county
Manufacturing-dependent county
Metro areas
Metropolitan Statistical Area (MSA)
Mining-dependent county
Nonmetro areas
Retirement-destination county
Rural
Rural Development Administration (RDA)
Rural Electrification Administration (REA)
Standard Metropolitan Statistical Area (SMSA)
Urban
Urbanized area

Science

See ***Animal and Plant Science***

Seeds

American Seed Trade Association (ASTA)
Certified seed
Damping off
Germplasm
Multiline
Pericarp
Scarified seed
Seed
Seed bank
Seed, certified
Seed, cotton
Seed, pure live
Seed, treated

See also ***Oilseeds***

Shipping and Transportation

Arrivals
Bulk cargo
Bulk carrier
Contract of affreightment
Cost, insurance, and freight (c.i.f.)
Dry-bulk carrier
Farm truck
F.o.b. (free-on-board) shipping point price
Forwarder
Forwarding agent
Free in (FI)
Free in and out (FIO)
Free of particular average (FPA)
Free-on-board
Free out (FO)
Freight fowarder
Freight tariff
Grain truck
Hold
Horsepower
Horsepower hour
Infrastructure
Interregional marketing costs
Liquid bulk carrier
Livestock truck
Longshoreman
Loud spout
Marine insurance
Open policy
Point-to-point rate
Shippers Export Declaration
Special policy
Stevedores
Utility vehicle
With particular average (WPA)

See ***Farm Equipment and Buildings***

Soil

See ***Natural Resources (Soil and Water)***

Soybeans

Defatted soy flakes
Defatted soy flour
National Soybean Crop Improvement Council

(NSCIC)
National Soybean Processors Association (NSPA)
Soybean
Soybean meal
Soybean millfeed
Soybean oil
Soy grits and/or flour
Soy protein concentrate

Sugar and Sweeteners

Acesulfame-K
American Sugarbeet Growers Association (ASGA)
Amylase
Amylopectin
Amylose
Anhydrous dextrose
Aspartame (APM)
Beet
Carbohydrate
Carbohydrate-based fat substitute
Coffee, Sugar, and Cocoa Exchange, Inc.
Corn sweeteners
Crystalline fructose
Dextrose
Dialdehyde starch
Direct-consumption sugar
Fructose
Glucose
Glucose isomerase
Glucose syrup
Glucosinolates
Gluten
High fructose corn syrup
Hydrate dextrose
International Sugar Agreement (ISA)
Invert or invert sugar
Isoglucose
Low-calorie sweeteners
National Sugarbeet Growers Association
Ninety-Six-Degree Basis
No. 11 Contract price
No. 12 Contract price
No. 14 Contract price
Noncentrifugal sugar
No net cost
Nonrecourse loans
Polariscope
Polarization
Polysaccharides
Price-support programs
Program crop
Protein-based fat substitute
Quota-exempt sugar
Ratoon
Raw sugar
Reexport sugar
Refined corn syrup
Refined sugar
Saccharide
Saccharimeter
Saccharin
Starch
Sucrose
Sugar
Sugar Association, Inc. (SAI)
Sugarbeets
Sugarcane
Sugar-containing products
Syrup
U.S. Beet Sugar Association (USBSA)
U.S. Cane Sugar Refiners' Association (USCSRA)

Sunflowers

Sunflower
Sunflower Oil Assistance Program (SOAP)
Sunflower seed

Supply and Demand

Apparent consumption
Beginning stocks
Buffer stock
Carryover
Demand
Disappearance
Elasticity of demand
Ending stocks
Free stocks
Glut
Inventories
Inventory (CCC)
Invisible stocks
Normal yield
Overproduction
Production (agricultural)
Reserves
Residual market
Stocks
Supply
Supply control
Surplus
Utilization

Tariffs and Trade

Absolute quotas
Access
Accession
Accession compensatory amount (ACA)
Across-the-board linear tariff negotiation
Admission temporaire
Ad referendum
Ad valorem equivalent
Ad valorem tariff
Agreement on Import Licensing Procedures
Agricultural attachee
Agricultural Information and Marketing Services (AIMS)
Agricultural trade and development missions
Agricultural Trade Development and Assistance Act of 1954
Agricultural trade office (ATO)
Antidumping duty
Antidumping law
Appreciation (depreciation) of currency
Balance of payments
Balance of trade
Barter
Bilateral trade agreement
Bilateral trade negotiations
Blended credit
Bonded warehouse
Brussels Tariff Nomenclature (BTN)
Bulk products
Buy-back
Cairns Group
Capital account
Caribbean Basin Economic Recovery Act (CBERA)
Caribbean Basin Initiative (CBI)
Cash Export Certificate Program

CCC commercial credit
Certificate of origin
Clearing accounts
Codes of Conduct
Codex Alimentarius
Codex Alimentarius Commission
Commercial treaty
Commodity Import Programs (CIPS)
Common Agricultural Policy (CAP)
Common external tariff
Common market
Comparative advantage
Compensatory duty
Competitive imports
Complementary imports
Compound tariff
Concession
Concessional sales
Conditional Most-Favored Nation
Conselho Nacional do Comercio Exterior (CONCEX)
Contracting party
Controlled exchange rates
Cooperator program
Cooperators
Counterpurchase
Countertrade
Countervailing duty (CVD)
Credit guarantees
Creditor nation
Current account
Current account surplus
Customs
Customs classification
Customs Cooperation Council Nomenclature (CCCN)
Customs valuation
Dairy Export Incentive Program (DEIP)
Debtor nation
Devaluation
Dillon Round
Direct (export) credit
Discrimination
Drawback
Dual exchange rate system
Dual pricing
Dumping
Duty
Economic sanctions
Embargo
Enterprise for the Americas Initiative (EAI)
Entrepot
European Free Trade Association (EFTA)
Exchange rate
Exchange rate management
Exchange restrictions
Eximbank
Ex (point of origin)
Export
Export allocation or quota
Export assistance programs
Export Credit Guarantee Program (GSM-102)
Export credit insurance
Export Credit Revolving Fund
Export declaration
Export deficit
Export Enhancement Program (EEP)
Exporter
Export-Import Bank of the United States
Export Incentive Program (EIP)
Exportkhleb
Export license
Export limitation
Export Marketing Certificate Program
Export payment
Export programs
Export promotion programs
Export quota agreement
Export restitutions
Export restraints
Exports
Export subsidies
Export surplus
Export tax
Ex (tariff number)
Fair value
Fast-track negotiating authority
Fixed exchange rates
Flexible exchange rates
Food, Agriculture, Conservation, and Trade Act of 1990 (P.L. 101-624)
Food balance
Food for Progress
Food reserves
Foreign
Foreign Agricultural Service (FAS)
Foreign Credit Insurance Association (FCIA)
Foreign exchange controls
Foreign exchange rate
Foreign trade zone
Formula approach
Free list
Free market
Free port
Free trade
Free trade area
Free trade zone
Functioning of the GATT System (FOGS)
GATT
GATT Rounds
GATT Secretariat
General Agreement on Tariffs and Trade (GATT)
Generalized System of Preferences (GSP)
Generic advertising
Global quota
Graduation
Grandfather clause
GSM
GSM-5
GSM-102
GSM-103
Guaranteed Export Credit
Hard currency
Harmonized Commodity Description and Coding System
Harmonized System (HS)
High-value products (HVP)
Import
Import barriers
Importer
Import license
Import Licensing Code
Import quota
Import substitution
Import value
Inconvertible currency
Intellectual Property Rights (IPR)
Intermediate Export Credit Guarantee Program (GSM-103)
International trade barriers
International Trade Centre

International Trade Commission (ITC)
Jackson-Vanik Amendment
Kennedy Round
Latin American Free Trade Association (LAFTA)
Latin American Integration Association
Levy
Liberalization
Liberal (trade)
Licensing
Linear reduction of tariffs
Lome Convention of 1975
Margin of preference
Market access
Market development programs
Market Promotion Program (MPP)
Meat Import Law
Meat Products Export Incentive Program payments
Most-favored nation (MFN)
MTN (multilateral trade negotiations) agreements and arrangements
Multifiber Arrangement (MFA)
Multilateral agreement
Multilateral trade negotiations
Multitier system
National Grain Trade Council (NGTC)
Net exporter
Net importer
Noncompetitive imports
Nonfarm foods
Nontariff trade barriers
North American Export Grain Association (NAEGA)
Office of the U.S. Trade Representative
Offset requirement
Orderly marketing agreement
Organization for Economic Cooperation and Development (OECD)
Phytosanitary
Phytosanitary certificate
Phytosanitary regulations
Plurilateral
Preferences
Preferential trade agreement
Principal supplier
Prior import deposit
Protectionism
Punta del Este Declaration
Quantitative restriction
Quarantine
Quota
Reexport
Residual supplier
Retaliation
Reverse preferences
Round
Safeguard(s)
Sanitary and phytosanitary regulations
Schedule B
Section 22
Section 32
Section 201
Section 301
Self-sufficiency
Shippers Export Declaration
Snap-back provision
Soft currency
Special drawing rights (SDR)
Specific tariff
STABEX
Standard International Trade Classification (SITC)
State marketing boards
State trading
Subsidy
Sufferance
Supplementary imports
Surcharge
Surveillance
Switch trading
Tariff
Tariff preference
Tariff quota
Tariff rate quota (TRQ) system
Tariff schedule
Tariff surcharge
Tariff union
Technical barrier to trade
Terms of sale
Terms of trade
Tokyo Round
Trade Act of 1974
Trade Agreements Act of 1979
Trade and Tariff Act of 1984
Trade balance
Trade barriers
Trade deficit
Trade liberalization
Trade negotiations
Trade policy committee (TPC)
Trade preference
Trade servicing
Trade surplus
Trade-weighted exchange rate
Treaties
Trigger
Trigger level
Two-price plan
Unfair trade practices
United Nations Conference on Trade and Development (UNCTAD)
Upstream subsidization
Uruguay Round
U.S.–Canada Free Trade Agreement (CFTA)
U.S. International Trade Commission (USITC)
U.S. Meat Import Law
U.S. Tariff Commission
U.S. Trade Representative (USTR)
Variable levy
Voluntary export agreement
Voluntary restraint agreement
Wheat and Feed Grain Export Certificate Program
Wheat Trade Convention

See also ***European Union; General Agreement on Tariffs and Trade***

Tobacco

Acreage allotment
Aging
Air-cured tobacco
Auction
Basic commodities
Binder tobacco
Bulk curing
Burley tobacco
Chewing tobacco
Cigar
Cigar classes of tobacco
Cigarette

Cropping
Cube cut (tobacco)
Cure
Curing
Dark air-cured tobacco
Dry weight
Farm sale weight (FSW)
Filler tobacco
Fire-cured tobacco
Flue-cured tobacco
Granulated tobacco
Leasing of quota
Licorice
Light air-cured tobacco
Long cut (tobacco)
Loose leaf (tobacco)
Maryland tobacco
Once-over harvester
Plug (tobacco)
Priming
Priming aid
Prizing
Renting quota
Smoking tobacco
Snuff
Stalk cutting
Stemming
Threshing
Tipping
Tobacco
Tobacco Institute (TI)
Topping
Twist (tobacco)
Tying machine
Warehouse (tobacco)
Wrapper tobacco

Trade

See ***Tariffs and trade***

Tropical, Root, Tuber, and Other Crops and Products

Arabica
Arracacha
Bambara groundnut
Banana
Boniato
Burdock
Cacao
Cactus pads
Cocoa
Coffee tree
Ginger
Ginseng
International Cocoa Agreement (ICCA)
International Coffee Agreement (ICA)
Jerusalem artichoke
Jicama
Malanga
New York Coffee, Sugar, and Cocoa Exchange, Inc.
Oca
Robusta
Tea
Tef
Tropical products
Tubers
Yucca

U.S. Department of Agriculture

Agricultural agent
Agricultural attachee
Agricultural Cooperative Service (ASC)
Agricultural Environmental Quality Council
Agricultural Information and Marketing Services (AIMS)
Agricultural Marketing Service (AMS)
Agricultural Research Service (ARS)
Agricultural Stabilization and Conservation Service (ASCS)
Animal and Plant Health Inspection Service (APHIS)
Animal Damage Control Program (ADC)
Bureau of Agricultural Economics
Cooperative Extension Service
Cooperative State Research, Education, and Extension Service (CSREES)
Cooperative State Research Service (CSRS)
County extension agent
Economic Research Service (ERS)
Extension agent
Extension home economist
Extension Service
Farmers Home Administration (FmHA)
Farm Service Agency
Federal Crop Insurance Corporation (FCIC)
Federal Grain Inspection Service (FGIS)
Federal-State Market news service
Food and Consumer Service (FCS)
Food and Nutrition Service (FNS)
Food Safety and Inspection Service (FSIS)
Foreign Agricultural Service (FAS)
Forest Service (FS)
4-H Youth Programs
Grain Inspection, Packers, and Stockyards Administration
Human Nutrition Information Service (HNIS)
Market news service
National Agricultural Library (NAL)
National Agricultural Statistics Service (NASS)
National Resource Conservation Service
Office of Agricultural Biotechnology
Office of Energy (OE)
Office of International Cooperation and Development (OICD)
Office of Small-Scale Agriculture
Office of the Consumer Advisor
Office of Public Affairs (OPA)
Office of Transportation (OT)
Packers and Stockyards Administration (PSA)
Rural Business and Coopera-

tive Development Service
Rural Development Administration (RDA)
Rural Electrification Administration (REA)
Rural Utilities Service
Rural Housing and Community Development Service
World Agricultural Outlook Board (WAOB)

U.S. Government

Bureau of Labor Statistics (BLS)
Bureau of the Census
Bureau of Reclamation
Council of Economic Advisors (CEA)
Environmental Protection Agency (EPA)
U.S. Customs Service
U.S. Department of Agriculture
U.S. Department of Commerce
U.S. Department of Defense
U.S. Department of Education
U.S. Department of Energy
U.S. Department of Health and Human Services
U.S. Department of Housing and Urban Development
U.S. Department of Interior
U.S. Department of Justice
U.S. Department of Labor
U.S. Department of State
U.S. Department of Transportation
U.S. Department of Treasury
U.S. Department of Veterans Affairs

See also ***U.S. Department of Agriculture***

Vegetables

Artichoke
Arugula
Asparagus
Baby vegetables
Beet
Bitter melon
Bok choy
Breadfruit
Broccoli
Brussel sprout
Cabbage
Calabaza
Cardoon
Carrot
Cauliflower
Celery
Chard
Chayote
Chinese cabbage
Chinese pea pod
Competitive imports
Dehydrated onion
Dehydrating onion
Dried onion
Dry onion
Eggplant
Fiddlehead fern
Fresh weight equivalent
Grade and size (produce)
Greens
Kohlrabi
Italian red lettuce
Leek
Lettuce
Mexican husk tomato
Mushroom
Mushroom spawn
Mustard green
Mycelium
Napa
National Onion Association (NOA)
National Potato Council (NPC)
Okra
Onion
Oysterplant
Pea
Pearl onion
Pepper, bell
Pimento
Potato
Potato eye
Processed fresh vegetables
Processed vegetables
Produce
Produce grower
Produce wholesaler
Radicchio
Radish
Red chicory
Rhubarb
Romaine
Rutabaga
Salsify
Scallion
Shallot
Snapback provision
Snow pea
Sorrel
Spinach
Squash
Sugar pea
Sunchoke
Sweet potato
Sweet Potato Coucil of the United States (SPCUS)
Swiss chard
Textured vegetable protein
Tomatillo
Tomato
Tomato repacker
Zucchini

Water

See ***Natural Resources***

Weights and Measures

Acre
Acre foot
Animal unit
Bale (cotton)
Bushel
Carcass weight
Carcass weight equivalent
Cargo ton
Conversion factors
cwt
Feed conversion rate (feed efficiency)
Fiscal year
Freight ton
Grain-consuming animal unit (GCAU)
Hectare
Hogshead

Hundredweight (cwt)
Liveweight
Long ton
Marketing year
Measurement ton
Metric ton
Pound
Product weight
Quintal
Running bale
Short ton
Staple length
Test weight
Ton
Universal density bale
Yarn size

Wheat

Basic commodities
Bleaching
Blending
Bran
Broken kernel
Bromus secalinus (cheat)
Bulgur
Cash grains
Cereals
Club wheat
Dark northern wheat
Deficiency payment
Durum wheat (*Triticum durum*)
Endosperm
Extraction rate
Farina
Farmer-owned Reserve Program
Food grains
Foreign material
Gliadin
Grading
Grain
Grain Reserve Program
Grain screenings
Hard red spring wheat
Hard red winter wheat
Hard wheat
Hard wheat flour
International Wheat Agreement (IWA)
Intrinsic quality
Karnal bunt
Meslin
Mill mix
National Association of Wheat Growers (NAWG)
National farm program acreage
National Program Acreage (NPA)
Near-infrared reflectance spectroscopy (NIRS)
Nonrecourse loans
Northern wheat
Prevented planting disaster payments
Protein premium
Red wheat
Release price
Semolina
Shorts
Soft red winter wheat (*Triticum aesturim*)
Soft wheat
Soft wheat flour
Spring wheat
Tempering
Test weight
Triticale
U.S. Wheat Associates
Vital wheat gluten
Vitreous
Wheat
Wheat and Feed Grain Export Certificate Programs
Wheat starch
Wheat Trade Convention
White wheat
Winter wheat

See also ***Feed and Food Grains***

Appendix 3

Major U.S. Agricultural and Trade Legislation, 1933-1994

Agricultural Adjustment Act of 1933 (P.L. 73-10) was signed into law May 12, 1933. The law introduced the price-support programs, including production adjustments, and incorporated the Commodity Credit Corporation (CCC) under the laws of the state of Delaware on October 17, 1933. The act also made price-support loans by the CCC mandatory for the designated "basic" (storable) commodities (corn, wheat, and cotton). Support for other commodities was authorized upon recommendation by the secretary of agriculture with the president's approval. Commodity loan programs carried out by the CCC for 1933-1937 included programs for cotton, corn, turpentine, rosin, tobacco, peanuts, dates, figs, and prunes. The provisions in the act for production control and tax processing were later declared unconstitutional.

Agricultural Adjustment Act Amendment of 1935 (P.L. 74-320) was signed into law August 24, 1935. The law gave the president authority to impose quotas when imports interfere with agricultural adjustment programs.

Soil Conservation and Domestic Allotment Act of 1936 (P.L. 74-461) was signed into law February 26, 1936. The law provided for soil-conservation and soil-building payments to participating farmers but did not include strong price- and income-support programs.

Agricultural Adjustment Act of 1938 (P.L. 75-430) was signed into law February 16, 1938. The law was the first to make price support mandatory for corn, cotton, and wheat to maintain a sufficient supply in low production periods along with marketing quotas to keep the supply in line with market demand. It also established permissive supports for butter, dates, figs, hops, turpentine, rosin, pecans, prunes, raisins, barley, rye, grain sorghum, wool, winter cover-crop seeds, mohair, peanuts, and tobacco for the 1938-1940 period. The 1938 Act is considered part of permanent legislation. Provisions of this law are often superseded by more current legislation. However, if the current legislation expires and new legislation is not enacted, the law reverts back to the 1938 Act (along with the Agricultural Act of 1949).

Federal Food, Drug, and Cosmetic Act of 1938 (P.L. 75-717) was signed into law June 25, 1938. The law is intended to ensure that foods are pure and wholesome, safe to eat, and produced under sanitary conditions; that drugs and devices are safe and effective for their intended uses; that cosmetics are safe and made from appropriate ingredients; and that all labeling and packaging is truthful, informative, and not deceptive. The Food and Drug Administration (FDA) is responsible for enforcing the Federal Food, Drug, and Cosmetic Act.

Steagall Amendment of 1941 (P.L. 77-144) was signed into law July 1, 1941. The law required support for many nonbasic commodities at 85 percent of parity or higher. In 1942, the minimum rate was increased to 90 percent of parity and was required to be continued for two years after the end of World War II. The "Steagall commodities" included hogs, eggs, chickens (with certain exceptions), turkeys, milk, butterfat, certain dry peas, certain dry edible beans, soybeans, flaxseed and peanuts for oil, American-Egyptian (ELS) cotton, potatoes, and sweet potatoes.

Agricultural Act of 1948 (P.L. 80-897) was signed into law July 3, 1948. The law made price support mandatory at 90 percent of parity for basic commodities in 1949. It also provided that, beginning in 1950, parity would be reformulated to take into consideration average prices of the previous 10 years, as well as those of the 1910-1914 base period.

Agricultural Act of 1949 (P.L. 89-439) was signed into law October 31, 1949. The law, along with the Agricultural Adjustment Act of 1938, makes up the major part of permanent agricultural legislation that is still effective in amended form. The 1949 Act designated mandatory support for the following nonbasic commodities: wool and mohair, tung nuts, honey, Irish potatoes (excluded in the Agricultural Act of 1954), and milk, butterfat, and their products.

Agricultural Trade Development and Assistance Act of 1954 (Food for Peace) (P.L. 83-480) was signed into law July 10, 1954. The law became the basis for sales and barter of surplus commodities overseas and for overseas relief. The program made U.S. agricultural commodities available through long-term credit sales at low interest rates, provided food relief, and authorized "food for development" projects.

Agricultural Act of 1954 (P.L. 83-690) was signed into law August 28, 1954. It established a flexible price support for basic commodities (excluding tobacco) at 82.5 to 90 percent of parity and authorized a Commodity Credit Corporation reserve for foreign and domestic relief.

> **National Wool Act of 1954** (Title VII of Agricultural Act of 1954) provided for a new price support program for wool and mohair to encourage a certain level of domestic production (set at 300 million pounds for 1955).

Agricultural Act of 1956 (P.L. 84-540) was signed into law May 28, 1956. This law began the Soil Bank Act that authorized the Acreage Reserve Program for wheat, corn, rice, cotton, peanuts, and several types of tobacco. It also provided for a 10-year Conservation Reserve Program.

Consolidated Farm and Rural Development Act (P.L. 87-128) was signed into law August 8, 1961. The law authorized USDA farm-lending activities through the Farmers Home Administration.

Food and Agricultural Act of 1962 (P.L. 87-703) was signed into law September 27, 1962. This law authorized an emergency wheat program with voluntary diversion of wheat acreage and continued the feed grain support program. It also included a marketing certificate program for wheat. The program, however, was rejected by wheat producers who were required to approve its marketing quota.

Agricultural Act of 1964 (P.L. 88-297) was signed into law April 11, 1964. This law authorized a two-year voluntary marketing certificate program for wheat and a payment-in-kind (PIK) program for cotton.

Food Stamp Act of 1964 (P.L. 88-525) was signed into law August 31, 1964. The law provided the basis for the Food Stamp Program. It was later replaced by the food stamp provisions (Title XIII) of the Food and Agricultural Act of 1977.

Food and Agricultural Act of 1965 (P.L. 89-321) was signed into law November 3, 1965. This law was the first multi-year farm legislation, providing for four-year commodity programs for wheat, feed grains, and upland cotton. It was extended for one more year through 1970 (P.L. 90-559). It authorized a Class I milk base plan for the 75 federal milk marketing orders and a long-term diversion of cropland under a Cropland Adjustment Program. The law also continued payment and diversion programs for feed grains and cotton and certificate and diversion programs for wheat.

Agricultural Act of 1970 (P.L. 91-524) was signed into law November 30, 1970. The law, in effect through 1973, established the cropland set-aside program and a payment limitation per producer (at $55,000 per crop). It also amended and extended the authority of the Class I Base Plan in milk marketing order areas.

Act of April 14, 1971 (P.L. 92-10) provided for poundage quotas for burley tobacco in place of farm acreage allotments.

Agriculture and Consumer Protection Act of 1973 (P.L. 93-86) was signed into law August 10, 1973. The law established target prices and deficiency payments to replace former price-support payments. It also set payment limitations at $20,000 for all program crops and authorized disaster payments and disaster reserve inventories to alleviate distress caused by a natural disaster.

Trade Act of 1974 (P.L. 93-618) provided the president with tariff and nontariff trade barrier negotiating authority for the Tokyo Round of multilateral trade negotiations. It also gave the president broad authority to counteract injurious and unfair foreign trade practices.

Section 201 of the act requires the U.S. International Trade Commission to investigate petitions filed by domestic industries or workers claiming injury or threat of injury due to expanding imports. Investigations must be completed within six months. If such injury is found, restrictive measures may be implemented. Action under Section 201 is allowed under the escape clause, GATT Article XIX.

Section 301 was designed to eliminate unfair foreign trade practices which adversely affect U.S. trade and investment in both goods and services. Under Section 301, the president must determine whether the alleged practices are unjustifiable, unreasonable, or discriminatory and burden or restrict U.S. commerce. If the president determines that action is appropriate, the law directs that all appropriate and feasible action within the president's power should be taken to secure the elimination of the practice.

Food and Agriculture Act of 1977 (P.L. 95-113) was signed into law September 9, 1977. The law increased price and income supports and established a farmer-owned reserve for grain. It also established a new two-tiered pricing program. Under the program, producers were given an acreage allotment on which a poundage quota was set. Growers could produce in excess of their quota, within their acreage allotment but would receive the higher of the two price-support levels only for the quota amount. Peanuts in excess of the quota are referred to as "additionals."

Food Stamp Act of 1977 (Title XIII) permanently amended the Food Stamp Act of 1964 by eliminating purchase requirements and simplifying eligibility requirements.

National Agricultural Research, Extension, and Teaching Policy Act (Title XIV) made USDA the leading federal agency for agricultural research, extension, and teaching programs. It also consolidated the funding for these programs.

Trade Agreements Act of 1979 (P.L. 96-39) was signed into law on July 26, 1979. This act provided the implementing legislation for the Tokyo Round of multilateral trade agreements in such areas as customs valuation, standards, and government procurement.

Federal Crop Insurance Act of 1980 (P.L. 96-365) was signed into law September 26, 1980. The law expanded crop insurance into a national program covering the majority of crops.

Agriculture and Food Act of 1981 (P.L. 97-98) was signed into law December 22, 1981. The law emphasized making U.S. commodities competitive abroad. It set specific target prices for four years, eliminated rice allotments and marketing quotas, and lowered dairy supports.

Omnibus Budget Reconciliation Act of 1982 (P.L. 97-253) was signed into law September 8, 1982. The law froze dairy price supports and mandated loan rates and acreage reserve programs for the 1983 crops.

Temporary Emergency Food Assistance Act of 1983 (P.L. 98-8) was signed into law March 24, 1983. The law authorized distribution of foodstuffs owned by the Commodity Credit Corporation to indigent persons.

Extra-Long Staple Cotton Act of 1983 (P.L. 98-88) was signed into law August 26, 1983. The law eliminated marketing quotas and allotments for extra-long staple cotton and tied its support to upland cotton through a formula that sets the loan rate at not less than 150 percent of the upland cotton loan level.

Dairy and Tobacco Adjustment Act of 1983 (P.L. 98-180) was signed into law November 29, 1983. The law froze tobacco price supports, launched a voluntary dairy diversion program, and established a dairy promotion order.

Agricultural Programs Adjustment Act of 1984 (P.L. 98-258) was signed into law April 10, 1984. The law froze target price increases provided in the 1981 Act; authorized paid land diversions for feed grains, upland cotton, and rice; and provided a wheat payment-in-kind program for 1984.

Trade and Tariff Act of 1984 (P.L. 98-573) was signed into law on October 30, 1984. The law clarified the conditions under which unfair trade cases under Section 301 of the Trade Act of 1974 can be pursued. It also provided bilateral trade negotiating authority for the U.S.–Israel Free Trade Area and set out procedures to be followed for congressional approval of future bilateral trade agreements.

Balanced Budget and Emergency Deficit Control Act of 1985 (P.L. 99-177) was signed into law December 12, 1985. Also known as the Gramm-Rudman-Hollings Act, the law was designed to eliminate the federal budget deficit by October 1, 1990. The law was declared unconstitutional in 1986.

As amended in 1987 (P.L. 100-119), the law mandated annual reductions in the federal budget deficit to eliminate it by 1993. Under the law, automatic spending cuts could have occurred for almost all federal programs if Congress and the president could not agree on a targeted budget package for any specific fiscal year.

Food Security Act of 1985 (P.L. 99-198) was signed into law December 23, 1985. The law allowed lower price and income supports, lowered dairy supports, established a dairy herd buyout program, and created a conservation reserve program targeted at erosive croplands.

Farm Credit Restructuring and Regulatory Reform Act of 1985 (P.L. 99-205) was signed into law December 23, 1985. The law implemented interest rate subsidies for farm loans and restructured the Farm Credit Administration.

Technical Corrections to Food Security Act of 1985 Amendments (P.L. 99-253) was signed into law February 28, 1986. The law gave the secretary of agriculture discretion to require cross-compliance for wheat and feed grains instead of mandating them, changed acreage base calculations, and specified election procedures for local Agricultural Stabilization and Conservation committees.

Food Security Improvements Act of 1986 (P.L. 99-260) was signed into law March 20, 1986. The law made further modifications to the 1985 Act, including limiting the nonprogram crops that can be planted under the 50/92 provision, permitting haying and grazing on diverted wheat and feed grain acreage during a set five-month period if requested by the State Agricultural Stabilization and Conservation Committee, and increasing deductions taken from the price of milk received by producers to fund the whole herd buyout program.

Consolidated Omnibus Budget Reconciliation Act of 1985 (P.L. 99-272) was signed into law April 7, 1986. This law cancelled the flue-cured and burley tobacco quotas announced for the 1986 programs, giving the secretary of agriculture discretion to set the quotas.

Omnibus Budget Reconciliation Act of 1986 (P.L. 99-509) was signed into law October 21, 1986. The law requires advance deficiency payments to be made to producers of 1987 wheat, feed grains, upland cotton, and rice crops at a minimum of 40 percent for wheat and feed grains and 30 percent for rice and upland cotton. It also amends the Farm Credit Act of 1971.

Making Continuing Appropriations for the Fiscal Year 1987, and for Other Purposes (P.L. 99-591) was signed into law October 30, 1986. The law, in addition to providing funding for federal programs, modified the 1985 farm bill by limiting program payments to $50,000 per person for deficiency and paid land diversion payments and included honey, resource adjustment (excluding land diversion), disaster, and Findley payments under a $250,000 payment limitation.

Futures Trading Act of 1986; Grain Quality Improvement Act of 1986; and Processed Products Inspection Improvement Act of 1986 (P.L. 99-641) was signed into law November 10, 1986. The law reauthorized appropriations to carry out the Commodity Exchange Act and made technical improvements to that act.

Farm Disaster Assistance Act of 1987 (P.L. 100-45) was signed into law May 27, 1987. The law provided assistance to producers who experienced crop losses from natural disasters in 1986.

Stewart B. McKinney Homeless Assistance Act (P.L. 100-77) was signed into law July 22, 1987. The law provided housing, food assistance, and job training for the homeless.

Uniform Cotton Classing Fees Act of 1987 (P.L. 100-108) was signed into law August 20, 1987. The law provided continuing authority to the secretary of agriculture to recover costs associated with cotton classing services.

Omnibus Budget Reconciliation Act of 1987 (P.L. 100-203) was signed into law December 22, 1987. The law set the 1988 fiscal year budget for agriculture and all federal agencies. It set target prices for 1988 and 1989 program crops, established loan rates for program and nonprogram crops, and required a voluntary paid land diversion for feed grains. The law also further defined who is eligible to receive program payments ("defining a person").

Agricultural Credit Act of 1987 (P.L. 100-233) was signed into law January 6, 1988. The law provided credit assistance to farmers, strengthened the Farm Credit System, and facilitated the establishment of secondary markets for agricultural loans.

Disaster Assistance Act of 1988 (P.L. 100-387) was signed into law August 6, 1988. The law provided assistance to farmers hurt by the drought and other natural disasters in 1988. Crop producers with losses greater than 35 percent of production were eligible for financial assistance, and feed assistance was available to livestock producers.

Agricultural Credit Technical Corrections Act (P.L. 100-399) was signed into law August 17,

1988. The law corrected the Agricultural Credit Act of 1987, restoring language that exempted mergers of the Farm Credit System institutions from state transfer taxes.

Omnibus Trade and Competitiveness Act of 1988 (P.L. 100-418) was signed into law August 23, 1988. The law revised statutory procedures for dealing with unfair trade practices and import damage to U.S. industries. It gave the secretary of agriculture discretionary authority to trigger marketing loans for wheat, feed grains, and soybeans, if it was determined that unfair trade practices existed. The secretary could extend export programs, such as the Export Enhancement Program and the Targeted Export Assistance Program, in response to unfair competition.

Hunger Prevention Act of 1988 (P.L. 100-435) was signed into law September 19, 1988. The law amended the Temporary Emergency Food Assistance Act of 1983 to require the secretary of agriculture to make additional types commodities available for the Temporary Emergency Food Assistance Program, to improve the child nutrition and food stamp programs, and to provide other hunger relief.

United States–Canada Free Trade Agreement Implementation Act of 1988 (P.L. 100-449) was signed into law September 28, 1988. The law implemented the bilateral trade agreement between the United States and Canada, including agricultural trade. The agreement phases out tariffs between the two countries over 10 years and revises other trade rules.

Disaster Assistance Act of 1989 (P.L. 101-82) was signed into law August 14, 1989. The law provided assistance to farmers hurt by drought or other natural disasters in 1988 or 1989. To qualify for financial assistance, crop producers must have lost at least 35 percent of production. The requirement was higher for farmers without crop insurance, as well as producers of nonprogram crops and those who did not participate in farm programs. Other assistance was similar to that provided by the Disaster Assistance Act of 1988 (P.L. 100-387).

Omnibus Budget Reconciliation Act 1989 (P.L. 101-239) was signed into law December 19, 1989. The law superseded the 10-25 planting provision of the Disaster Assistance Acts of 1988 and 1989. The act allowed program crop producers to plant up to 25 percent of their permitted acreage to soybeans, sunflowers, and safflowers for the 1990 crop.

Omnibus Budget Reconciliation Act 1990 (P.L. 101-508) was signed November 5, 1990. The law amended the Food, Agriculture, Conservation, and Trade Act of 1990 (P.L. 101-624) to reduce agricultural spending for 1991-1995. It included a mandatory 15 percent planting flexibility and assessments on nonprogram crop producers. The law also requires the U.S. Department of Agriculture (USDA) to calculate deficiency payments for 1994 and 1995 wheat, feed grain, and rice crops using a 12-month average market price instead of the 5-month average required under previous law.

Under the Omnibus Budget law, USDA was also directed to take specified actions to improve the competitiveness of U.S. agricultural exports if the negotiations in the Uruguay Round of the General Agreement on Tariffs and Trade (GATT) failed to result in the signing and implementation of a trade agreement.

Food, Agriculture, Conservation, and Trade Act of 1990 (P.L. 101-624) was signed November 28, 1990. The five-year farm bill continues to move agriculture in a market-oriented direction. It freezes target prices and allows more planting flexibility. New titles include rural development, forestry, organic certification, and commodity promotion programs. The law establishes a Rural Development Administration (RDA) in the U.S. Department of Agriculture to administer programs relating to rural and small community development. P.L. 101-624 also extends and improves the Food Stamp Program and other domestic nutrition programs and makes major changes in the operation of P.L. 480.

Food, Agriculture, Conservation, and Trade Act Amendments of 1991 (P.L. 102-237) was signed into law on December 12, 1991. The law amends the Food, Agriculture, Conservation, and Trade Act of 1990 (P.L. 101-624) to correct errors and alleviate problems in implementing the law. The law also allows the Farm Credit Bank for Cooperatives to make loans for agricultural exports and establishes a new regulatory scheme and capital standards for the Federal Agricultural Mortgage Corporation ("Farmer Mac"). The law also establishes new handling requirements for eggs to help prevent food-borne illness.

WIC Farmers' Market Nutrition Act of 1992 (P.L. 102-314) was signed into law on July 2, 1992. The law establishes a program that provides participants in the Special Supplemental Food Program for Women, Infants, and Children (WIC) with supplmental food coupons that can be used to purchase fresh, unprocessed foods, such as fruits and vegetables at farmers' markets.

Futures Trading Practices Act of 1992 (P.L. 102-546) was signed into law on October 28, 1992. The law amends the Commodity Exchange Act to improve the regulation of futures and options traded under rules and regulations of the Commodity Futures Trading Commission (CFTC), to establish registration standards for all exchange floor traders, and to restrict practices that may lead to fraud and abuse. The law also reauthorizes the CFTC through fiscal year 1994.

Farm Credit Banks and Associations Safety and Soundness Act of 1992 (P.L. 102-552) was signed into law on October 28, 1992. The law is designed to enhance the financial safety and soundness of the banks and associations of the Farm Credit System by establishing new mechanisms to ensure repayment of Farm Credit System debt resulting from federal financial assistance provided to the system under the Agricultural Credit Act of 1987 (P.L. 100-233) and to make other changes. The law also directs the Department of Agriculture to purchase, process, and distribute additional agricultural commodities for the emergency food assistance program established under the Temporary Emergency Food Assistance Act of 1983 (P.L. 98-8).

Agricultural Credit Improvement Act of 1992 (P.L. 102-554) was signed into law on October 28, 1992. The law establishes new Farmers Home Administration (FmHA) loan programs to assist beginning farmers and ranchers. The law establishes FmHA operating and equipment loan and loan guarantee programs for beginning farmers and ranchers and a program to provide 10-year loans for beginning farmers and ranchers to purchase their own farm or ranch in return for a down payment equivalent to 10 percent of the purchase price of the land. The law revises farm credit program requirements to improve women farmers' access to FmHA assistance. The law also limits the total number of years any borrower may participate in the agency's farm ownership and operating loan programs.

Bankruptcy, Title II U.S.C. Extension (P.L. 103-65) was signed into law on August 6, 1993. The law extends the Chapter 12 provision of the Bankruptcy Code through October 1, 1998. Chapter 12, which would have expired in October 1993, establishes special provisions governing bankruptcy proceedings for family farmers.

Omnibus Budget Reconciliation Act of 1993 (P.L. 103-66) was signed into law on August 10, 1993. The law makes changes in the federal farm programs and related programs to reduce federal spending by $3 billion over five years, including eliminating the U.S. Department of Agriculture's authority to waive minimum acreage set-aside requirements for wheat and corn, reducing deficiency payments to farmers participating in the 0/92 and 50/92 programs from 92 percent to 85 percent of the normal payment level, reducing the acreage to be enrolled in the Conservation Reserve Program and Wetlands Reserve Program, and requiring improvement in the actuarial soundness of the federal crop insurance program. The measure also provides for a temporary moratorium on sales of synthetic bovine growth hormone, reduces the loan rate for soybeans, and requires changes in the Market Promotion Program.

The law provides for the designation of three empowerment zones and 30 enterprise communities for rural areas and six empowerment zones and 65 enterprise communities for urban areas. These designated areas will receive special consideration for various federal programs and other assistance and qualify for specific tax credits. The law also provides $1 billion in spending under the federal block grant program, with each rural empowerment grant totaling $20 million.

Emergency Supplemental Appropriations for Relief from the Major Widespread Flooding in the Midwest Act of 1993 (P.L. 103-75) was signed into law on August 16, 1993. This supplemental appropriations bill provided $1.35 billion for Commodity Credit Corporation (CCC) disaster payments to farmers who lost their crops due to natural disasters in 1993. Disaster payments are provided under the formula in the Food, Conservation, and Trade Act of 1990 (P.L. 101-624). The Emergency Supplemental Appropriations law authorized the use of other CCC funds if the bill's appropriations are insufficient to make full disaster payments to farmers.

Agriculture, Rural Development, Food and Drug Administration, and Related Agencies Appropriations Act, 1994 (P.L. 103-111) was signed into law on October 21, 1993. The law prohibits the use of funds made available under this legislation to provide price supports for honey in the 1994 crop year.

National Wool Act of 1954, Amendment (P.L. 103-130) was signed into law on November 1, 1993. The law provides for reductions in the federal incentive payments to wool and mohair producers for the 1994 and 1995 marketing years. The wool and mohair price support program is terminated beginning in 1996.

Federal Crop Insurance Reform and Department of Agriculture Reorganization Act of 1994 (P.L. 100-354) was signed into law on October 4, 1994. The act, in effect beginning with 1995 crops, supplements the federal crop insurance program with a new catastrophic coverage level (CAT) available to farmers for a processing fee of $50 per crop with a cap of $200 per farmer per county and $600 per farmer total. Farmers may purchase additional insurance coverage providing higher yield or price protection levels. The law stipulates producers must purchase crop insurance coverage at the CAT level or above to participate in federal commodity support programs, Farmers Home Administration loans, and the Conservation Reserve Program.

The act also creates the Noninsured Assistance Program (NAP), a permanent aid program for crops not covered by crop insurance.

The law also authorized a major restructuring of the U.S. Department of Agriculture.

Appendix 4

North American Free Trade Agreement (NAFTA) and U.S.–Canada Free Trade Agreement Provisions Affecting Agriculture

The North American Free Trade Agreement (NAFTA) sets out separate bilateral undertakings on cross-border trade in agricultural products, one between the United States and Mexico and the other between Canada and Mexico. In general, the rules of the U.S.–Canada Free Trade Agreement on tariff and nontariff barriers will continue to apply to agricultural trade between the United States and Canada.

Key NAFTA Provisions

Elimination of Nontariff Barriers: Under NAFTA, the United States and Mexico eliminated all nontariff barriers to agricultural trade, generally through their conversion to tariff rate quotas (TRQs) or ordinary tariffs.

> No tariffs will be imposed on imports within the TRQ amount. Tariffs on imports above the TRQ amount will be phased out during a 10- or 15- year transition period, depending on the product. The quantities to enter duty free under the TRQ are based on recent average trade levels and will increase generally at 3 percent per year. TRQs will be eliminated at the end of the transition period.

Elimination of Tariffs: The United States and Mexico eliminated tariffs on a broad range of agricultural products, and most tariffs will be eliminated within 10 years. Duties on a few highly sensitive products will be phased out over 15 years.

Special Safeguard Provisions: During the first 10 years that NAFTA is in effect, a special safeguard provision will apply to certain products. A designated quantity of imports will be allowed at a NAFTA preferential tariff rate. Once imports exceed the designated quantity, the importing country may apply the tariff rate in effect at the time NAFTA went into effect or the then current most-favored-nation rate, whichever is lower. The United States can apply the special safeguard provisions to seasonal imports of fresh tomatoes, eggplants, chili peppers, squash, watermelons, and onions.

Country-of-Origin Rules: NAFTA increases incentives for buying within the NAFTA region and ensures that Mexico will not serve as a platform for exports from other countries to the United States. Under NAFTA, only North American producers can obtain benefits from tariff preferences. Non-Mexican origin commodities must be transformed or processed significantly in Mexico so that they become Mexican goods before they can receive the lower NAFTA duties for shipment to the United States.

Key Provisions Of The U.S.–Canada Free Trade Agreement

Article 401: Tariffs on all agricultural goods will be removed over 10 years. Agricultural products that were duty-free before the imposition of the U.S.–Canada Free Trade Agreement (CFTA) will continue to be so. Most agricultural products fall under Category C (see below). However, a provision (401.5) permits accelerated elimination of tariffs, provided both governments agree. The phase-out categories in which products are placed are:

A: Immediate tariff elimination, effective January 1, 1989;
B: Five-year tariff elimination in five equal annual stages, duty-free on January 1, 1993;
C: Ten-year tariff elimination in 10 equal annual stages, duty-free on January 1, 1998.

Article 701: Public organizations cannot export agricultural goods to the other country at a price below the acquisition price, plus storage and handling costs. Neither country can use export subsidies in bilateral trade, and each will consider the other when using export subsidies to other countries. Canadian goods (eligible grains and oilseeds) shipped to the United States through west coast ports are excluded from receiving transport subsidies.

Article 702: Tariffs on fresh fruits and vegetables can snap back to the tariff that was in effect at the time the agreement was signed if certain price and acreage restrictions are met.

Article 703: Both countries shall work toward improving market access by removing trade barriers.

Article 704: Each country excludes the other from its meat import law.

Article 705: Canada will remove its import license on wheat, barley, and oats when U.S. government support levels for those grains are equal to or less than Canadian government support.

Article 706: Canada enlarged its global import quotas for chicken, turkey, and shell eggs.

Article 707: The United States will not restrict Canadian imports of products containing 10 percent or less sugar.

Article 708: The United States and Canada will work toward harmonizing their technical regulations and standards.

Chapter 8: This chapter reduced trade barriers in wine and distilled spirits arising from measures related to their internal sale and distribution.

Chapter 19: Binational dispute settlement panels will rule on cases involving countervailing and antidumping duties.

Sources: U.S. Department of Agriculture. *Effects of the North American Free Trade Agreement on U.S. Agricultural Commodities.* Office of Economics. March 1993. pp. 1–2. Carol Goodloe and Mark Simone. *A North American Free Trade Area for Agriculture: The Role of Canada and the U.S. Agreement.* U.S. Dept. Agr., Econ. Res. Serv. AIB-644. March 1992. p. 10.

Appendix 5

Where Major U.S. Crops Are Grown: The Top Producing States, 1991–1993

The list of states below each crop are in descending order of average annual production, 1991–1993

Corn	Cotton	Rice
Iowa	Texas	Arkansas
Illinois	California	Louisiana
Nebraska	Mississippi	Texas
Indiana	Arkansas	Mississippi
Minnesota	Louisiana	Missouri
Ohio	Arizona	
Wisconsin	Georgia	
Michigan	Tennessee	
Missouri	Alabama	
South Dakota	North Carolina	

Soybeans	Wheat	Peanuts
Illinois	Kansas	Georgia
Iowa	North Dakota	Texas
Indiana	Montana	Alabama
Minnesota	Oklahoma	North Carolina
Ohio	Washington	Oklahoma
Missouri	Texas	Virginia
Arkansas	South Dakota	Florida
Nebraska	Idaho	New Mexico
Kansas	Minnesota	South Carolina
South Dakota	Colorado	
Michigan	Nebraska	

Tobacco

North Carolina
Kentucky
Tennessee
South Carolina
Virginia
Georgia
Ohio
Pennsylvania
Indiana
Florida

Oats

South Dakota
North Dakota
Wisconsin
Minnesota
Iowa
Nebraska
Pennsylvania
Ohio
Texas

Rye

South Dakota
Georgia
North Dakota
Nebraska
Oklahoma
Minnesota
South Carolina
North Carolina
Pennsylvania
Michigan
Wisconsin

Appendix 6

Standard Weights and Measures with Equivalents (U.S. to Metric; Metric to U.S.)

U.S. Standard Units

U.S. Customary Unit	*U.S. Equivalent*	*Metric Equivalent*
Linear Measure		
inch	= 0.083 foot	= 2.540 centimeters
foot	= 12 inches	= 0.305 meter
yard	= 36 inches	= 0.914 meter
rod, pole, or perch	= 16.5 feet	= 5.029 meters
	= 5.5 yards	
furlong	= 660 feet	
	= 40 rods	
	= 10 chains	= 201.168 meters
U.S. statute mile	= 8 furlongs	
	= 5,280 feet	= 1.609 kilometers
international nautical mile	= 6,076.11549 feet (approx.)	
	= 1.151 statute miles	= 1.852 kilometers
Area measure		
square inch	= 0.007 square foot	= 6.452 square centimeters
square foot	= 144 square inches	= 929.030 square centimeters
square yard	= 1,296 square inches	
	= 9 square feet	= 0.836 square meters
square rod	= 272.25 square feet	= 25.289 square meters
acre	= 43,560 square feet	
	= 4,840 square yards	
	= 160 square rods	= 4,047 square meters
square mile	= 640 acres	= 2,590 square kilometers
1 section of land	= 1 square mile	= 2,590 square kilometers
1 township	= 6 square miles	= 15,540 square kilometers
Cubic measure		
cubic inch	= 0.00058 cubic foot	= 16.387 cubic centimeters
cubic foot	= 1,728 cubic inches	= 0.028 cubic meter
cubic yard	= 27 cubic feet	= 0.765 cubic meter

U.S. Customary Unit	*U.S. Equivalent*	*Metric Equivalent*
Liquid measure		
fluid ounce	= 8 fluid drams	
	= 1.804 cubic inches	= 29.573 milliliters
pint	= 16 fluid ounces	
	= 28.875 cubic inches	= 0.473 liter
quart	= 2 pints	
	= 57.75 cubic inches	= 0.946 liter
gallon	= 4 quarts	
	= 231 cubic inches	= 3.785 liters
barrel	= from 31 to 42 gallons, established by law or usage	
Dry measure		
pint	= 1/2 quart	
	= 33.6 cubic inches	= 0.551 liter
quart	= 2 pints	
	= 67.2006 cubic inches	= 1.101 liters
peck	= 8 quarts	
	= 537.605 cubic inches	= 8.810 liters
bushel	= 32 quarts	
	= 4 pecks	
	= 2,150.42 cubic inches	= 35.239 liters
Avoirdupois weight		
grain	= 0.036 dram	
	= 0.002285 ounce	= 64.798 milligrams
dram	= 27.344 grains	
	= 0.0625 ounce	= 1.772 grams
ounce	= 16 drams	
	= 437.5 grains	= 28.35 grams
pound	= 256 drams	
	= 7,000 grains	
	= 16 ounces	= 453.592 grams
hundredweight	= 100 pounds	
ton (short)	= 2,000 pounds	
	= 20 hundredweights	= 0.907 metric ton
		= 1,000 kilograms
Values in gross or long measure		
gross or long hundredweight	= 112 pounds	= 50.4 kilograms
gross or long ton	= 1.12 short tons	
	= 2,240 pounds	
	= 20 gross or long hundredweights	= 1.016 metric tons

Conversion: U.S. Unit to Metric Equivalent

Basic formula: *Multiply (U.S. unit) by (approximate factor) to convert to (metric equivalent)*

Length

U.S. Unit	*Approximate factor*	*Metric equivalent*
inches	= 25.40	= millimeters
inches	= 2.54	= centimeters
feet	= 30.48	= centimeters
yards	= 0.91	= meters
miles	= 1.61	= kilometers
Area		
square inches	= 6.45	= square centimeters
square feet	= 0.09	= square meters
square yards	= 0.84	= square meters
square miles	= 2.60	= square kilometers
acres	= 0.40	= hectares
Weight		
ounces	= 28.35	= grams
pounds	= 0.45	= kilograms
short tons	= 0.91	= tons
Volume		
teaspoons	= 4.93	= milliliters
tablespoons	= 14.78	= milliliters
fluid ounces	= 29.57	= milliliters
cups	= 0.24	= liters
pints	= 0.47	= liters
quarts	= 0.95	= liters
gallons	= 3.79	= liters
Temperature		
Fahrenheit	= -32, 5/9 x remainder	= Celsius

The Metric System

Linear Measure

Metric Unit	*Number of Meters*	*U.S. Equivalent*
millimeter	= 0.001	= 0.039 inch
centimeter	= 0.01	= 0.394 inch
decimeter	= 0.1	= 3.937 inches
meter	= 1	= 39.370 inches
decameter	= 10	= 32.808 feet
hectometer	= 100	= 109.361 yards
kilometer	= 1,000	= 0.621 mile
myriameter	= 10,000	= 6.214 miles

Area measure

Metric Unit	*Number of square meters*	*U.S. Equivalent*
square centimeter	= 100 square millimeters = 0.0001 square meter	= 0.115 square inch
square decimeter	= 100 square centimeters = 0.01 square meter	= 11.5 square inch
square meter	= 100 square decimeters	= 10.764 square feet
centare	= 1 square meter	= 10.764 square feet
deciare	= 10 square meters	= 11.960 square yards
are	= 100 square meters	= 119.599 square yards
hectare	= 10,000 square meters	= 2.477 acres
square kilometer	= 1,000,000 square meters	= 0.386 square mile

Volume

Metric Unit	*Number of Liters*	*U.S. Equivalent*	
		Liquid	*Dry*
milliliter	= 0.001	= 0.271 fluid dram	
centiliter	= 0.01	= 0.338 fluid ounce	
deciliter	= 0.10	= 0.211 pint	= 0.182 pint
liter	= 1	= 1.057 quarts	= 0.908 quart
decaliter	= 10	= 2.642 gallons	= 1.135 pecks
hectoliter	= 100	= 2.838 bushels	
kiloliter	= 1,000		

Weight

Metric Unit	*Number of Grams*	*U.S. Equivalent*
milligram	= 0.001	= 0.015 grain
centigram	= 0.01	= 0.154 grain
decigram	= 0.10	= 1.543 grains
gram	= 1	= 0.035 ounce
decagram	= 10	= 0.3523 ounce
hectogram	= 100	= 3.527 ounce
kilogram	= 1,000	= 2.205 pounds
quintal	= 100,000	= 220.462 pounds
metric ton	= 1,000,000	= 1.102 tons

Conversion: Metric Unit to U.S. Equivalent

Basic formula: *Multiply (metric unit) by (approximate factor) to convert to (U.S. equivalent)*

Length

Metric Unit	*Approximate factor*	*U.S. equivalent*
millimeters	= 0.04	= inches
centimeters	= 0.39	= inches
meters	= 3.28	= feet
meters	= 1.09	= yards
kilometers	= 0.62	= miles
Area		
square centimeters	= 0.16	= square inches
square meters	= 1.20	= square yards
square kilometers	= 0.39	= square miles
hectares	= 2.47	= acres
Weight		
grams	= 0.035	= ounce
kilograms	= 2.21	= pounds
tons	= 1.10	= short tons
Volume		
milliliters	= 0.20	= teaspoons
milliliters	= 0.06	= tablespoons
milliliters	= 0.03	= fluid ounces
liters	= 4.23	= cups
liters	= 2.12	= pints
liters	= 1.06	= quarts
liters	= 0.26	= gallons
cubic meters	= 35.32	= cubic feet
cubic meters	= 1.35	= cubic yards
Temperature		
Celsius	= (9/5) + 32	= Fahrenheit

Source: U.S. Department of Agriculture. *Weights, Measures, and Conversion Factors for Agricultural Commodities and Their Products.* Econ. Res. Serv. Ag. Handbook No. 697 (June 1992), pp. 4-5.

Appendix 7

Individual Commodity Weights and Measures

Commodity	Unit	Approximate net weight Kilograms	Pounds
Alfalfa seed	Bushel	27.2	60
Apples	Bushel basket or carton	18.1	40
	Carton, tray, or cell pack	18.1	40
Apricots	Lug, loose	10.9	24
Western	4-basket crate	11.8	26
Artichokes	Carton	10.4	23
Globe	1/2 box	9.1	20
Jerusalem	Bushel	22.7	50
Asparagus	Crate	13.6	30
Avocados	Lug	5.4–6.8	12–15
	Flat or carton, 2-layer	11.8	26
Bananas	Carton	18.1	40
Barley	Bushel	21.8	48
Beans:			
Lima, dry	Bushel	25.4	56
Other, dry	Bushel	27.2	60
	Sack	45.4	100
Lima, unshelled	Bushel	12.7–14.5	28–32
Snap	Bushel	12.7–14.5	28–32
Beets:			
Topped	Sack	11.3	25
Bunched	Crate or carton	17.2	38
Berries, frozen pack:			
Without sugar	50-gallon barrel	172	380
3 + 1 pack	50-gallon barrel	193	425
2 + 1 pack	50-gallon barrel	204	450
Blackberries	12, 1/2-pint baskets	2.7	6
Bluegrass seed	Bushel	6.4–13.6	14–30
Broccoli	Carton	10.4	23
Broomcorn (6 bales per ton)	Bale	151	333
Broomcorn seed	Bushel	20.0–22.7	44–50
Brussels sprouts	Carton	11.3	25
Buckwheat	Bushel	21.8	48
Butter	Box	30.9	68

Commodity	Unit	Approximate net weight Kilograms	Pounds
Cabbage:	Open mesh bag, sack	22.7	50
	Wirebound crate	22.7	50
	Western crate	36.3	80
Chinese cabbage	15 1/2-inch wirebound crate	22.7–24.0	50–53
	1 1/9-bushel wirebound crate	18.1–20.4	40–45
Cantaloupes	1/2 carton or crate	18.1	40
Carrots, without tops	Sacks, 48 1-pound and 24 2-pound	21.8	48
	Sacks	22.7	50
Castor beans	Bushel	18.6	41
Castor oil	Gallon	3.6	8
	Western Grower's Association crate	22.7–27.2	50–60
Cauliflower	Carton, film-wrapped, trimmed	11.3	25
	LI wirebound crate	27.2	60
Celery	Carton or crate	27.2	60
Cherries	Lug, California	8.2	18
	Lug, Northwest	9.1	20
Chives	Flat of 12 pots	4.5	10
Clover seed	Bushel	27.2	60
Coffee	Bag	60	132.3
Corn:			
Ear, husked	Bushel	31.8	70
Shelled	Bushel	25.4	56
Meal	Bushel	22.7	50
Oil	Gallon	3.5	7.7
Syrup	Gallon	5.3	11.72
Sweet	Carton	22.7	50
	Wirebound crate	19.1	42
Cotton	Bale, gross	227	500
	Bale, net	218	480
Cottonseed	Bushel	14.5	32
Cottonseed oil	Gallon	3.5	7.7
Cowpeas	Bushel	27.2	60
Cranberries	Barrel	45.4	100
	Carton, 24 12-ounce filmbags	8.2	18
Cream, 40 percent butterfat	Gallon	3.80	8.38
Cucumbers	1 1/9-bushel, carton/crate	24.9	55
Dewberries	Flat, 12 1/2-pint baskets	2.7	6
Eggplant	1 1/9-bushel, carton/crate	15.0	33
Eggs, average size	Case, 30 dozen	21.3	47.0
Escarole	1 1/9-bushel, carton/crate	11.3	25
Figs, fresh	Flat, 1-layer, tray pack	2.7	6
Flaxseed	Bushel	25.4	56
Flour, various	Bag	45.4	100
Garlic	Carton or crate, bulk	13.6	30

Commodity	*Unit*	*Approximate net weight* Kilograms	Pounds
	Carton of 12-tube or 12-film bag package (2 cloves each)	4.5	10
Grapefruit:			
Florida and Texas	1/2-box mesh bag	18.1	40
Florida	4/5-bushel carton	18.1	40
Texas	7/10-bushel carton	18.1	40
California and Arizona	Carton	15.4	34
Grapes	Carton or lug	10.0–10.4	22–23
Eastern	12-quart basket	9.1	20
Western	Lug	12.7	28
	4-basket crate	9.1	20
Hempseed	Bushel	20.0	44
Hickory nuts	Bushel	22.7	50
Honey	Gallon	5.4	11.84
Honeydew melons	2/3 carton	13.6	30
Hops	Bale, gross	90.7	200
Horseradish roots	Sack	22.7	50
Hungarian millet seed	Bushel	21.8–22.7	48–50
Kale	Carton or crate	11.3	25
Kapok seed	Bushel	15.9–18.1	35–40
Kiwifruit:			
California	1-layer flat	1.8–2.7	4–6
New Zealand	1-layer carton	3.2–4.1	7–9
Leeks	4/5-bushel crate	9.1	20
Lemons:			
Florida	4/5-bushel crate	19.1	42
California and Arizona	Carton	17.2	38
Lentils	Bushel	27.2	60
Lettuce	Carton	22.7	50
Lettuce, hothouse	24-quart basket	4.5	10
Limes	Carton	17.2	38
Linseed oil	Gallon	3.5	7.7
Malt	Bushel	15.4	34
Mangoes:			
Florida	Flat	6.4	14
Mexico	Lug	4.5–5.0	10–11
Maple syrup	Gallon	5.00	11.02
Meadow fescue seed	Bushel	10.9	24
Milk	Gallon	3.90	8.62
Millet	Bushel	21.8–22.7	48–60
Molasses, edible	Gallon	5.3	11.74
Molasses, inedible	Gallon	5.3	11.74
Mustard seed	Bushel	26.3–27.2	58–60
Nectarines	Los Angeles lug, 2-layer, tray pack	10.0	22
	Lug or carton, tight-fill	11.3	25
Oats	Bushel	14.5	32

Commodity	*Unit*	*Approximate net weight* *Kilograms*	*Pounds*
Okra	Bushel hamper or crate	13.6	30
	5/9-bushel crate	8.2	18
	Carton	8.2	18
	12-quart basket, crate, or carton	6.8–8.2	15–18
Olives	Lug	11.3–13.6	25–30
Olive oil	Gallon	3.5	7.6
Onions, dry	Sack	22.7	50
Onions, green bunched	Carton	5.9	13
Onion sets	Bushel	12.7–14.5	28–32
Oranges:			
Florida	4/5-bushel carton	19.5	43
Texas	7/10-bushel carton	19.1	42
California and Arizona	Carton	17.2	38
Orchardgrass seed	Bushel	6.4	14
Palm oil	Gallon	3.5	7.7
Papayas	Carton	4.5	10
Parsley	Carton, bushel basket, or crate, 5-dozen bunches	9.1–11.3	20–25
Parsnips	Bushel	22.7	50
Peaches	3/4-bushel, carton/crate	17.2	38
	2-layer carton or lug	10	22
Peanut oil	Gallon	3.5	7.7
Peanuts, unshelled:			
Virginia type	Bushel	7.7	17
Runners, southeastern	Bushel	9.5	21
Spanish	Bushel	11.3	25
Pears:			
California	Carton	16.3	36
	4/5-bushel carton	20.9	46
Northwest	4/5-bushel carton	20.4	45
Peas, green			
Unshelled	Bushel	12.7–13.6	28–30
Dry	Bushel	27.2	60
Peppers, green	Bushel, 1 1/9-bushel carton/crate	12.7	28
Perilla seed	Bushel	16.8–18.1	37–40
Persimmons	2-layer tray pack, lug or carton	9.1–11.3	20–25
	1-layer tray pack, flat or carton	4.5–5.4	10–12
Pineapples	Carton	18.1	40
Plantains	Carton	22.7	50
Plums	1/2-bushel carton	12.7	28
Prunes	1/2-bushel carton	13.6	30
Pomegranates	2-layer carton or lug	10.0–11.8	22–26
Popcorn:			
On ear	Bushel	31.8	70
Shelled	Bushel	25.4	56

Commodity	*Unit*	*Approximate net weight* Kilograms	Pounds
Poppy seed	Bushel	20.9	46
Potatoes	Carton	45.4	100
	Sack	45.4	100
Prickly pears	Box, wrapped pack	8.2	18
Quinces	Carton/lug, 2-layer	10.0	22
Radishes, topped	Carton of 24, 8-ounce, film bags	5.4	12
	Carton of 30, 6-ounce, film bags	5.0–5.4	11–12
	40-pound, film bag	18.1	40
Rapeseed	Bushel	22.7–27.2	50–60
Raspberries	Flat, 12 1/2-pint baskets	2.7	6
Redtop seed	Bushel	22.7–27.2	50–60
Refiners' syrup	Gallon	5.2	11.45
Rice:			
Rough	Bushel	20.4	45
	Bag	45.4	100
	Barrel	73.5	162
Milled	Pocket or bag	45.4	100
Rosin	Drum, net	236	520
Rhubarb	Carton or lug	9.1	20
	5-pound carton	2.3	5
Rutabagas	Sack	22.7	50
Rye	Bushel	25.4	56
Savory	Sack, crate, or carton	16.8	37
Sesame seed	Bushel	20.9	46
Shallots	Sacks of 8, 5-pound bags	18.1	40
Sorgo:			
Seed	Bushel	22.7	50
Syrup	Gallon	5.2	11.55
Sorghum grain	Bushel	25.4	56
Soybeans	Bushel	27.2	60
Soybean oil	Gallon	3.5	7.7
Spelt	Bushel	18.1	40
Spinach	Bushel	11.3	25
Strawberries	12, 1-pint	5.4	12
Sudangrass seed	Bushel	18.1	40
Sugarcane:			
Syrup (sulfured or unsulfured)	Gallon	5.2	11.45
Sunflower seed	Bushel	10.9–14.5	24–32
Sweetpotatoes	Carton	18.1	40
Tangerines:			
California and Arizona	Carton	11.3	25
Florida	4/5-bushel carton/crate	19.5	43
Timothy seed	Bushel	20.4	45

Commodity	*Unit*	*Approximate net weight* *Kilograms*	*Pounds*
Tobacco:			
Maryland	Hogshead	352	775
Flue-cured	Hogshead	431	950
Burley	Hogshead	442	975
Dark air-cured	Hogshead	522	1,150
Virginia fire-cured	Hogshead	612	1,350
Kentucky and Tennessee fire-cured	Hogshead	680	1,500
Cigar-leaf	Case	113–166	250–365
	Bale	68.0–79.4	150–175
	Crate	27.2	60
Tomatoes	Carton	11.3	25
	2-layer flat	9.1	20
Tomatoes, hothouse	12-quart basket	9.1	20
Tung oil	Gallon	3.5	7.8
Turnips:			
Without tops	Sack	11.3	25
Bunched	Carton	17.2	38
Turpentine	Gallon	3.3	7.23
Velvetbeans (hulled)	Bushel	27.2	60
Vetch	Bushel	27.2	60
Walnuts	Sacks	22.7	50
Watermelon	Carton	38.6	85
	Bin	476.3	1,050
Watercress	Carton, 25 bunches	3.6	8
Wheat	Bushel	27.2	60

Note: Much of this table on individual commodity weights and measures was taken from *Agricultural Statistics, 1990,* prepared by the U.S. Department of Agriculture's National Agricultural Statistics Service, Agricultural Statistics Board. Some of the weights were suggested by the Agricultural Marketing Service, U.S. Department of Agriculture. The table covers many important agricultural products, but it does not cover all farm products or all containers for any one product.

The information was assembled from State schedules of legal weights, various sources with the U.S. Department of Agriculture, and materials from other Government agencies. For most products, especially fruits and vegetables, there is considerable variation in weight per unit of volume because of differences in variety, size, condition, tightness of pack, degree to which the container is heaped, and other factors. An effort was made to select representative averages for listed products. For commodities for which there is considerable shrinkage, the point of origin weight or weight at harvest is used.

The approximate or average weights given in this table do not necessarily have official standing as a basis for packing or as grounds for settling disputes. Nor are they all recognized as legal weights.

Considerable information is available on dimensions of the various units and containers listed in *Agricultural Statistics.*

References

Angelo, Luigi, Robert Barry, Peter Buzzanell, Fred Gray, David Harvey, Ron Lord, and William Moore. *U.S. Sugar Statistical Compendium.* U.S. Dept. Agr., Econ. Res. Serv., Stat. Bull. No. 830., Aug. 1991.

Animal Health Institute. *Food Safety Expert List.* 1992.

Ash, Mark S. *An Animal Feeds Compendium.* U.S. Dept. Agr., Econ. Res. Serv., AER-656, May 1992, pp. 151–158.

Barry, Robert D., Luigi Angelo, Peter J. Buzzanell, and Fred Gray. *Sugar: Background for the 1990 Farm Legislation.* U.S. Dept. Agr., Econ. Res. Serv., AGES 90-06, Feb. 1990.

Becker, Geoffrey S. *Glossary of Agricultural Terms.* Library of Congress, Cong. Res. Serv, May 9, 1989.

Bennett, William F., Billy B. Tucker, and A. Bruce Maunder. *Modern Grain Sorghum Production.* Ames: Iowa State Univ. Press, 1990, pp. 157–162.

Brown, Jr., Richard N., and Nydia R. Suarez. *U.S. Markets for Caribbean Basin Fruits and Vegetables: Selected Characteristics for 17 Fresh and Frozen Imports, 1975-87.* U.S. Dept. Agr., Econ. Res. Serv., Stat. Bull. 821, March 1991.

Central Intelligence Agency. *The World Factbook 1991.* Washington, D.C.: GPO, 1991, pp. 361, 366, 384.

Cliver, Dean O. *Eating Safely: Avoiding Foodborne Illness.* New York: Am. Council on Science and Health, Jan. 1993.

Council for Agricultural Science and Technology. *Genetic Engineering in Food and Agriculture.* Rept. No. 110, Sept. 1986.

Crowder, Brad, and Cecil Davison. *Soybeans: Background for 1990 Farm Legislation.* U.S. Dept. Agr., Econ. Res. Serv., AGES 89-41, Sept. 1989.

Dicks, Michael R., and Katharine C. Buckley, eds. *Alternative Opportunities in Agriculture: Expanding Output Through Diversification.* U.S. Dept. Agr., Econ. Res. Serv., AER-633, May 1989.

Dowdy, George T., Sr., and Leroy W. Garnett. *Dictionary of Agricultural Economics and Related Terms.* Tuskegee, AL: Tuskegee Institute, 1966.

Duewer, Lawrence A. *Changing Trends in the Red Meat Distribution System.* U.S. Dept. Agr., Econ. Res. Serv., AER-509, Feb. 1984.

Elving, Phyllis. *Fresh Produce A to Z: How to Select, Store, and Prepare.* Menlo Park, CA: Lane Publishing, Jan. 1987.

Fuell, Lawrence D., David C. Miller, and Merritt Chesley. *Dictionary of International Agricultural Trade.* U.S. Dept. Agr., For. Agr. Serv., Ag. Hndbk. No 411, June 1988.

Gilliam, Henry C. Jr., *U.S. Beef Cow-Calf Industry.* U.S. Dept. Agr., Econ. Res. Serv., Econ. Rept. No. 515, Sept. 1984.

Glynn, Priscilla. "New Crops, and Old, Offer Alternative Opportunities." *Farmline* XII, no. 1 (Dec. 1990–Jan. 1991), pp. 12–16.

Gollehon, Noel. *Chemigation: A Technology for the Future?* U.S. Dept. Agr., Econ. Res. Serv., AIB-608, July 1990.

Goodloe, Carol, and Mark Simone. *A North American Free Trade Area for Agriculture: The Role of Canada and the U.S.–Canada Agreement.* U.S. Dept. Agr., Econ. Res. Serv., AIB-644, March 1992.

Grise, Verner N. *Tobacco: Background for 1990 Farm Legislation.* U.S. Dept. Agr., Econ. Res. Serv., Staff Report AGES 89-48, Oct. 1989, pp. 37–42.

Hampe, Edward C., Jr., and Merle Wittenberg. *The Life Line of America. Development of the Food Industry.* New York: McGraw-Hill, 1964, pp. 271–273.

Harwood, Joy L., Mack N. Leath, and Walter G. Heid, Jr. *The U.S. Milling and Baking Industries.* U.S. Dept. Agr., Econ. Res. Serv., AER- 611, Dec. 1989.

Heid, Walter G., Jr. *U.S. Wheat Industry.* U.S. Dept. Agr., Econ. Res. Serv., AER-432, Aug. 1979, pp. 110–117.

Heifner, Richard G., Bruce H. Wright, and Gerald E. Plato. *Using Cash, Futures, and Options Contracts in the Farm Business.* U.S. Dept. Agr., Econ. Res. Serv., AIB-665, April 1993.

Hoff, Frederic L., and Jane K. Phillips. *Honey: Background for 1990 Farm Legislation.* U.S. Dept. Agr., Econ. Res. Serv. Staff Report AGES-89- 43, Sept. 1989, pp. 40–42.

Johnson, Doyle, and Tarra M. Johnson. *Financial Performance of U.S. Floriculture and Environmental Horticulture Farm Businesses, 1987-91.* U.S. Dept. of Agr., Econ. Res. Serv., SB-862, Sept. 1993, pp. 3–4.

Krause, Kenneth R. *The Beef Cow-Calf Industry, 1964-87; Location and Size.* U.S. Dept. Agr., Econ. Res. Serv., AER-659, June 1992, pp. 100–103.

Lasley, Floyd A. *The U.S. Poultry Industry: Changing Economics and Structure.* U.S. Dept. Agr., Econ. Res. Serv., AER-502, July 1983, p. iii.

Lasley, Floyd A., William L. Henson, and Harold B. Jones, Jr. *The U.S. Turkey Industry.* U.S. Dept. Agr., Econ. Res. Serv., AER-525, March 1985.

Lasley, Floyd A., Harold B. Jones, Jr., Edward H. Easterling, and Lee A. Christensen. *The U.S. Broiler Industry.* U.S. Dept. Agr., Econ. Res. Serv., AER-591, Nov. 1988.

Lipton, Kathryn L. *Agriculture, Trade, and the* GATT: *A Glossary of Terms.* U.S. Dept. Agr., Econ. Res. Serv., AIB-625, June 1991.

Lipton, Kathryn L., and Susan L. Pollack. *A Glossary of Food and Agricultural Policy Terms, 1989.* U.S. Dept. Agr., Econ. Res. Serv., AIB-573, Nov. 1989.

———. "Glossary of Food and Agricultural Terms." *Agricultural-Food Policy Review: U.S. Agricultural Policies in a Changing World.* U.S. Dept. Agr., Econ. Res. Serv. AER- 620. pp. 383–401.

Lord, Ron, and Robert D. Barry. *The World Sugar Market—Government Intervention and Multilateral Policy Reform.* U.S. Dept. Agr., Econ. Res. Serv., AGES-9062, Sept. 1990.

Love, John M. *U.S. Onion Statistics, 1960–93.* U.S. Dept. Agr., Econ. Res. Serv., Stat. Bull. No. 880, April 1994.

Lucier, Gary, Agnes Chesley, and Mary Ahearn. *Farm Income Data: A Historical Perspective.* U.S. Dept. Agr., Econ. Res. Serv., Stat. Bull. No. 740, May 1986, pp. 2–5.

Manchester, Alden C. *Paying for Marketwide Services in Fluid Milk Markets.* U.S. Dept. Agr., Econ. Res. Serv., AIB-514, April 1987.

———. *Rearranging the Economic Landscape: The Food Marketing Revolution, 1950-91.* U.S. Dept. Agr., Econ. Res. Serv., Ag. Econ. Report No. 660, Sept. 1992.

———. *The Structure of Wholesale Produce Markets.* U.S. Dept. Agr., Econ. Res. Serv., AER-45, April 1964, pp. 121–123.

———. *Transition in the Farm and Food Sector.* U.S. Dept. Agr., Econ. Res. Serv. Unpublished paper for the National Planning Association Committee on Agriculture, April 1992.

McDowell, Howard, Ann M. Fleming, and Richard F. Fallert. *Federal Milk Marketing Orders: An Analysis of Alternative Policies.* U.S. Dept. Agr., Econ. Res. Serv., AER-598, Sept. 1988, pp. 60–61.

Mercier, Stephanie. *Corn: Background for 1990 Farm Legislation.* U.S. Dept. Agr., Econ. Res. Serv., Staff Report No. AGES 89-47, pp. 48–50.

Morse, Roger A., and Kim Flottum. *ABC and XYZ of Bee Culture.* A.I. Root Co., 1990, pp. 489–500.

National Research Council. *Alternative Agriculture*. Washington, D.C.: National Academy Press, 1989, pp. 419–426.

Nelson, Kenneth E., Lawrence A. Duewer, and Terry L. Crawford. *Reevaluation of the Beef Carcass-to-Retail Weight Conversion Factor*. U.S. Dept. Agr., Econ. Res. Serv., AER-623, 1989, pp. iv–v.

Office of Technology Assessment. *Grain Quality in International Trade: A Comparison of Major U.S. Competitors*. 1989, pp. 148–149.

Paul, Allen B., Richard G. Heifner, and John W. Helmuth. *Farmers' Use of Forward Contracts and Futures Markets*. U.S. Dept. Agr., Econ. Res. Serv., AER-320, 1976, p. iv.

Pollack, Susan L., and Lori Lynch. *Provisions of the Food, Agriculture, Conservation, and Trade Act of 1990*. U.S. Dept. Agr., Econ. Res. Serv., AIB-624, June 1991.

Powers, Nicholas J. *Federal Marketing Orders for Fruits, Vegetables, Nuts, and Specialty Crops*. U.S. Dept. Agr., Econ. Res. Serv., Ag. Econ. Report No. 629, March 1990.

———. *Marketing Practices for Vegetables*. U.S. Dept. Agr., Econ. Res. Serv., Ag. Info. Bull. No. 702. Aug. 1994.

Powers, Nicholas J. and Richard G. Heifner. *Federal Grade Standards for Fresh Produce: Linkages to Pesticide Use*. U.S. Dept. Agr., Econ. Res. Serv., Ag. Inf. Bull. No. 675, Aug. 1993.

Runyan, Jack L. *A Summary of Federal Laws and Regulations Affecting Agricultural Employers*. U.S. Dept. Agr., Econ. Res. Serv., AIB-550, Jan. 1989.

Russell, James R., and James C. Hanson. *Producers; Guide to Grain Marketing Terminology*. Univ. of Maryland, Coop. Exten. Serv., Fact Sheet 498, 1988.

Soil Conservation Society of America. *Resource Conservation Glossary*. Ankeny, Iowa:, 1976.

Storck, John, and Walter Dorwin Teague. *Flour for Man's Bread: A History of Milling*. St. Paul: North Central Pub., 1952.

Stults, Harold, Edward H. Glade, Jr., Scott Sanford, and Leslie A. Meyer. *Cotton: Background for 1990 Farm Legislation*. U.S. Dept. Agr., Econ. Res. Serv., Staff Report No. AGES 89-42, pp. 51–65.

U.S. Department of Agriculture. *1989 Fact Book of Agriculture*. Off. of Public Affairs, Misc. Pub. 1063, August 1989.

———. *A Glossary of Farm Terms*. Off. of Sec., May 1988.

———. *Agricultural Outlook*. Econ. Res. Serv., Selected issues.

———. *Changes in the Market Structure of the Breakfast Foods Industry*. Econ. Res. Serv., MRR-623, 1972, p. 25.

———. *Dictionary of International Trade*. For. Ag. Serv., Ag. Handbook No. 411, April 1971.

———. *Effects of the North American Free Trade Agreement on U.S. Agricultural Commodities*. Office of Economics, March 1993.

———. *Enhancing the Quality of U.S. Grain for International Trade*. 1989, pp. 274–279.

———. *Farm Bill Issues: Background Facts*. Off. of Pub. Affairs, May 1990.

———. *Food Marketing Review, 1991*. Econ. Res. Serv., AER-657, March 1992.

———. *Industrial Uses of Agricultural Materials: Situation and Outlook*. Econ. Res. Serv., IUS-1, June 1993.

———. *National Food Review*. Econ. Res. Serv., selected issues.

———. *Synthetics and Substitutes for Agricultural Products: Projections for 1980*. Econ. Res. Serv., MRR-947, 1972, pp. 44–45.

———. *The Formula Feed Industry, 1969: A Statistical Summary*. Econ. Res. Serv., SB-485, May 1972, pp. 2–3.

———. *The 1990 Farm Act and the Budget Reconciliation Act: How U.S. Farm Policy Mechanisms Will Work Under New Legislation*. Econ. Res. Serv., Misc. Pub. 1489, Dec. 1990.

———. *U.S. Fats and Oils Statistics, 1961–76*. Econ. Res. Serv., Stat. Bull. No. 574, June 1977.

———. *Weights, Measures, and Conversion Factors for Agricultural Commodities and Their Products*. Ag. Handbook 697, June 1992.

———. "Where Major U.S. Crops Are Grown." *Farmline* XIII, no. 9 Sept. 1992), pp. 8–21.

———. *Yearbook of Agriculture*. Off. of Pub. and Visual Comm, Selected years.

Van Arsdall, Roy. *Structural Characteristics of the U.S. Hog Production Industry*. U.S. Dept. Agr., Econ. Stat. and Coop. Serv. AER-415. Dec. 1978, pp. iii–iv.

Van Arsdall, Roy, and Kenneth E. Nelson. *Characteristics of Farmer Cattle Feeding*. U.S. Dept. Agr., Econ. Res. Serv., AER-503, Aug. 1983, pp. iii–iv.

———. *U.S. Hog Production Industry*. U.S. Dept. Agr., Econ. Res. Serv., AER-511, June 1984, p. iv.

Williams, Sheldon W., David A. Vose, Charles E. French, Hugh L. Cook, and Alden C. Manchester. *Organization and Competition in the Midwest Dairy Industries*. Ames: Iowa State Univ. Press, pp. 327–334.

Wood, Marcia. "Exotic Fruits Promise Taste of the Tropics," *Agricultural Research* 39, no. 10 (October 1991), pp. 18–20.

Zellner, James A., and Rosanna Mentzer Morrison. *How Do Government Programs and Policies Influence Consumers' Food Choices?* U.S. Dept. Agr., Econ. Res. Serv., AIB-553, Nov. 1988.

About the Book and the Author

This practical, comprehensive dictionary provides an invaluable source of information for everyone from undergraduates to seasoned policymakers interested in the disciplines related to food and agriculture. With more than 3,000 detailed entries, it covers a wide array of topic areas, including:

- Agricultural policies and programs
- Animal and plant science
- Conservation and environmental protection
- Commodities
- Credit and banking
- Farm and agricultural-related labor
- Farm equipment and buildings
- Fibers and textiles
- Food marketing, wholesaling, and retailing
- Food safety and animal and human nutrition and health
- Futures and options markets
- Industrial crops and products
- Natural resources
- Tariffs and trade

The dictionary is alphabetically arranged, but also contains an index of terms by topic area. The definitions include cross-references to guide the reader easily to related terms.

Appendixes provide descriptions of U.S. agricultural legislation from 1933 through mid-1994, an outline of the NAFTA provisions affecting agriculture, as well as a listing of individual commodity weights and measures and related conversions.

Kathryn L. Lipton has worked extensively in the field of food and agriculture. An agricultural economist with the U.S. Department of Agriculture, she has published more than 60 articles and monographs. She served as economics editor of the *National Food Review* from 1983–1987.